Robust Control System Design

Robust Control System Design

Advanced State Space Techniques

Third Edition

Chia-Chi Tsui

CRC Press is an imprint of the
Taylor & Francis Group, an **informa** business

Third edition published 2022
by CRC Press
6000 Broken Sound Parkway NW, Suite 300, Boca Raton, FL 33487-2742

and by CRC Press
4 Park Square, Milton Park, Abingdon, Oxon, OX14 4RN

CRC Press is an imprint of Taylor & Francis Group, LLC

Second edition published by CRC Press 2004

Library of Congress Cataloging-in-Publication Data
Names: Tsui, Chia-Chi, 1953- author.
Title: Robust control system design : advanced state space techniques /
Chia-Chi Tsui.
Description: Third edition. | Boca Raton, FL : CRC Press, 2022. | Includes
bibliographical references and index.
Identifiers: LCCN 2021054036 (print) | LCCN 2021054037 (ebook) |
ISBN 9781032195193 (hbk) | ISBN 9781032195223 (pbk) |
ISBN 9781003259572 (ebk)
Subjects: LCSH: Feedback control systems—Design and construction. |
Control theory.
Classification: LCC TJ216 .T79 2022 (print) | LCC TJ216 (ebook) |
DDC 629.8/3—dc23/eng/20220110
LC record available at https://lccn.loc.gov/2021054036
LC ebook record available at https://lccn.loc.gov/2021054037

ISBN: 978-1-032-19519-3 (hbk)
ISBN: 978-1-032-19522-3 (pbk)
ISBN: 978-1-003-25957-2 (ebk)

DOI: 10.1201/9781003259572

Typeset in Palatino
by codeMantra

Contents

Preface ... ix
Preface (Second Edition) ... xiii
Author ... xvii

1. System Mathematical Models ... 1
1.1 Two Kinds of Mathematical Models ... 1
1.2 Eigenstructure Decomposition of the State Space Model ... 8
1.3 System Order, Controllability, and Observability ... 11
1.4 System Poles and Zeros ... 18
Exercises ... 20

2. Single-System Performance and Sensitivity ... 25
2.1 System Performance ... 25
2.2 System Sensitivity and Robustness ... 34
2.2.1 The Sensitivity of Eigenvalues (Robust Performance) ... 37
2.2.2 The Sensitivity of System Stability (Robust Stability) ... 41
2.3 Conclusion ... 46
Exercises ... 46

3. Feedback System Sensitivity ... 49
3.1 Sensitivity and Loop Transfer Function of the Feedback Systems ... 49
3.1.1 Sensitivity to System Model Uncertainty ... 51
3.1.2 Sensitivity to Control Input Disturbance ... 53
3.2 Sensitivity of Feedback Systems of the Modern Control Theory ... 56
3.2.1 State Feedback Control Systems ... 56
3.2.2 Static Output Feedback Control Systems ... 60
3.2.3 Observer Feedback System – Loop Transfer Recovery ... 62
3.3 Summary ... 70

4. A New Feedback Control Design Principle/Approach ... 73
4.1 Basic Observer Design Concept – Generating State Feedback Signal Directly Without Generating Explicit System States ... 74
4.2 Performance of the Observer Feedback System – Separation Property ... 77
4.3 Eight Drawbacks and Irrationalities of the Modern Control Design and Separation Principle ... 79
4.3.1 Drawback 1 of Separation Principle: Invalid Basic Assumption ... 79

4.3.2 Drawback 2 of Separation Principle: Ignor Key Parameters 80
4.3.3 Drawback 3 of Separation Principle: Wrong Design Priority 80
4.3.4 Drawback 4 of Separation Principle: Unnecessary Design Requirement 81
4.3.5 Drawback 5 of Separation Principle: Abandon Existing Control Structure 81
4.3.6 Drawback 6 of Separation Principle: Failed Robust Realization 82
4.3.7 Drawback 7 of Separation Principle: Two Extreme Controls 84
4.3.8 Drawback 8 of Separation Principle: Two Extreme Control Structures 84
4.4 A New Design Principle That Guarantees the General and Full Realization of Robustness of the Generalized State Feedback Control 85
Exercises 92

5. Solution of Matrix Equation $TA-FT=LC$ 97
5.1 Computation of System's Observable Hessenberg Form 97
5.1.1 Single-Output Systems 97
5.1.2 Multiple-Output Systems 99
5.2 Computation of the Solution of Matrix Equation $TA-FT=LC$ 105
5.2.1 Eigen-Structure Case A 106
5.2.2 Eigen-Structure Case B 109
Exercises 115

6. Observer Design for Robust Realization 117
6.1 Solution of Matrix Equation $TB=0$ 118
6.2 Analysis and Examples of This Design Solution 120
6.3 Complete Unification of Two Existing Basic Modern Control System Structures 133
6.4 Observer Order Adjustment to Tradeoff between Performance and Robustness 135
Exercises 138

7. Observer Design for Other Special Purposes 143
7.1 Minimal-Order Linear Functional Observer Design 144
7.1.1 Simplest Possible Design Formulation – Most Significant Theoretical Development 144
7.1.2 Really Systematic Design Algorithm and Guaranteed Observer Order Upper Bound 146

7.1.3 The Lowest Possible Observer Order Upper Bound – The Best Possible Theoretical Result – The Whole Design Problem Is Essentially Solved 155
7.2 Fault Detection, Isolation, and Control Design 157
7.2.1 Fault Models and Design Formulation of Fault Detection and Isolation 157
7.2.2 Design Algorithm and Examples of Fault Detection and Isolation 160
7.2.3 Adaptive Fault Control and Accommodation (Tsui, 1997) 163
7.2.4 The Treatment of Model Uncertainty and Measurement Noise (Tsui, 1994b) 167
Exercises 171

8. Control Design for Eigenvalue Assignment 177
8.1 Eigenvalue (Pole) Selection 177
8.2 Eigenvalue Assignment by State Feedback Control 179
8.3 Eigenvalue Assignment by Generalized State Feedback Control 181
8.4 Modifications of Generalized State Feedback Control for Eigenstructure Assignment (Tsui, 2004b,c, 2005) 193
8.5 Summary of Eigenstructure Assignment Designs 197
Exercises 199

9. Control Design for Eigenvector Assignment 203
9.1 Numerical Iterative Methods (Kautsky et al., 1985) 204
9.2 Analytical Decoupling Method 209
9.3 Summary of Eigenstructure Assignment of Chapters 8 and 9 219
Exercises 220

10. Control Design for LQ Optimal Control 223
10.1 Direct State Feedback Control Design 225
10.2 Design of Generalized State Feedback Control 227
10.3 Comparison and Conclusion of Feedback Control Designs 231
Exercises 234

Appendix A: Linear Algebra & Numerical Linear Algebra 237

Appendix B: Design Projects and Problems 263

References 271

Index 281

Preface

The third edition of *Robust Control System Design* makes more assertion, articulation, and demonstration, on the advantages of the main contribution of this book – a fundamentally novel design principle of the modern control theory. This new design principle designs a generalized state feedback control, based on the parameters of observer/feedback controller and of output measurement, of each system. Therefore, this new design principle can be called a "synthesized design principle".

Existing modern control design has always followed the well-known "separation (design) principle" for 60 years since the start. This design principle first assumes that all system's internal states are measurable and known, designs a state feedback control based on this assumption, and only then designs an observer to realize this state feedback control. Therefore, under separation principle, the state feedback control and its realizing observer are designed independently and separately.

Because almost all real systems cannot satisfy that basic assumption of separation principle, the robustness properties, or the loop transfer function, of the direct state feedback control *cannot* be actually realized by an observer for a great majority of real systems, even though the signals of state feedback control can be estimated.

Here the robustness is defined as the system's low sensitivity against system model uncertainty, control input disturbance, and output measurement noise, or is defined as the system's ability to maintain high performance (at least stability) against these undesirable effects. Therefore, robustness is also the reliability of the system.

It is well known that robustness/reliability has always been the foremost purpose of feedback control. It is also known that high performance and high robustness are contradictory to each other, although both are required in practice.

Therefore, the above situation of "*cannot* realize the robustness properties of state feedback control" is the fatal drawback of the existing modern control design theory and separation principle. This fatal and pivotal drawback must be overcome.

Because of this fatal and pivotal drawback, the modern control theory has not found many successful applications in the past 60 years, even less than the successful applications of the classical control theory. As a result, the control research community has largely turned back since the 1980s to the classical control theory such as the "neo-classical control theory".

However, it is also well known that the transfer function/loop transfer function model of the classical control theory is far less direct, detailed, and simple than the state space model of the modern control theory. It is also well known that as a result, the classical control theory is far less effective in both analysis and design than the modern control theory.

For example, its gain/phase margins are far less generally accurate in measuring robust stability than the measure which is based on the real part of system poles and their sensitivities of the modern control theory. Gain and phase margins are far less effective either in guaranteeing the system performance than the system poles. This is also the reason that the control research community turned from the classical control theory to the modern control theory in the 1960s and 70s in the first place. To summarize, because of the above fatal and pivotal drawback of the modern control design theory, the whole control theory has been essentially stagnant without real progress for six decades.

This situation of "without real progress" certainly does not mean "without new design formulations and re-formulations" – there have been many. However, if a new design formulation or reformulation cannot guarantee higher system performance and robustness, or cannot have a solution that is really practical, then such a new formulation or re-formulation cannot be considered a "real progress".

For example, an optimal control, whether the quadratic optimal control or the other recently formulated and sensitivity function/loop transfer function-based optimal control, cannot really optimize the actual system performance and robustness, as its name suggested. For example, the quadratic optimal criterion is defined by too many parameters (in the order of n^2) whose selection has no general and explicit rules. Therefore, the optimization of this criterion cannot imply the direct improvement of actual system performance and robustness, at least far less direct than optimizing this book's robust stability criterion that is based on system poles and their sensitivities.

The design solutions of these optimal controls are not practical either. For example, the quadratic criterion can only be optimized by state feedback control. Because the robustness of state feedback control cannot be actually realized by observers generally, the quadratic optimal control cannot be actually practical if not all internal system states are measurable. In addition, the solutions of these optimal controls are numerical. The existence of a numerical solution is often unpredictable, and numerical-solution-based design results cannot be reversely adjusted based on simulation, even though such adjustment of theoretically guided designs is necessary in practice.

To summarize, it has been the consensus that optimal control results are not really practical and are unable to develop further besides reformulation to another new optimal criterion.

Modern control theoreticians only formulated the feedback control design problem in various ways. The point however is to solve this design problem really satisfactorily.

The new design principle of this book can design feedback controllers that *can* fully realize the robustness properties of their generalized state feedback control, for a great majority of system conditions! Therefore, this new design principle really overcomes the fatal and pivotal drawback of the existing modern control design theory of the past 60 years since the start, and therefore is the real and decisive progress of the modern control design theory for 60 years.

This new design principle *can* also adjust effectively the tradeoff between the strength and the degree of robust realization of its generalized state feedback control. This adjustment is based on the actual system conditions, design requirements, and simulation results, and is achieved simply by adjusting the controller order! These features are also breakthroughs of the modern control design theory for 60 years.

The generalized state feedback control of this new design principle is designed based on much better information of the state space model of the system and of its realizing observer, and is therefore far more effective than any other basic forms of feedback control. This control also unifies the existing state feedback control and static output feedback control as its extremes. With the guaranteed full realization of robustness of this control, the entire control systems theory is now revived!

Another advantage of this book's new design principle lies in its simplicity in both analysis and design.

It is well known that only the simple and clear analytical result is the result of really thorough understanding. For example, linear system performance is most directly determined by system poles. Therefore, this book's robust stability/robust performance criterion M_3, which is based on system poles and their sensitivities, is proven to be exemplarily generally accurate (see Subsection 2.2.2). Another example is that the condition to realize the robustness/loop transfer function of generalized state feedback control is zero input feedback. This condition is formulated in this book in the simplest possible way as $TB=0$ (Theorem 3.4).

It is also well known that only simple design methods can really be learned and grasped by practical control system designers. Furthermore, only after being really grasped, a design method can be applied in practice with real system conditions and real design requirements, and can really be reversely adjusted (including the original design requirement) by the data of the final simulation.

The design computation of this book is very simple and is mostly in closed form (not numerical). As a result, unlike other books of a similar scope, this book demonstrated far more MIMO system examples with orders ranging from 3 to 9. The design of most examples is carried out by hand computation only, so that the result of each design step can be clearly revealed. Only this kind of design methods can be really learned and grasped by the readers and be applied and adjusted in real practice. There are a total of 12 computational algorithms in this book. These algorithms are simple and explicit enough to be coded directly into computer programs.

Finally, the main improvement of this third edition is at more detailed, sufficient, and complete explanation and illustration of the new design principle of this book. The main improvement is in the following two parts.

(A) Chapter 7 (Observer Design for Other Special Purposes) is completely re-written:

First, the conclusion of "the best possible theoretical result of minimal order function observer design" is more formally and rigorously proved. Based on this strengthened conclusion, more conclusive claims were made that the theoretical part of this design problem is solved, and that because further complication of the design computation algorithm for further reduced observer order is generically not worthwhile, the computational part of the problem and the whole design problem, are also solved. These conclusions were presented in the new Section 7.1.

Second, the result of the 2nd edition's Chapter 10 (fault detection, isolation, and control) is now placed in the new Section 7.2, as another special purposed observer. The whole result is presented in a more general way in this edition. This topic has been treated exclusively in some other monograph books, and can complement the normal feedback observers/controllers of this book.

(B) Chapter 8 (Eigenstructure Assignment) is much improved and expanded and is now divided in this edition into two chapters, covering respectively eigenvalue assignment and eigenvector assignment. The total number of chapters of this edition remains at 10.

Because system state matrix eigenvalues are system poles, which determine most directly system performance, and because the sensitivity of an eigenvalue is determined by its associated eigenvectors, these two assignments can most directly and therefore effectively improve system performance and robustness. Eigenstructure assignment can only be achieved by state/generalized state feedback control. Therefore, the improvement, development even illustration, of the design algorithms of this assignment, is never enough.

For example, two third-order examples using design methods of the entire book are added to Section 8.3. The feedback system poles assigned in these two examples are complex conjugates, which are common in real practical designs. The second edition examples of Section 8.3 only assign distinct and real poles. These two new examples can demonstrate the power and advantages of the new design principle of this book more clearly, completely, and sufficiently. Ten more such examples are added as exercise problems at the end of the chapter, while in the other comparable books there usually is at most one example of such complete design. Furthermore, the existing classical control theory cannot even analyze accurately the 2×2 loop transfer function matrix of this example, while the existing separation principle cannot guarantee robustness at all for this example, which has more inputs than outputs and whose control achieved arbitrary pole assignment and one-pole-sensitivity assignment.

This new edition has a most important reference published in 2015: https://www.researchgate.net/publication/273352547_Observer_design_-_A_survey.

by this author,
April 3, 2021 in New York

Preface (Second Edition)

This second edition of Robust Control System Design introduces a new design approach to modern control systems. This design approach guarantees, for the first time, the full realization of robustness properties of generalized state feedback control for most open-loop system conditions. State and generalized state feedback control can achieve feedback system performance and robustness far more effectively than other basic forms of control. Performance and robustness (versus model uncertainty and control disturbance) are mutually contradictory, yet they are the key properties required by practical control systems. Hence, this design approach not only enriches the existing modern control system design theory but also makes possible its wide application.

Modern (or state space) control theory was developed in the 1960s. The theory has evolved such that the state feedback control and its implementing observer are designed separately (following the so-called separation principle (Willems, 1995)). With this existing design approach, although the direct state feedback system can be designed to have good performance and robustness, almost all the actual corresponding observer feedback systems have entirely different robustness. In the new design approach presented here, the state feedback control and its implementing observer are designed together. More explicitly, the state feedback control is designed based on the results of its implementing observer. The resulting state feedback control is the generalized state feedback control (Tsui, 1999b).

This fundamentally new approach guarantees – for all open-loop systems with more outputs than inputs or with at least one stable transmission zero – the same loop transfer function and therefore the same robustness of the observer feedback system and the corresponding direct state feedback system. Most open-loop systems satisfy either of these two conditions. For all other open-loop systems, this approach guarantees that the difference between the loop transfer functions of the above two feedback systems is kept minimal in a simple least-square sense.

The modern and classical control theories are the two major components of control systems theory. Compared with classical control theory, modern control theory can describe a single system's performance and robustness more accurately, but it lacks a clear concept of feedback system robustness, such as the loop transfer function of classical control theory. By fully using the concept of loop transfer functions, the approach exploits the advantages of both classical and modern control theories. This approach guarantees the robustness and loop transfer function of classical control theory while designing this loop transfer function much more effectively (though indirectly) using modern control design techniques. Thus, it achieves both good robustness and performance for feedback control systems.

If the first edition of this book emphasized the first of the above two advantages (i.e., the true realization of robustness properties of feedback control), then this second edition highlights the second of the above two advantages – the far more effective design of high performance and robustness feedback control itself.

A useful control theory should provide general and effective guidance on complicated control system design. To achieve this, the design formulation must fully address both performance and robustness. It must also exploit fully the existing design freedom and apply a general, simple, and explicit design procedure. The approach presented here truly satisfies these requirements. Since this book concentrates on this new design approach and its relevant analysis, other analytical control theory results are presented with an emphasis on their physical meanings, instead of their detailed mathematical derivations and proofs.

The following list shows several of the book's most important results. With the exception of the third item, these results are not presented in any other books:

The first general dynamic output feedback compensator can implement state or generalized state feedback control, and its design procedure. The feedback system of this compensator is the first general feedback system that has the same robustness properties as its corresponding direct state feedback system (Chapters 3–6).

A systematic, simple, and explicit eigenvalue assignment procedure using static output feedback control or generalized state feedback control (Section 8.1). This procedure enables the systematic eigenvector assignment procedures of this book and is general to most open-loop system conditions if based on the generalized state feedback control of this book.

Eigenvector assignment procedures that can fully use the freedom of this assignment. Both numerical algorithms and analytical procedures are presented (Section 8.2).

A general failure detection, isolation, and accommodation compensator that is capable of considering system model uncertainty and measurement noise, and its systematic design procedure (Chapter 10).

The simplest possible formulation, and a truly systematic and general procedure, of minimal order observer design (Chapter 7).

Solution of the matrix equation $TA-FT=LC$ [matrix pair (A, C) is observable and eigenvalues of matrix F are arbitrarily assigned]. This solution is general and has all eigenvalues of F and all rows of T completely decoupled (F is in Jordan form). This solution uniquely enables the full use of the remaining freedom of this matrix equation, which is fundamentally important in most of the basic design problems of modern control theory (Chapters 5–8, 10).

The basic design concept of generating a state feedback control signal without estimating all state variables, and the generalization of this design concept from function observers only to all feedback compensators (Chapters 3–10).

The complete unification of two existing basic feedback structures of modern control theory – the zero-input gain state observer feedback structure and the static output feedback structure (Section 6.3).

A more generally accurate robust stability measure is expressed in terms of the sensitivities of each system pole. This analytical measure can be used to guide systematic feedback system design (Sections 2.2.2 and 8.2).

Comparison of computational complexity and therefore trackability (ability to adjust the original design formulation based on the final and numerical design results) of all feedback control design techniques (Section 9.3).

Emphasis on the distinct advantages of high performance/robustness control design using eigenstructure assignment techniques over the techniques for the direct design of loop transfer functions (Chapters 2, 3, 8, 9).

The concept of adaptive control and its application in failure accommodation and control (Section 10.2).

The first five of the above results are actual design results. The last seven are new theoretical results and concepts that have enabled the establishment of the first five results. In other words, the main new result (result 1, the full realization of robustness properties of state/generalized state feedback control) is enabled by some significant and fundamental developments (such as results 6 to 8), and is validated by the distinct effectiveness of state/generalized state feedback control (results 2 to 3 and 9 to 11).

This book also addresses the computational reliability of its analysis and design algorithms. This is because practical control problems usually require a large amount of computation, and unreliable computation can yield totally unreliable results. Every effort has been made to use reliable computational methods in design algorithms, such as the computation of Hessenberg form (instead of the canonical form) and orthogonal matrix operation (instead of elementary matrix operation).

As a result, the computation required in this book is slightly more complicated, but the more reliable results thus obtained make the effort worthwhile. It should be noted that the computation of polynomials required by the classical control theory is usually unreliable. The development of computational software has also eased considerably the complexity of computation. Each design procedure is presented in algorithm form, and each step of these algorithms can be implemented directly by the existing computational software.

This book will be useful to control system designers and researchers. Although a solid background in basic linear algebra is required, it requires remarkably less mathematical sophistication than other books similar in scope. This book can also be used as a textbook for students who have had a first course (preferably including state space theory) in control systems. Multi-input and multi-output systems are discussed throughout. However, readers will find that the results have been substantially simplified to be quite easily understandable, and that the results have been well unified with the single-input and single-output system results. In addition, this book is comprehensive and self-contained, with every topic introduced at the most basic level. Thus, it could also be used by honor program students with a background in signals and systems only.

An overview of each chapter follows. Chapter 1 introduces basic system models and properties. Chapter 2 analyzes the performance and sensitivity of a single overall system. Chapter 3 describes the critical role of loop transfer functions on the sensitivity of feedback systems, including the observer feedback systems. Chapter 4 proposes the new design approach and analyzes its advantages. Chapter 5 presents the solution of a basic matrix equation. This solution is used throughout the remaining chapters (except Chapter 9). Chapter 6 presents the design of the dynamic part of the observer such that for any state feedback control signal generated by this observer, the loop transfer function of this control is also fully realized. Chapter 7 presents the design of the function observer, which generates an arbitrarily given state feedback control signal, with minimized observer order. Chapter 8 presents the eigenvalue/vector assignment control design methods. Chapter 9 introduces the linear-quadratic optimal control design methods. Both designs of Chapters 8 and 9 will determine the output part of the observer of Chapter 6, as well as the "target" closed-loop system loop transfer function. Comparison of various designs reveals two distinct advantages of eigenstructure assignment design. Chapter 10 deals with the design of general failure detection, isolation, and (adaptive) accommodation compensator that is capable of considering system model uncertainty and measurement noise. This compensator has the compatible structure of—and can be implemented in coordination with—the normal (free of major failure) robust control compensator of this book. There is a set of simple exercises at the end of each chapter.

To make the book self-contained, Appendix A provides a simple introduction to the relevant mathematical background material. Appendix B lists the mathematical models of eight real-world systems for synthesized design practice.

I would like to thank everyone who helped me, especially during my student years. I also thank my former student Reza Shahriar, who assisted with some of the computer graphics.

Chia-Chi Tsui

Author

Chia-Chi Tsui was born in 1953, Shanghai, China. He worked at a state farm in northeast China between 1969 and 1975. He received Bachelor of Computer Science degree from Concordia University, Montreal, Canada in 1979. He received his Masters and Ph.D. degrees from the Electrical Engineering Department, State University of New York at Stony Brook in 1980 and 1983, respectively. He has held teaching positions at Northeastern University, City University of New York Staten Island College, and DeVry University New York. His research interest is linear feedback control system design, including robust control design.

1

System Mathematical Models

Unlike other engineering specialties whose subject of study is a specific engineering system such as an engine system or an airborne system, the control systems theory deals only with a general mathematical model of engineering systems. This chapter introduces two basic mathematical models and some basic system properties revealed by those models. There are four sections in this chapter.

Section 1.1 introduces the state space model and the transfer function model of linear time-invariant multi-input and multi-output systems and the derivation of the two models.

Section 1.2 describes the eigenstructure decomposition of the state space model, where the state matrix of the model is in Jordan form.

Section 1.3 introduces two basic system properties – controllability and observability. These two properties can be simply and clearly described based on the eigenstructure decomposition of the state space model. The corresponding canonical form of state space model realization of a transfer function model in a polynomial matrix fraction description form can also be simply and clearly derived.

Section 1.4 introduces two basic system parameters – system poles and zeros.

1.1 Two Kinds of Mathematical Models

This book deals with linear time-invariant systems, which have also been the main subject of the control systems theory. A linear time-invariant system can be represented by two kinds of mathematical models – the state space model and the transfer function model. The control theory that is based on the state space model is called the "state space control theory" or the "modern control theory", and the control theory that is based on the transfer function model is called the "classical control theory".

We will first introduce the state space model and its derivation.

A state space model is formed by a set of first-order linear differential equations with constant coefficients (1.1a) and a set of linear equations (1.1b).

DOI: 10.1201/9781003259572-1

$$dx(t)/dt = A\mathbf{x}(t) + B\mathbf{u}(t) \tag{1.1a}$$

$$\mathbf{y}(t) = C\mathbf{x}(t) + D\mathbf{u}(t) \tag{1.1b}$$

where

$\mathbf{x}(t) = [x_1(t), \ldots, x_n(t)]'$ is the system state vector (the symbol "′" stands for transpose),

$x_i(t)$, $i = 1, \ldots, n$, represents the system state variables,

$\mathbf{u}(t) = [u_1(t), \ldots, u_p(t)]'$ is the system input,

$\mathbf{y}(t) = [y_1(t), \ldots, y_m(t)]'$ is the system output,

and the system matrices (A, B, C, D) are real and constant and have dimensions $n \times n$, $n \times p$, $m \times n$, and $m \times p$ respectively.

In the above model (1.1), Eq. (1.1a) is called the "dynamic equation" and it describes the "dynamic part" of the system and how the initial system state $\mathbf{x}(0)$ and system input $\mathbf{u}(t)$ will determine the system state. Hence, matrix A is called the "state matrix" of the system. Equation (1.1b) describes how the system state $\mathbf{x}(t)$ and system input $\mathbf{u}(t)$ will instantly determine the system output $\mathbf{y}(t)$. This is the "output part" of the system and is static and memoryless as compared with the dynamic part of the system (1.1a).

From the Definition of (1.1), parameters p and m represent the number of system inputs and outputs, respectively. If $p > 1$, then we call the corresponding system "multi-input". If $m > 1$, then we call the corresponding system "multi-output". A multi-input and multi-output system is called a "MIMO" system. A single-input and single-output system is called a "SISO" system.

From (1.1), the physical meaning of system state $\mathbf{x}(t)$ is that it can be used to describe completely the energy distribution of the system at time t, especially at $t = 0$.

For example, in electrical circuit systems with linear time-invariant circuit elements (inductors, resistors, and capacitors), the system state is formed by all independent capacitor voltages and inductor currents. Thus, its initial condition $\mathbf{x}(0)$ can completely describe the initial electrical charges and initial magnetic fluxes stored in the circuit system.

Another example is in linear motion mechanical systems with time-invariant elements (springs, dampers, and masses), in which the system state is formed by all independent mass velocities and spring forces. Thus, its initial state $\mathbf{x}(0)$ completely describes the initial dynamic energy and initial potential energy stored in the mechanical system.

Because of this reason, the number (n) of state variables is also the number of the system's independent energy storage devices.

Example 1.1

The following electrical circuit system is a linear time-invariant system (Figure 1.1).

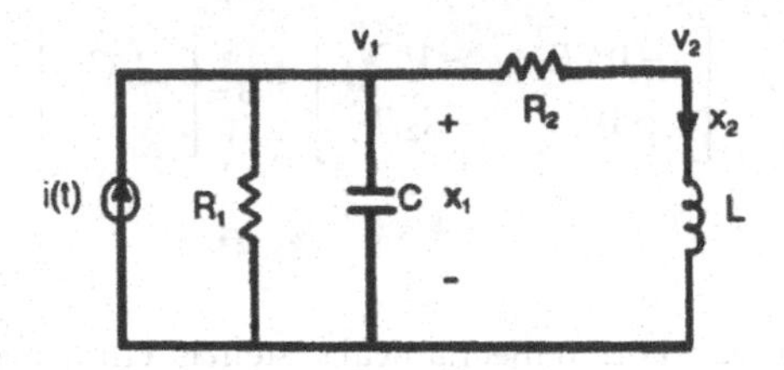

FIGURE 1.1
A linear time-invariant circuit system.

Let $v_1(t)$ and $v_2(t)$ be the node voltages of the circuit and the capacitor voltage and inductor current be the two system states $x_1(t)$ and $x_2(t)$, respectively, we have

$$v_1(t) = x_1(t), \text{ and } v_2(t) = x_1(t) - R_2 x_2(t) \tag{1.2}$$

In other words, all node voltages and branch currents can be expressed in terms of system states and inputs. Hence, the output part of the system (1.1b) can be directly derived. For example, if the output $\mathbf{y}(t)$ is designated as $[v_1(t)\ v_2(t)]'$, then from (1.2),

$$\mathbf{y}(t) = \begin{bmatrix} y_1(t) \\ y_2(t) \end{bmatrix} = \begin{bmatrix} 1 & 0 \\ 1 & -R_2 \end{bmatrix} \begin{bmatrix} x_1(t) \\ x_2(t) \end{bmatrix} + 0 = C\mathbf{x}(t) + D\mathbf{u}(t)$$

The dynamic equation (1.1a) of this circuit system can also be derived from standard circuit analysis. Applying Kirchoff's current law at each node of the circuit, we have

$$i(t) = C\dot{v}_1(t) + \frac{v_1(t)}{R_1} + \frac{[v_1(t) - v_2(t)]}{R_2} \tag{1.3a}$$

$$0 = \frac{[v_2(t) - v_1(t)]}{R_2} + \frac{\left[\int v_2(t)\,dt\right]}{L} \tag{1.3b}$$

By substituting (1.2) into (1.3) and after simple mathematical manipulation, we obtain the form of (1.1a) as

$$dx_1(t)/dt = (-1/CR_1)x_1(t) + (-1/C)x_2(t) + (1/C)i(t)$$

$$dx_2(t)/dt = (1/L)x_1(t) + (-R_2/L)x_2(t)$$

Thus, compared to (1.1a), the system matrices are

$$A = \begin{bmatrix} -1/CR_1 & -1/C \\ 1/L & -R_2/L \end{bmatrix}, \quad B = \begin{bmatrix} 1/C \\ 0 \end{bmatrix}$$

Example 1.2

The following linear motion mechanical system is a linear time-invariant system (Figure 1.2).

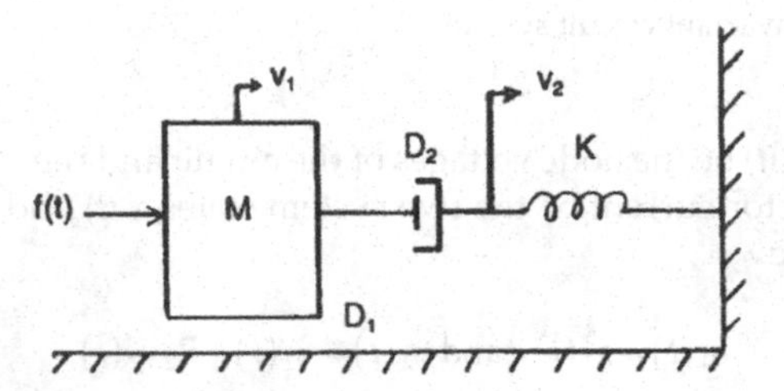

FIGURE 1.2
A linear time-invariant mechanical system.

Let $v_1(t)$ and $v_2(t)$ be the node velocities in the system and the mass velocity and spring force be the system states $x_1(t)$ and $x_2(t)$, respectively, then

$$v_1(t) = x_1(t), \text{ and } v_2(t) = x_1(t) - D_2^{-1} x_2(t) \tag{1.4}$$

In other words, all velocities and forces within the mechanical system can be expressed in terms of system states and the applied input force. Hence the system's output part (1.1b) can be directly derived. For example, if the output $\mathbf{y}(t)$ is designated as $[v_1(t)\ v_2(t)]'$, then from (1.4),

$$\mathbf{y}(t) = \begin{bmatrix} v_1(t) \\ v_2(t) \end{bmatrix} = \begin{bmatrix} 1 & 0 \\ 1 & -D_2^{-1} \end{bmatrix} \begin{bmatrix} x_1(t) \\ x_2(t) \end{bmatrix} + 0 = C\mathbf{x}(t) + D\mathbf{u}(t)$$

The dynamic equation of this mechanical system can also be derived using the standard dynamic analysis. By balancing the forces at each node of this system, we have

$$f(t) = M\,dv_1(t)/dt + D_1 v_1(t) + D_2\left[v_1(t) - v_2(t)\right] \tag{1.5a}$$

$$0 = D_2\left[v_2(t) - v_1(t)\right] + K\left[\int v_2(t)dt\right] \tag{1.5b}$$

After substitution of (1.4) into (1.5) and simple mathematical manipulation, we obtain the following form of (1.1a)

$$dx_1(t)/dt = (-D_1/M)x_1(t) + (-1/M)x_2(t) + (1/M)f(t)$$

$$dx_2(t)/dt = K\,x_1(t) + (-K/D_2)x_2(t)$$

Thus, compared to (1.1.a), the system matrices are

$$A = \begin{bmatrix} -D_1/M & -1/M \\ K & -K/D_2 \end{bmatrix}, \quad B = \begin{bmatrix} 1/M \\ 0 \end{bmatrix}$$

In the above two systems, the forms and derivations of the state space model are very similar to each other. Different physical systems that are similar in terms of mathematical models are termed as "analogs". This property enables the simulation of the behavior of one physical system (such as a mechanical system) by comparison with an analog but different physical system (such as a circuit system) or by comparison with the numerical solution of the mathematical model of that system. We call the former "analog simulation" and the latter "digital simulation".

The use of analogs can be extended to a wide range of linear time-invariant physical systems, such as rotational mechanical systems, thermodynamic systems, and fluid dynamic systems. Therefore, although the mathematical models and the control theory, which is based on those models, are abstract, they can have very general applications.

A linear time-invariant system can have another kind of mathematical model – the transfer function model, which can be derived directly from its corresponding state space model.

Taking the Laplace transforms on both sides of (1.1) and after simple algebraic manipulation,

$$X(s) = (sI - A)^{-1}\mathbf{x}(0) + (sI - A)^{-1}BU(s) \tag{1.6a}$$

$$Y(s) = CX(s) + DU(s) \tag{1.6b}$$

where $X(s)$, $U(s)$, and $Y(s)$ are the Laplace transforms of $\mathbf{x}(t)$, $\mathbf{u}(t)$, and $\mathbf{y}(t)$, respectively, and I stands for an n-dimensional identity matrix such that $s\,I\,X(s) = s\,X(s)$.

By substituting (1.6a) into (1.6b), we have

$$Y(s) = \underbrace{C(sI - A)^{-1}\mathbf{x}(0)}_{\substack{\text{Zero input response} \\ Yzi(s)}} + \underbrace{\left[C(sI - A)^{-1}B + D\right]U(s)}_{\substack{\text{Zero state response} \\ Yzs(s)}} \tag{1.6c}$$

From the superposition principle of linear systems and Laplace transforms, among the two terms of Eqs. (1.6a) and (1.6c), the first term is due to the system's initial state $\mathbf{x}(0)$ only, and the second term is due to the system input $U(s)$ only. For example, in (1.6c), the system output (also called the system "response") $Y(s)$ equals the first term if the system input $U(s)$ is zero, we therefore define the first term of (1.6c) as "zero-input response $Yzi(s)$". Similarly, $Y(s)$ equals the second term of (1.6c) if the system initial state $\mathbf{x}(0)$ is zero, and we therefore define the second term of (1.6c) as "zero-state response $Yzs(s)$".

The transfer function model $G(s)$ of a system is defined from the zero-state response of the system as

$$Yzs(s) = G(s)U(s) \tag{1.7}$$

Therefore from (1.6c),

$$G(s) = C(sI - A)^{-1}B + D \tag{1.8}$$

The definition of $G(s)$ shows that the operation of (1.8) compressed and simplified the operation of (1.1) and (1.6a and b), which involves explicitly the system state $\mathbf{x}(t)$ and $X(s)$. Furthermore, $G(s)$ reflects only the relationship between system output $Y(s)$ and input $U(s)$, but not the initial state $\mathbf{x}(0)$, even though the response to $\mathbf{x}(0)$, the zero-input response, is as important as the zero-state response.

Example 1.3

Consider the above RC circuit system (a) and mechanical system (b) with mass M and friction D (Figure 1.3).

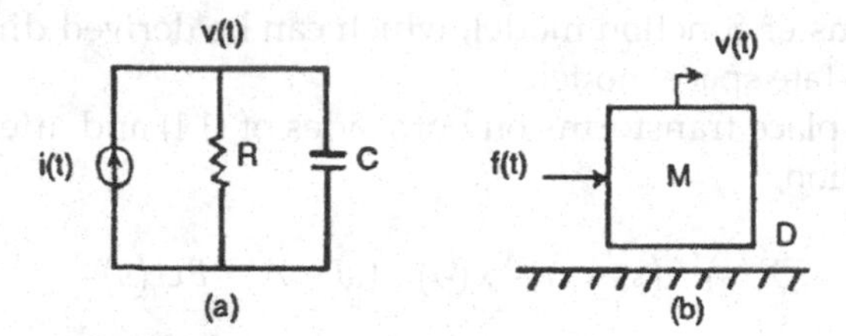

FIGURE 1.3
First-order circuit and mechanical systems.

Upon balancing the currents of system (a) and the forces of system (b), we have

$$i(t) = C\ dv(t)/dt + [v(t) - 0]/R$$

and

$$f(t) = M dv(t)/dt + D[v(t) - 0]$$

Compared to (1.1a), the system matrices ($A \equiv \lambda$, B) equals ($-1/RC$, $1/C$) and ($-D/M$, $1/M$) for these two systems, respectively. We let the system output equal the system state in this example, so that $C = I$ and Eqs. (1.6a) and (1.6c) are the same, in this example.

Taking Laplace transforms on these two equations and after simple manipulation, we have the form of (1.6a) or (1.6c) as

$$V(s) = [1/(s - \lambda)]v(0) + [B/(s - \lambda)]U(s)$$

where $V(s)$ and $U(s)$ are the Laplace transforms of $v(t)$ and the system input signal [$i(t)$ or $f(t)$], respectively.

Letting $U(s) = F/s$ (or step function) and taking the inverse Laplace transforms of the above equation, we have, for $t \geq 0$,

$$v(t) = e^{\lambda t}v(0) + F(-B/\lambda)\left[1 - e^{\lambda t}\right]$$

$$\equiv v_{zi}(t) + v_{zs}(t)$$

The waveforms of the two terms of $v(t)$, $v_{zi}(t)$, and $v_{zs}(t)$, are shown in Figure 1.4.

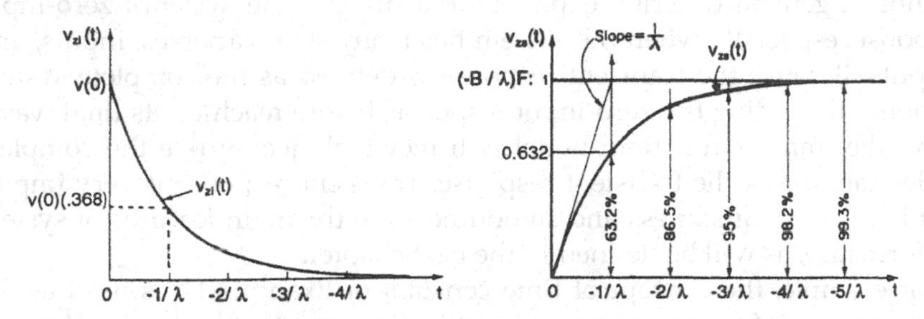

FIGURE 1.4
Waveforms of zero-input response and zero-state response of a first-order system.

The first waveform of Figure 1.4 shows the zero-input response. It starts at its initial condition $v(0)$ and then decays exponentially to zero with a time constant $|1/\lambda|$. In other words, $v_{zi}(t)$ decays to 36.8% of its initial value at time $t = |1/\lambda|$.

This waveform has a very clear physical meaning. In the circuit system (a), this waveform shows (when the input current is zero) how the initial capacitor charge [$=C\ v(t)$] is discharged to zero, through resistor R and with current $v(t)/R$, and with the time constant RC: the larger the capacitor C or resistor R, the slower the discharge. In the mechanical system (b), this waveform shows with zero-input force how the momentum [$= M\ v(t)$] slows down to zero, by the frictional force $Dv(t)$, and with the time constant M/D: the larger the mass M or the smaller the friction D, the longer the time for the velocity $v(t)$ to slow to 36.8% of its initial value.

The second waveform of Figure 1.4 shows the zero-state response. It starts at zero and then reaches exponentially to its final level which is specified by the input level F. This process also has a time constant $|1/{-\lambda}|$, which means that the response reaches 100% − 36.8% = 63.2% of its final level at time $t = |1/{-\lambda}|$.

This waveform also has a very clear physical meaning. In the circuit system (a), this waveform shows how the capacitor is charged from zero until $v(t) = F(-B/{-\lambda}) = F\ R$, by a constant current source $Fu(t)$. The final value of $v(t)$ equals the source side voltage $F\ R$, which means that the capacitor is fully charged. The time constant RC of this charging process means that the larger the capacitor C or the resistor R, the slower the charging process. In the mechanical system (b), this waveform shows how the mass is accelerated from zero speed to F/D by a constant force $Fu(t)$. This acceleration process has a time constant M/D, which means that the larger the mass M or the smaller the friction D (which means a higher final speed F/D), the longer the time for the mass to accelerate to 63.2% of its final speed.

Example 1.3 shows a very fitting analogy between the two different physical systems, their mathematical models, and their responses and behavior. This example also shows the importance of the zero-input response – the effects of the initial capacitor charge and the initial mass momentum.

Definitions (1.7) and (1.8) of the transfer function model $G(s)$ imply that $G(s)$ cannot in general describe explicitly and directly the system's zero-input response, especially when the system has many state variables, inputs, and outputs. Because the transient response is defined as the complete system response (including the zero-input response) before reaching its final waveform, the transfer function model will inevitably jeopardize the complete understanding of the transient response. Transient response is very important because its quickness and smoothness are the main features of system performance, as will be defined in the next chapter.

For example, the concept of time constant of Example 1.3 is used as the main measure of transient response, while time constant is closely related to the zero-input response.

In both the state space model (1.1) and the transfer function model (1.8), the system parameter D reflects only an independent and static relationship between system inputs and outputs. This relationship can be easily measured and canceled in analysis and design. For this reason, we will assume $D = 0$ in the rest of this book. Hence, the transfer function model of (1.8) becomes

$$G(s) = C(sI - A)^{-1} B \tag{1.9}$$

Finally, because of (1.6a) and (1.6c), the new transfer function model (1.9) can be decomposed into a series of two subblocks, as shown in Figure 1.5.

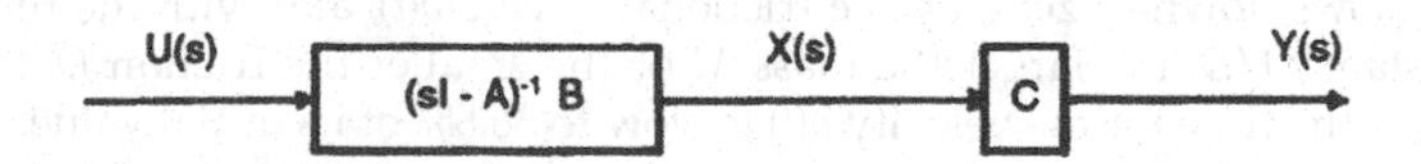

FIGURE 1.5
Decomposed block diagram representation of a system's transfer function model.

1.2 Eigenstructure Decomposition of the State Space Model

To gain a simpler yet deeper understanding of the system structure and properties, we partition the system state matrix

$$A = V\Lambda V^{-1} \equiv \begin{bmatrix} | & & | \\ V_1 : & \ldots : & V_q \\ | & & | \end{bmatrix} \begin{bmatrix} \Lambda_1 & & \\ & \ddots & \\ & & \Lambda_q \end{bmatrix} \begin{bmatrix} -T_1- \\ \vdots \\ -T_q- \end{bmatrix}$$

$$\equiv T^{-1}\Lambda T \tag{1.10a}$$

where $\Lambda = \text{diag}\{\Lambda_1, \ldots, \Lambda_q\}$ is called a "Jordan form matrix", whose diagonal matrix blocks Λ_i $(i = 1, \ldots, q)$ are called the "Jordan blocks" and are formed by the eigenvalues λ_i $(i = 1, \ldots, n)$ of matrix A according to the following rules:

$\Lambda_i = \lambda_i$, if λ_i is real and distinct.

$\Lambda_i = \begin{bmatrix} \sigma_i & \omega_i \\ -\omega_i & \sigma_i \end{bmatrix}$ if the eigenvalues are a complex conjugate pair $\sigma_i \pm j\omega_i$

$\Lambda_i = \text{diag}\{\Lambda_{i1}, \ldots, \Lambda_{iqi}\}$ if λ_i repeats n_i times, where the diagonal blocks

$$\Lambda_{i,j} = \begin{bmatrix} \lambda_i & 1 & & \\ & \lambda_i & \ddots & \\ & & \ddots & 1 \\ & & & \lambda_i \end{bmatrix} \tag{1.10b}$$

(blank entries are all 0's) and this is called a "bidiagonal form" (1.10b).

Where the sum of the dimensions of these q_i bidiagonal blocks equals n_i.

Finally, the sum of the dimensions of all q Jordan blocks equals n.

When matrix A is in the Jordan form (1.10), the corresponding state space model is said to be in the "Jordan canonical form". Any real square matrix A can have the eigenstructure decomposition as (1.10), and therefore any state space model can have its Jordan canonical form.

Because (1.10) implies $AV-V\Lambda = 0$, we call matrix V the "right eigenvector matrix" of matrix A and each column of matrix V, $\mathbf{v}_i$ $(i = 1, \ldots, n)$, the "right eigenvector" of matrix A corresponding to λ_i. Similarly, because $TA-\Lambda T = 0$, we call matrix T the "left eigenvector matrix" of matrix A and each row of matrix T, $\mathbf{t}_i$ $(i = 1, \ldots, n)$, the "left eigenvector" of matrix A corresponding to λ_i.

All but the first eigenvector of the bidiagonal blocks (1.10b) are derived based on other eigenvectors of the same block and are called the "generalized eigenvector".

From (1.10),

$$(sI-A)^{-1} = \left[V(sI-\Lambda)V^{-1}\right]^{-1} = V(sI-\Lambda)^{-1}V^{-1}$$

Therefore, from (1.9) and the inverse matrix rules,

$$G(s) = CV(sI - \Lambda)^{-1} V^{-1}B \tag{1.11a}$$

$$= \frac{CV \operatorname{adj}(sI - \Lambda)V^{-1}B}{\det(sI - \Lambda)} \tag{1.11b}$$

$$= \frac{CV \operatorname{adj}(sI - \Lambda)V^{-1}B}{(s - \lambda_1)\dots(s - \lambda_n)} \tag{1.11c}$$

where adj(.) and det(.) stand for the adjoint and the determinant of the corresponding matrix, respectively.

From (1.11c), the transfer function $G(s)$ is a rational polynomial matrix. It has an n-th order denominator polynomial which is called the "characteristic polynomial" of the system and whose n roots equal the n eigenvalues of the system state matrix A.

Compared to (1.11a) with (1.9), a new system matrix triple (Λ, $V^{-1}B$, CV) has the same transfer function as that of system matrix triple (A, B, C), provided that $A = V \Lambda V^{-1}$. We call these two state space models and their corresponding systems "similar" to each other and call the transformation between the two similar state space models "similarity transformation". This property can be extended to any system matrix triple ($Q^{-1}AQ$, $Q^{-1}B$, CQ) for a nonsingular matrix Q.

The physical meaning of similar systems can be interpreted as follows. Let $\mathbf{x}(t)$ and $\underline{\mathbf{x}}(t)$ be the state vectors of state models (A, B, C) and ($Q^{-1}AQ$, $Q^{-1}B$, CQ), respectively. Then from (1.1),

$$d\underline{\mathbf{x}}(t)/dt = Q^{-1}AQ\underline{\mathbf{x}}(t) + Q^{-1}B\mathbf{u}(t) \tag{1.12a}$$

$$\mathbf{y}(t) = CQ^{-1}\underline{\mathbf{x}}(t) + D\mathbf{u}(t) \tag{1.12b}$$

It is clear from the comparison between (1.1) and (1.12) that

$$\mathbf{x}(t) = Q\underline{\mathbf{x}}(t) \text{ and } \underline{\mathbf{x}}(t) = Q^{-1}\mathbf{x}(t) \tag{1.13}$$

From the Definitions A.3 and A.4 of Appendix A, (1.13) implies that the only difference between the state space models (1.1) and (1.12) is that their state vectors are based on different basis vector matrices I and Q, respectively.

Similarity transformation, especially when the state space model is transformed to the "Jordan canonical form" where the state matrix is in Jordan form, is a very effective and very commonly used scheme that can substantially simplify the understanding of the system, as will be shown in the next section.

1.3 System Order, Controllability, and Observability

Definition 1.1

The order n of a system equals the order of the system's characteristic polynomial. It is clear from (1.1c) and (1.11c) that a system order also equals the number of state variables of the system.

Let us discuss the situation of the existence of common factors between the transfer function's numerator and denominator polynomials. Because this denominator polynomial is the characteristic polynomial of the system, and because common factors can cancel out each other, the above situation implies that the corresponding system order is reducible. We call this kind of system "reducible". Otherwise, the system is said to be "irreducible".

The situation of reducible systems can be more explicitly described by their corresponding state space models. Definition 1.1 implies that in reducible systems, some of the system states are not involved with the system's input and output relationship $G(s)$. In other words, in reducible systems, some of the system states either cannot be influenced by any of the system inputs or cannot influence any of the system outputs. We will define these two situations separately in the following.

Definition 1.2

If there is at least one system state that cannot be influenced by any of the system inputs, then the system is uncontrollable; otherwise, the system is controllable. Among many existing criteria of controllability, the simplest is that "a system is controllable if and only if there exists no constant λ such that the rank of matrix $[\lambda I–A: B\ ]$ is less than n".

Definition 1.3

If there is at least one system state that cannot influence any of the system output, then the system is unobservable; otherwise, the system is observable. Among many existing criteria of observability, the simplest is that "a system is observable if and only if there exists no constant λ such that the rank of matrix $[\lambda I–A': C']'$ is less than n".

Because the rank of matrix $\lambda I–A$ always equals n if λ is not an eigenvalue of matrix A, the above criteria can be checked only for the n values of λ which equal the eigenvalues of matrix A.

An irreducible system must be both controllable and observable. Any uncontrollable or unobservable system is also reducible.

Up to this point, we can see a common and distinct phenomenon of linear systems – duality. For example, in linear systems, current and voltage, force and velocity, charge and flux, dynamic energy and potential energy, capacitance and inductance, and mass and spring are dual pairs. In linear algebra and linear control theory which describe linear systems, matrix columns and rows, right and left eigenvectors, inputs and outputs, and controllability and observability are also dual pairs.

The phenomenon of duality can not only help us understand linear systems more comprehensively, but also help us solve some specific analysis and design problems. For example, the determination of whether a system (A, B) is controllable can be replaced by the determination of whether a system ($A = A'$, $C = B'$) is observable instead.

Because matrix $[\lambda I-Q^{-1}AQ\colon Q^{-1}B] = Q^{-1}[(\lambda I-A)Q\colon B]$ has the same rank as that of matrix $[\lambda I-A\colon B]$, similarity transformation will not change the controllability property of the original system. Similarity transformation changes only the basis vector matrix representation of the system's state vector, and therefore cannot change the system's basic properties such as controllability. From duality, similarity transformation cannot change the observability property of the system either. It is therefore valid to determine a system's controllability and observability after similarity transformation.

The following three examples show the relative simplicity of determining the controllability and observability, when the system matrices are in special forms, in particular, the Jordan canonical form. These special forms can be computed generally by similarity transformation.

Example 1.4

Determine whether the system

$$(A,B,C)=\left(\begin{bmatrix} -1 & 0 & 0 \\ 0 & -2 & 0 \\ 0 & 0 & -3 \end{bmatrix}, \begin{bmatrix} -\mathbf{b}_1- \\ -\mathbf{b}_2- \\ -\mathbf{b}_3- \end{bmatrix}, [\mathbf{c}_1 : \mathbf{c}_2 : \mathbf{c}_3]\right)$$

is controllable and observable.

From Definition 1.2, it is clear that if any row of matrix B equals zero, say $\mathbf{b}_i = \mathbf{0}$, ($i$ = 1–3), then there exist a constant $\lambda = -i$ such that the i-th row of matrix $[\lambda I-A\colon B]$ equals zero (rank$<n$). Only when every row of matrix B is nonzero, the rank of matrix $[\lambda I-A\colon B]$ equals n, for all three eigenvalues of A ($\lambda = -1, -2, -3$). Thus, the necessary and sufficient condition for this system to be controllable is that every row of matrix B is nonzero.

Similarly, from duality, the necessary and sufficient condition for this system to be observable is that every column of matrix C is nonzero.

From (1.9), the transfer function of this system is

$$G(s) = C(sI - A)^{-1}B$$
$$= \frac{\mathbf{c}_1(s+2)(s+3)\mathbf{b}_1 + \mathbf{c}_2(s+1)(s+3)\mathbf{b}_2 + \mathbf{c}_3(s+1)(s+2)\mathbf{b}_3}{(s+1)(s+2)(s+3)}$$

It is clear that if any $\mathbf{b}_i$ or any $\mathbf{c}_i$ equals zero (i = 1–3), then there will be common factors between the numerator and denominator polynomials of $G(s)$ or a reducible system. However, a reducible transfer function $G(s)$ cannot indicate the converse: whether a row of matrix B is zero or a column of matrix C is zero, or whether the system is uncontrollable or unobservable or both. Therefore, the information provided by the transfer function model is less complete and explicit than the state space model.

Controllability and observability conditions can also be clearly revealed from the block diagram of the system in the Jordan canonical form (Figure 1.6).

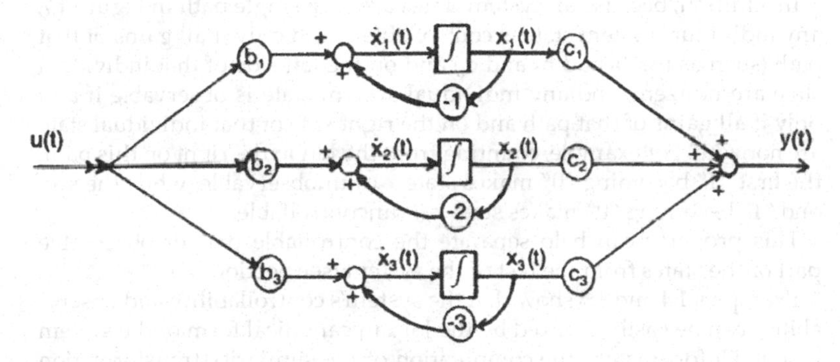

FIGURE 1.6
Block diagram of the system of Example 1.4.

Figure 1.6 shows clearly that any system state $x_i(t)$ is influenced by the system input $\mathbf{u}(t)$ if and only if the corresponding gain $\mathbf{b}_i \neq 0$, and that any state $x_i(t)$ can influence system output $\mathbf{y}(t)$ if and only if the corresponding gain $\mathbf{c}_i \neq 0$.

Example 1.5

Example 1.4 is a Jordan canonical form system with distinct and real eigenvalues. The present example studies the same system with multiple eigenvalues (see (1.10b)). Let

$$(A,B,C) = \left(\begin{bmatrix} \lambda & 1 & 0 \\ 0 & \lambda & 1 \\ 0 & 0 & \lambda \end{bmatrix}, \begin{bmatrix} -0- \\ -0- \\ -b_3- \end{bmatrix}, \begin{bmatrix} c_1 & 0 & \dots & 0 \end{bmatrix} \right)$$

The rank of matrix $[\lambda I\text{–}A: B]$ equals n if and only if $\mathbf{b}_3 \neq 0$, and the rank of matrix $[\lambda I\text{–}A': C']$ equals n if and only if $\mathbf{c}_1 \neq 0$.

In examining the block diagram of this system below (Figure 1.7), it is clear that $\mathbf{b}_3$ and $\mathbf{c}_1$ are the only links between all system states and the system's inputs and outputs, respectively. Therefore, this system is controllable if and only if $\mathbf{b}_3 \neq 0$ and observable if and only if $\mathbf{c}_1 \neq 0$.

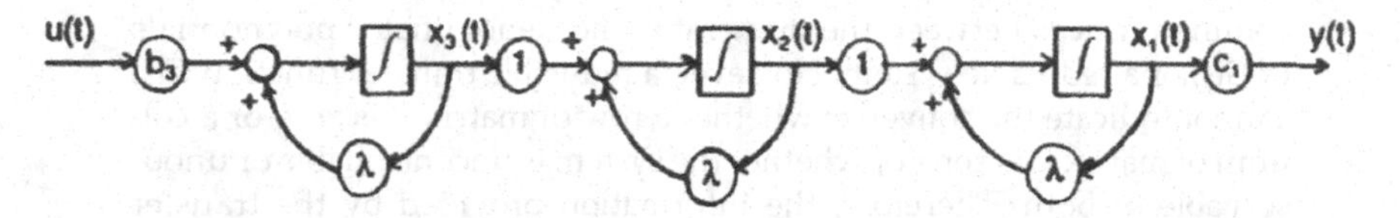

FIGURE 1.7
Block diagram of the system of Example 1.5 Jordan canonical form with multiple λ values.

In addition, because all system states are on a single path in Figure 1.7, any individual system state is controllable if and only if all gains of that path (such as the 1s and $\mathbf{b}_3$ and $\mathbf{c}_1$) and on the left side of that individual state are nonzero, and any individual system state is observable if and only if all gains of that path and on the right side of that individual state are nonzero. For example, counting from the left to the right on this path, the first "1" becoming "0" makes state $x_3(t)$ unobservable, while the second "1" becoming "0" makes state $x_1(t)$ uncontrollable.

This property can help separate the controllable part or observable part of the states from the rest of the system (see Section 5.1).

Examples 1.4 and 1.5 show that the system's controllability and observability can be easily checked by the Jordan canonical form of the system model. Unfortunately, the computation of the similarity transformation to the Jordan canonical form is difficult and is often very sensitive to the initial data variation.

On the other hand, the form of matrix A of Example 1.5 is a special case of the so-called Hessenberg form, and similarity transformation to the Hessenberg form matrix can be computed reliably. In the next two examples, we will study another special form of the Hessenberg form state space model. A very important property of that special form is that it has a one-to-one parametric match with the corresponding transfer function model (in the polynomial fraction description form).

Example 1.6

The observable canonical form of the state space model

$$(A,B,C)=\left(\begin{bmatrix} -a_1 & 1 & 0 & \dots & 0 \\ -a_2 & 0 & 1 & & \\ \vdots & \vdots & & & 0 \\ -a_{n-1} & 0 & & & 1 \\ -a_n & 0 & & \dots & 0 \end{bmatrix}, \begin{bmatrix} -\mathbf{b}_1- \\ -\mathbf{b}_2- \\ \vdots \\ -\mathbf{b}_{n-1}- \\ -\mathbf{b}_n- \end{bmatrix}, [c_1,0,\dots,0]\right) \tag{1.14}$$

This is a single-output (although it can be multi-input) system. The above system matrices are said to be in the "observable Hessenberg form". In addition, the above state matrix A of (1.14) is represented in the "companion form" or the "canonical form". Let us examine the block diagram (Figure 1.8) of this system.

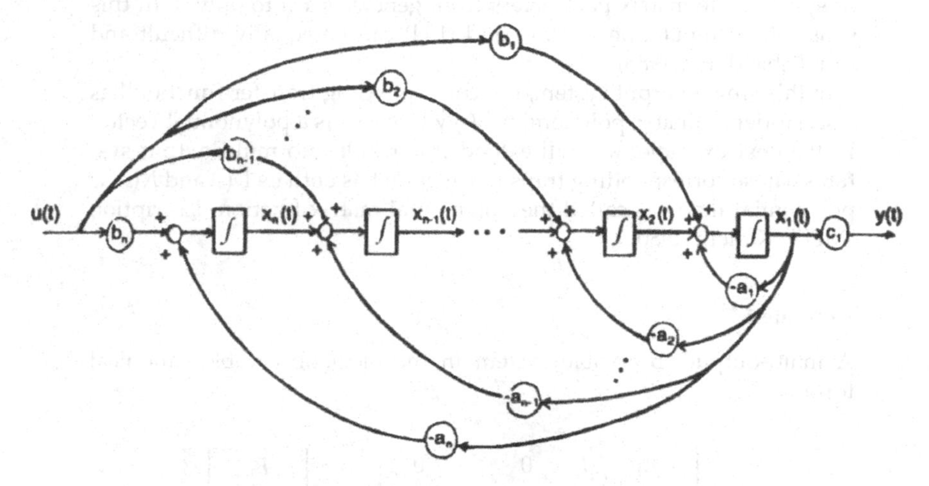

FIGURE 1.8
Block diagram of a single-output system in an observable canonical form.

Figure 1.8 shows that all system states can influence the system output (observable) if and only if parameter $c_1 \neq 0$. Furthermore, if any "1" entry of matrix A becomes "0", then all states subsequent to this "1" will become unobservable. For example, if the "1" entry at the $(n-1)$th row of matrix A, or the "1" after the first integration of the path of Figure 1.8, becomes 0, then state $x_n(t)$ becomes unobservable; if the "1" entry at the first row of matrix A, or the "1" after the last integration of the path of Figure 1.8, becomes 0, then all states from $x_2(t)$ to $x_n(t)$ become unobservable.

It has been proven that any single-output observable system is similar to the form (1.14) (Luenberger, 1967; Chen, 1984).

From duality, the system model (A', C', B') from (1.14) is said to be in the "controllable canonical form" and is controllable if and only if $c_1 \neq 0$. Any single-input controllable system is similar to the system model (A', C', B') from (1.14).

Controllable and observable canonical form state space models have a one-to-one parametric match with their corresponding transfer function $G(s)$. By substituting (1.14) into (1.9), we get

$$G(s) = C(sI - A)^{-1} B$$

$$= \frac{c_1\left(\mathbf{b}_1 s^{n-1} + \mathbf{b}_2 s^{n-2} + \cdots + \mathbf{b}_{n-1} s + \mathbf{b}_n\right)}{s^n + a_1 s^{n-1} + a_2 s^{n-2} + \cdots + a_{n-1} s + a_n} \equiv \frac{N(s)}{D(s)} \tag{1.15}$$

All unknown parameters of (1.15) match with that of (1.14). Furthermore, all n unknown coefficients of characteristic polynomial $D(s)$ of (1.15) match with the n unknown parameters of matrix A of (1.14). For this reason, we also call all canonical forms (Jordan, observable, or controllable) of the state space model the "minimal parameter" model.

The computation of the similarity transformation from a general state space model to canonical forms (1.10) and (1.14) implies the compression of system state matrix parameters from general n x n to only n. In this sense, the computation of (1.10) and (1.14) can be equally difficult and unreliable (Laub, 1985).

In this single-output system, the corresponding transfer function has a scalar denominator polynomial $D(s)$ while $N(s)$ is a polynomial vector. In the next example, we will extend this result into multi-output systems whose corresponding transfer function has both its $D(s)$ and $N(s)$ as polynomial matrices called the polynomial matrix fraction description (MFD) (Kaileth, 1980).

Example 1.7

A multi-output observable system in the block-observable canonical form:

$$A = \begin{bmatrix} A_1 & I_2 & 0 & \dots & 0 \\ A_2 & 0 & I_3 & & \vdots \\ \vdots & \vdots & & & 0 \\ A_{v-1} & 0 & & & I_v \\ A_v & 0 & & \dots & 0 \end{bmatrix}, \quad B = \begin{bmatrix} B_1 \\ B_2 \\ : \\ B_{v-1} \\ B_v \end{bmatrix}$$

$$C = C_1 [I_1 : 0 \;\dots\;\dots\;\dots\;\dots\; 0] \tag{1.16}$$

where the matrix blocks I_i $(i = 1, \dots, \nu)$ have dimensions $m_{i-1} \times m_i$ $(m_0 = m)$ and equal an m_{i-1}-dimensional identity matrix with the $m_{i-1}-m_i$ columns removed. Here, $m_1 + \dots + m_\nu = n$. C_1 is a lower triangular matrix.

For example, if $m_{i-1} = 3$, then all seven possible corresponding I_i matrix blocks are:

$$\begin{bmatrix} 1 & 0 & 0 \\ 0 & 1 & 0 \\ 0 & 0 & 1 \end{bmatrix}, \begin{bmatrix} 1 & 0 \\ 0 & 1 \\ 0 & 0 \end{bmatrix}, \begin{bmatrix} 1 & 0 \\ 0 & 0 \\ 0 & 1 \end{bmatrix},$$

$$\begin{bmatrix} 0 & 0 \\ 1 & 0 \\ 0 & 1 \end{bmatrix}, \begin{bmatrix} 1 \\ 0 \\ 0 \end{bmatrix}, \begin{bmatrix} 0 \\ 1 \\ 0 \end{bmatrix}, \text{and} \begin{bmatrix} 0 \\ 0 \\ 1 \end{bmatrix} \tag{1.17}$$

Without loss of generality, we assume $m_1 = m$ (all m outputs are linearly independent (Chen, 1984)). Hence I_1 will be an m-dimensional identity matrix. The m columns of I_1 will be removed gradually at subsequent matrix blocks I_i ($i = 2, 3, \ldots$). Once the j-th column of I_i is removed, this column and its corresponding row will also be removed at subsequent matrix blocks $I_{i+1}, \ldots$. We can therefore assign a constant parameter $\nu_j = i$.

From Example 1.6, the removal of the j-th column indicates that the j-th output is no longer influenced by any subsequent system states. In other words, parameter ν_j ($j = 1, \ldots, m$) equals the number of observable states that are associated with the j-th output and is called the j-th "observability index". It is apparent from (1.16) that the largest possible value of the m observability indices is ν.

It is proven that all observable systems are similar to (1.16) (Luenberger, 1967; Chen, 1984). Form (1.16) is called the "block-observable canonical form". The block diagram of (1.16) is shown in Figure 1.9.

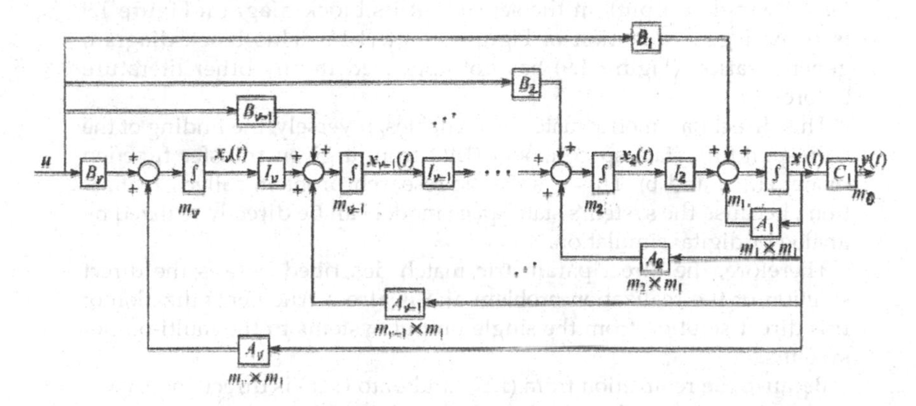

FIGURE 1.9
Block diagram of a multi-output system in a block-observable canonical form.

To match all unknown parameters of (1.16) with that of the MFD of

$$G(s) = C_1 D(s)^{-1} N(s) \tag{1.17a}$$

or into that of the $m \times m$ dimensional polynomial matrix $D(s)$ and the $m \times p$ dimensional polynomial matrix $N(s)$, we need to perform the following two simple preliminary steps.

Step 1: Fill $m - m_i$ zero rows into each matrix block A_i and B_i ($i = 1, \ldots, \nu$). The rows will be filled at the positions corresponding to all the removed columns of matrix block I_i. For example, if $m = 3$ and for the last six of the seven possible forms of I_i ($i = 2, \ldots, 6$) of (1.17), the zero rows will be filled at the third, second, first, second and third, first and third, and first and second positions respectively. At the end of this step, all matrix blocks of A_i and B_i will become $m \times m$ and $m \times p$ dimensional, respectively, and will be denoted as $\underline{A}_i$ and $\underline{B}_i$ ($i = 1, \ldots, \nu$), respectively.

Step 2: Form matrices $[I\text{: } -\underline{A}_1\text{: } -\underline{A}_2\text{: } \ldots\text{: } -\underline{A}\nu]$ and $[\underline{B}_1\text{: } \ldots\text{: } \underline{B}\nu]$. Then shift to the right each row (say the j-th row), of these two matrices, by $m(\nu - \nu_j)$ positions and $p(\nu - \nu_j)$ positions, respectively, $j = 1, \ldots, m$. Denote the resulting two matrices $[I_0\text{: } A_1\text{: } A_2\text{: } \ldots\text{: } A_\nu]$ and $[B_1\text{: } \ldots\text{: } B_\nu]$, respectively (for this procedure only).

Now to match all parameters of $D(s)$ and $N(s)$ of (1.17a):

$$D(s) = I_0 s^{\nu} + A_1 s^{\nu-1} + \cdots + A_{\nu-1} s + A_{\nu},$$

$$\text{and } N(s) = B_1 s^{\nu-1} + B_2 s^{\nu-2} + \cdots + B_{\nu-1} s + B_{\nu} \qquad (1.17b)$$

It can be verified that by substituting (1.16) into (1.9), the result matches that of (1.17a and b) (Tsui and Chen, 1983a). This parametric matching does not change, add, or eliminate any parameter of (1.16) and is therefore direct. Please try numerical Exercises 1.4–1.8.

Equation (1.16) ($m \geq 1$, multi-output) is truly a generalization of (1.14) ($m = 1$, single-output), in the sense that its block diagram Figure 1.9 is truly a generalization of Figure 1.8 of (1.14). This block diagram generalization (Figure 1.9) has not appeared in any other literature before.

This direct parametric match also enables, reversely, the finding of the corresponding state space model (1.16) from a given transfer function model (1.17a and b). This is a classic research problem called "realization", because the system's state space model can be directly realized by analog or digital simulation.

Therefore, the direct parametric match described here is the direct solution of the realization problem and is also a true generalization of this direct solution from the single-output systems to the multi-output systems.

Because the realization from (1.17a and b) to (1.16) is direct and easy, a transfer function (1.17a and b) based analysis and design result can easily find its counterpart in the state space theory. On the other hand, the computation of similarity transformation of (1.16) from a general state space model (1.1) is very difficult and unreliable as discussed in Example 1.6. Thus the analysis and design result based on the state space model (1.1) cannot find easily its counterpart in the classical control theory. This is another clear advantage of the modern control theory over the classical control theory.

The analysis and design of this book are for the controllable and observable part (or the irreducible part) of systems only, although the determination and separation of this part of a system from the rest is also discussed in Example 1.6 and in Section 5.1.

1.4 System Poles and Zeros

Definition 1.4

A system pole is a constant λ such that $G(s = \lambda) = \infty$. From (1.10), a system pole is a root of the system's characteristic polynomial and is also an eigenvalue of the system's state matrix A. Thus, the number of poles of an irreducible system is n.

Definition 1.5

For SISO irreducible systems, a system zero is a finite constant z such that $G(s = z) = 0$. From (1.11), a system zero is a root of the numerator polynomial $CV\,\mathrm{adj}(sI-\Lambda)\,V^{-1}B$.

For MIMO irreducible systems, $CV\,\mathrm{adj}(sI-\Lambda)\,V^{-1}B$ is a polynomial matrix, so the definition of system zeros is more complicated (Rosenbrock, 1973). We define any finite constant z such that $G(s = z) = 0$ a "blocking zero". A system with blocking zero z has zero response to the input $\mathbf{u}_0 e^{zt}$ for any $\mathbf{u}_0$.

We also define any finite constant z such that the rank of $G(s = z)$ is less than $\min\{m, p\}$ as a "transmission zero". A system with transmission zero z and with more outputs than inputs ($m > p$) has zero response to the input $\mathbf{u}_0 e^{zt}$ for at least one $\mathbf{u}_0$ such that $G(s = z)\mathbf{u}_0 = 0$.

Therefore, a blocking zero is a special case of transmission zero in MIMO systems. There is no difference between blocking zeros and transmission zeros in SISO systems.

Example 1.8

Let the transfer function of a system of three outputs and two inputs be

$$G(s)=\begin{bmatrix} 0 & (s+1)/(s^2+1) \\ s(s+1)/(s^2+1) & (s+1)(s+2)/(s^2+2s+3) \\ s(s+1)(s+2)/(s^4+2) & (s+1)(s+2)/(s^2+2s+2) \end{bmatrix}$$

From Definition 1.5, this system has a blocking zero –1 and two transmission zeros –1 and 0. But –2 is not a transmission zero.

Although transmission zero is defined mainly from the transfer function model $G(s)$, there is a simple and clear relationship between transmission zero and the general state space model (A, B, C, D) (Chen, 1984). Because

$$\begin{bmatrix} I & 0 \\ C(zI-A)^{-1} & -I \end{bmatrix}\begin{bmatrix} zI-A & B \\ C & -D \end{bmatrix}=\begin{bmatrix} zI-A & B \\ 0 & G(s=z) \end{bmatrix}$$

Therefore,

$$\mathrm{rank}[S] \equiv \mathrm{rank}\begin{bmatrix} zI-A & B \\ C & -D \end{bmatrix} = \mathrm{rank}\begin{bmatrix} zI-A & B \\ 0 & G(s=z) \end{bmatrix}$$

$$= \mathrm{rank}[\,zI-A] + \mathrm{rank}[G(s=z)] = n + \min\{m,p\} \tag{1.18}$$

In other words, transmission zero z must make the rank of matrix S of (1.18) (which is formed by the state space model parameters (A, B, C, D)) less than $n+\min\{m, p\}$.

The above definition is based on the assumption of irreducible systems, so that transmission zero z cannot be a system pole, so that rank $[zI - A]$ is always n.

Example 1.8 shows that when a system has different numbers of outputs and inputs ($m \neq p$), its number of transmission zeros is usually much less than $n-m$. However, when a system has the same number of outputs and inputs, its number of transmission zeros is generically $n-m$. Furthermore, if such a system satisfies the condition that the matrix CB is nonsingular (rank $= m = p$), then that system always has $n-m$ transmission zeros. These properties were derived based on the determinant of matrix S of (1.18) (Davison and Wang, 1974).

The property of transmission zeros is as follows. Suppose there are r transmission zeros of the system (A, B, C). Then as K approaches infinity, among the n eigenvalues of matrix $A-BKC$ for any nonsingular matrix K, r of them will approach each of the r transmission zeros, and the rest $n-r$ of them will approach infinity (Davison, 1978).

Another property of transmission zeros is that when a system $G(s)$ is connected with a dynamic feedback compensator system $H(s)$, the set of transmission zeros of the overall feedback system equals the union of the transmission zeros of $G(s)$ and the poles of $H(s)$ (Patel, 1978).

In addition, we need to assign all stable transmission zeros of $G(s)$ to match an equal number of poles of the feedback compensator $H(s)$, if that $H(s)$ is to realize (including the loop transfer function of) a generalized state feedback control (see Conclusion 6.1). Hence it is important to compute accurately the transmission zeros of $G(s)$.

There are several methods for computing transmission zeros (Davison and Wang, 1974; Davison 1976, 1978; Kouvaritakis and Macfarlane, 1976; MacFarlane and Karcaniar, 1976; Sinswat et al., 1976). The QZ method of (Laub and Moore, 1978) adapts a numerically stable algorithm (Moler and Stewart, 1973) for computing the generalized eigenvalues and computes all finite values z such that there exists an $n+p$ dimensional vector $\mathbf{w}$ such that $S\mathbf{w} = 0$, where matrix S is defined in (1.18).

We have discussed the properties of system zeros. The properties of system poles will be discussed in the next chapter, which shows that the system poles are the most important parameters in determining the performance of that system.

Exercises

1.1 For a linear time-invariant circuit system shown in Figure 1.10,

   a. Let the currents of the two resistors be the two outputs of this system, respectively. Find the state space model (1.1) of this system.

   b. Derive the transfer function model (1.9) of this system.

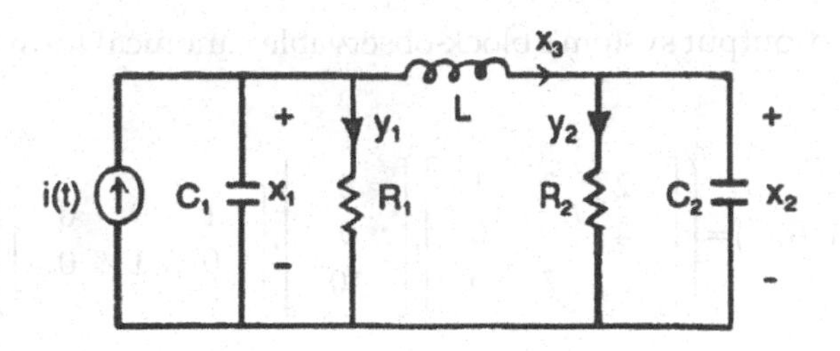

FIGURE 1.10
A linear time-invariant circuit system.

c. Plot a linear motion mechanical system which is analogous to this circuit system. Indicate all signals and elements of the mechanical system in terms of the signals and elements of the circuit system. Hint: Figure 1.11 below.

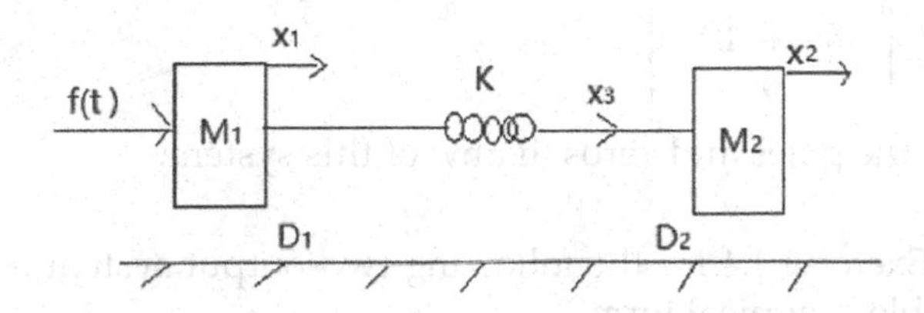

FIGURE 1.11
A linear time-invariant mechanical system.

1.2 For a linear time-invariant mechanical system shown in Figure 1.11,

a. Let the forces (same direction as force x_3) of the two dampers be the two outputs of this system, respectively. Find the state space model (1.1) of this system.

b. Derive the transfer function model (1.9) of this system.

c. Plot the circuit system which is analogous to this mechanical system. Indicate all signals and elements of the circuit system in terms of the signals and elements of the mechanical system. Hint: Figure 1.10 above.

1.3 Let a controllable canonical form state space model be the dual of (A, B, C) of Example 1.6.

$$(\underline{A}, \underline{B}, \underline{C}) = \left(A', C' = [c_1,\ 0 \ldots 0]', B' = \left[\mathbf{b}_1' : \ \ldots \ : \mathbf{b}_n'\right]\right)$$

a. Plot the block diagram similar to Figure 1.8.

b. Prove that this system is controllable if and only if $c_1 \neq 0$.

c. Prove that the transfer function of this system equals the transpose of that of (1.14).

1.4 Let a two-output system's block-observable canonical form model be

$$(A,B,C)=\left(\begin{bmatrix} 2 & 3 & 1 \\ 4 & 5 & 0 \\ 6 & 7 & 0 \end{bmatrix},\begin{bmatrix} 8 \\ 9 \\ 10 \end{bmatrix},\begin{bmatrix} 1 & 0 & 0 \\ 0 & 1 & 0 \end{bmatrix}\right)$$

a. From the description of Example 1.7, find the observability indices ν_i, $i = 1, 2$.

b. Following the two-step procedure of Example 1.7, derive the $D(s)$ and $N(s)$ of this system's polynomial MFD transfer function (1.17a and b).

Answer: $\nu_1 = 2$, $\nu_2 = 1$, $D(s) = \begin{bmatrix} s^2-2s-6 & : & -3s-7 \\ -4 & : & s-5 \end{bmatrix}$, and $N(s) = \begin{bmatrix} 8s + 10 \\ 9 \end{bmatrix}$.

c. Find the poles and zeros (if any) of this system.

1.5 Repeat Exercise 1.4 for the following two-output system in a block-observable canonical form

$$(A,B,C)=\left(\begin{bmatrix} a & b & 0 \\ c & d & 1 \\ e & f & 0 \end{bmatrix},\begin{bmatrix} g & h \\ i & j \\ k & l \end{bmatrix},\begin{bmatrix} 1 & 0 & 0 \\ 0 & 1 & 0 \end{bmatrix}\right)$$

a. From the description of Example 1.7, find the observability indices ν_i, $i=1, 2$.

b. Following the two-step procedure of Example 1.7, derive the $D(s)$ and $N(s)$ of this system's polynomial MFD transfer function (1.17a and b).

Answer: $\nu_1 = 1, \nu_2 = 2$, $D(s) = \begin{bmatrix} s-a & : & -b \\ -cs-e & : & s^2-ds-f \end{bmatrix}$, and

$N(s) = \begin{bmatrix} g & : & h \\ is+k & : & js+l \end{bmatrix}$.

1.6 Repeat Exercise 1.4 for the following two-output system in a block-observable canonical form

$$(A,B,C)=\left(\begin{bmatrix} a & b & 1 & 0 \\ c & d & 0 & 1 \\ e & f & 0 & 0 \\ g & h & 0 & 0 \end{bmatrix}, \begin{bmatrix} i \\ j \\ k \\ l \end{bmatrix}, \begin{bmatrix} 1 & 0 & 0 & 0 \\ 0 & 1 & 0 & 0 \end{bmatrix}\right)$$

a. From the description of Example 1.7, find the observability indices ν_i, $i = 1, 2$.

b. Following the two-step procedure of Example 1.7, derive the $D(s)$ and $N(s)$ of this system's polynomial MFD transfer function (1.17a and b).

Answer: $\nu_1 = \nu_2 = 2$,

$$D(s)=\begin{bmatrix} s^2-as-e & : & -bs-f \\ -cs-g & : & s^2-ds-h \end{bmatrix}, \text{ and}$$

$$N(s)=\begin{bmatrix} is+k \\ js+l \end{bmatrix}$$

1.7 Repeat Exercise 1.4 for the following two-output system in a block-observable canonical form

$$(A,B,C)=\left(\begin{bmatrix} a & b & 1 & 0 \\ c & d & 0 & 0 \\ e & f & 0 & 1 \\ g & h & 0 & 0 \end{bmatrix}, \begin{bmatrix} i \\ j \\ k \\ l \end{bmatrix}, \begin{bmatrix} 1 & 0 & 0 & 0 \\ 0 & 1 & 0 & 0 \end{bmatrix}\right)$$

Answer: $\nu_1 = 3$, $\nu_2 = 1$,

$$D(s)=\begin{bmatrix} s^3-as^2-es-g & : & -bs^2-fs-h \\ -c & : & s-d \end{bmatrix}, \text{ and}$$

$$N(s)=\begin{bmatrix} is^2+ks+l \\ j \end{bmatrix}$$

1.8 Repeat Exercise 1.4 for the following two-output system in a block-observable canonical form

$$(A,B,C)=\left(\begin{bmatrix} a & b & 0 & 0 \\ c & d & 1 & 0 \\ e & f & 0 & 1 \\ g & h & 0 & 0 \end{bmatrix},\begin{bmatrix} i \\ j \\ k \\ l \end{bmatrix},\begin{bmatrix} 1 & 0 & 0 & 0 \\ 0 & 1 & 0 & 0 \end{bmatrix}\right)$$

Answer: $\nu_1 = 1, \nu_2 = 3,$

$$D(s)=\begin{bmatrix} s-a & : & -b \\ -cs^2-es-g & : & s^3-ds^2-fs-h \end{bmatrix}, \text{ and}$$

$$N(s)=\begin{bmatrix} i \\ js^2+ks+l \end{bmatrix}$$

1.9 Verify the two polynomial matrices $N(s)$ of (1.17a and b) of Examples 6.1 and 6.2.

1.10 Derive the 4×4 polynomial matrix $D(s)$ from the ninth order system matrices of (5.8b).

1.11 Verify the polynomial MFD's (1.17a and b) of the two systems of Examples 8.3 and 8.4.

1.12 Compute the eigenstructure decomposition (1.10) of the following two state matrices:

$$A_1=\begin{bmatrix} -1 & 1 & 0 \\ 0 & -1 & -1 \\ 0 & 0 & -2 \end{bmatrix}, A_2=\begin{bmatrix} -1 & 0 & 0 \\ 0 & -1 & 1 \\ 0 & 0 & -2 \end{bmatrix}.$$

2

Single-System Performance and Sensitivity

High system performance and low sensitivity are the two required properties of control systems. Here low sensitivity is defined with respect to the system's mathematical model uncertainty and terminal disturbance, and is called "robustness".

Unfortunately, high performance and robustness are usually contradictory to each other – high-performance systems usually have high sensitivity and worse robust properties. Yet, both high performance and high robustness are essential to most practical engineering systems. Usually, only high-performance systems have high sensitivity, yet only such systems are worthy of controlling. Robustness, which can be considered as reliability, is also essential in almost all practical applications.

Therefore, both performance and sensitivity properties must be analyzed. This chapter deals with these two properties in two sections.

Section 2.1 is about system properties such as stability, quickness, and smoothness of the system transient response. These performance properties are the most important and most difficult to achieve. This section explains how these properties are most directly and explicitly determined by the system poles. An example also shows convincingly that the bandwidth (BW) of the classical control theory cannot reflect system performance correctly in general.

Section 2.2 deals with the system sensitivity via a novel perspective of the sensitivities of system poles. A basic numerical linear algebra result and an often-neglected result, yet a result that is directly useful in analyzing the pole sensitivity, is that the sensitivity of an eigenvalue is determined by its left and right eigenvectors. A far more generally accurate robust stability (or robust performance) criterion, which is based on the system poles and their sensitivities and which can be generally and effectively optimized by eigenvalue and eigenvector assignment, is proposed.

2.1 System Performance

The reason that the system control theory has concentrated mainly on linear time-invariant systems is that only the mathematical models of this kind of systems can have general and explicit solutions. Furthermore, only the

DOI: 10.1201/9781003259572-2

general and explicit understanding of the system can be used to guide generally, systematically, and effectively the complicated control system design.

The analytical solution of the state space model (1.1a) is, for $t \geq 0$,

$$\mathbf{x}(t) = e^{At}\mathbf{x}(t=0) + \int_0^t e^{A(t-\tau)}B\,\mathbf{u}(\tau)d\tau \tag{2.1}$$

where $\mathbf{x}(t=0)$ and $\mathbf{u}(\tau)$ are the given system initial state and system input, respectively. One way of deriving the result of (2.1) is by making the inverse Laplace transform on (1.6a). We call (2.1) the "complete response" of system state $\mathbf{x}(t)$.

By substituting (1.10) into (2.1) and using the Cayley–Hamilton theorem,

$$\mathbf{x}(t) = Ve^{\Lambda t}V^{-1}\mathbf{x}(t=0) + \int_0^t Ve^{\Lambda(t-\tau)}V^{-1}B\mathbf{u}(\tau)d\tau \tag{2.2}$$

$$= \sum_{i=1}^{q} V_i e^{\Lambda_i t}T_i\mathbf{x}(t=0) + \int_0^t \sum_{i=1}^{q} V_i e^{\Lambda_i(t-\tau)}T_i B\mathbf{u}(\tau)d\tau \tag{2.3}$$

Therefore, $e^{\Lambda t}$ is the only time function related to system parameters (Λ), in the system response (2.2)–(2.3). In other words, the eigenvalues Λ of the system's state matrix A are the system parameters that most directly and explicitly determine the system response (2.1)–(2.3).

Let us analyze the waveforms of all possible time functions of $e^{\Lambda_i t}$, for all different Jordan blocks of Λ_i of (1.10).

$$e^{\Lambda_i t} = e^{\lambda_i t} \quad \text{if } \Lambda_i = \lambda_i \tag{2.4a}$$

$$e^{\Lambda_i t} = \begin{bmatrix} e^{\sigma t}\cos(\omega t) & e^{\sigma t}\sin(\omega t) \\ -e^{\sigma t}\sin(\omega t) & e^{\sigma t}\cos(\omega t) \end{bmatrix} \quad \text{if } \Lambda_i = \begin{bmatrix} \sigma & \omega \\ -\omega & \sigma \end{bmatrix} \tag{2.4b}$$

The linear combination of the elements of this matrix can be simplified as

$$Ae^{\sigma t}\cos(\omega t) + Be^{\sigma t}\sin(\omega t) = \left[A^2 + B^2\right]^{1/2} e^{\sigma t}\cos\left(\omega t - \tan^{-1}(B/A)\right)$$

where A and B are real constants.

$$e^{\Lambda_i t} = \begin{bmatrix} 1 & t & t^2/2 & \cdots & \cdots & \cdots & t^{n-1}/(n-1)! \\ 0 & 1 & t & \cdots & \cdots & \cdots & t^{n-2}/(n-2)! \\ 0 & 0 & 1 & \cdots & \cdots & \cdots & t^{n-3}/(n-3)! \\ \vdots & & & & & & \vdots \\ 0 & 0 & 0 & \cdots & \cdots & 0 & 1 \end{bmatrix} e^{\lambda_i t} \tag{2.4c}$$

if Λ_i is an n-dimensional bidiagonal matrix of (1.10b).

Figure 2.1 plots all different waveforms of (2.4). In this figure, an eigenvalue is marked by the symbol "×", and its value is indicated by its position coordinates. The waveform of $e^{\Lambda_i t}$ for the eigenvalue Λ_i is plotted near its mark.

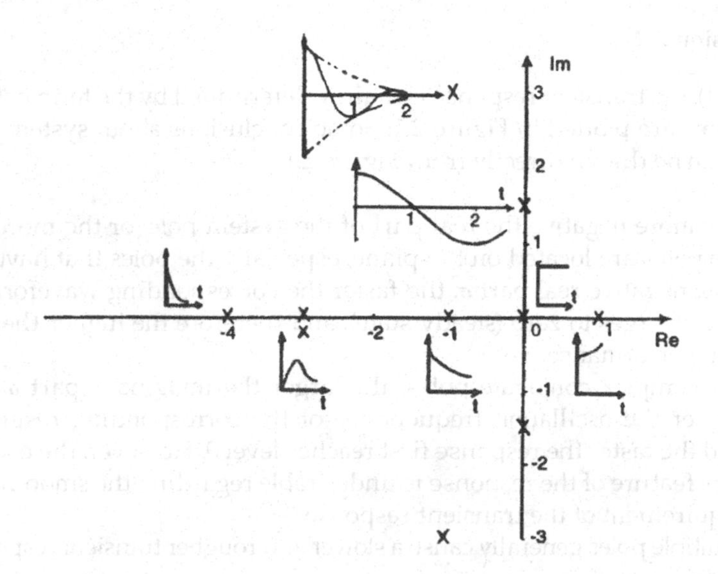

FIGURE 2.1
Possible system poles and waveforms of their corresponding system responses.

The following important conclusions of system performance can be derived directly from Figure 2.1.

Definition 2.1

A system is asymptotically stable if and only if for any initial state $\mathbf{x}(0)$, the system's zero-input response $e^{At}\,\mathbf{x}(0)$ converges to zero.

Conclusion 2.1

From Figure 2.1, a system is asymptotically stable if and only if every system pole (or every state matrix eigenvalue) has negative real parts, or every system pole is located at the open left half plane of Figure 2.1. We will refer to "asymptotically stable" as "stable" in the rest of this book.

Definition 2.2

The response (2.1) of an asymptotically stable system always reaches a steady state, which is called "steady-state response" and which is usually the desired state of response. The system response (2.1) before reaching its steady state is called the "transient response". Therefore, the faster and smoother the transient response, the better (higher) the performance of the system.

Conclusion 2.2

From (2.1), the transient response is mainly determined by the term $e^{\Lambda i t}$ whose waveforms are plotted in Figure 2.1. Some conclusions about system performance can be drawn directly from Figure 2.1.

a. The more negative the real part of the system pole, or the more left the poles are located on the plane, especially the poles that have the least negative real part σ, the faster the corresponding waveform of $e^{\sigma t}$ converges to zero (steady state), and therefore the higher the system performance.
b. For complex conjugate poles, the larger the imaginary part ω, the higher the oscillation frequency ω of the corresponding response, and the faster the response first reaches level 0. However, the oscillatory feature of the response is undesirable regarding the smoothness requirement of the transient response.
c. Multiple poles generally cause a slower and rougher transient response.

We define stability and fastness and smoothness of the system transient response, as the main measures of system performance. Conclusions 2.1 and 2.2 indicate that the system poles determine the system performance most directly, accurately, and comprehensively.

For example, the first-order systems of Example 1.3 are stable, because their only pole λ is negative. Furthermore, the more negative the λ, the smaller the time constant $|1/\lambda|$, and the faster the zero-input response and zero-state response to reach their respective steady-state value, as depicted in Figure 1.4. In addition, the first-order systems do not have multiple poles, and hence their responses are smooth.

In the classical control theory, the system performance is measured by bandwidth (BW). Assume that a second-order SISO system has complex conjugate poles $\sigma \pm j\omega_0$

$$G(s) = \frac{\omega_n^2}{\left[s-(\sigma+j\omega_0)\right]\left[s-(\sigma-j\omega_0)\right]}$$

$$G(s) = \frac{\omega_n^2}{s^2+(-2\sigma)s+\left[\sigma^2+\omega_0^2\right]} \tag{2.5a}$$

$$G(s) = \frac{\omega_n^2}{s^2+2\zeta\omega_n s+\omega_n^2}$$

where natural frequency $\omega_n = (\sigma^2 + \omega_0^2)^{1/2}$ and damping ratio

$$\zeta = -\sigma/\omega_n \tag{2.5b}$$

System (2.5) is called an underdamped system when the damping ratio ζ is between 0 and 1.

The magnitude of frequency response $|G(j\omega)|$ of underdamped system (2.5) is depicted in Figure 2.2.

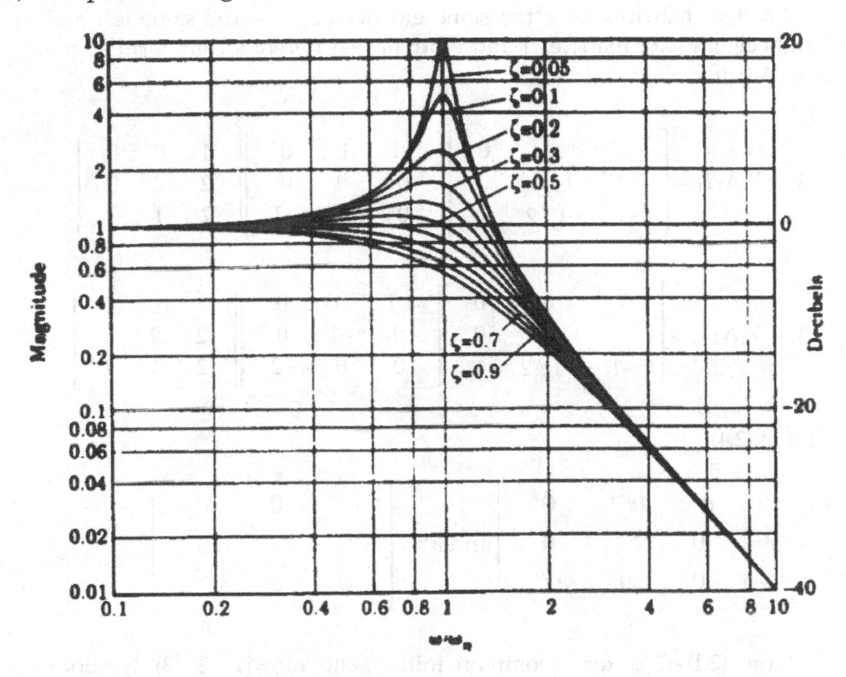

FIGURE 2.2
Frequency response of an underdamped system.

Figure 2.2 shows that as the frequency ω increases from 0 to infinity, the function $|G(j\omega)|$ starts at 1 and eventually decays to 0. BW is defined as the frequency at which $|G(j\omega)|$ reaches $1/\sqrt{2} \approx 0.707$. Figure 2.2 shows that (Chen, 1993)

$$\text{BW} \approx 1.6\omega_n \text{ to } 0.6\omega_n, \quad \text{when } \zeta \text{ is from } 0.1 \text{ to } 1 \tag{2.6a}$$

or BW is proportional to ω_n or proportional to $|\sigma|$ and ω_0. Therefore, from Conclusion 2.2, the wider the BW, the higher the performance of the system generally.

However, relationship (2.6a) is based on a rather strict class of system (2.5), and the relationship between BW and performance is indirectly derived from Conclusion 2.2. Therefore, the BW is far less generally accurate than the system poles, in indicating the system performance. The following Example 2.2 is a convincing proof of this inaccuracy.

Example 2.1: Zero-Input Response of Two Third-Order Systems

Let the state matrices of the two systems be

$$A_1=\begin{bmatrix} 1 & 2 & 0 \\ -2 & -3 & 0 \\ 0 & 1 & -2 \end{bmatrix} \text{ and } A_2=\begin{bmatrix} -1 & 0 & 0 \\ 0 & -1 & 0 \\ 2 & -1 & -2 \end{bmatrix}$$

The two matrices have the same eigenvalues and the same left and right eigenvector matrices T and V but have different Jordan form matrices Λ (1.10):

$$A_1 = V_1\Lambda_1 T_1 = \begin{bmatrix} 1 & 0 & 0 \\ -1 & 1/2 & 0 \\ -1 & -1/2 & 1 \end{bmatrix}\begin{bmatrix} -1 & 1 & 0 \\ 0 & -1 & 0 \\ 0 & 0 & -2 \end{bmatrix}\begin{bmatrix} 1 & 0 & 0 \\ 2 & 2 & 0 \\ 2 & 1 & 1 \end{bmatrix}$$

$$A_2 = V_2\Lambda_2 T_2 = \begin{bmatrix} 1 & 0 & 0 \\ -1 & 1/2 & 0 \\ -1 & -1/2 & 1 \end{bmatrix}\begin{bmatrix} -1 & 0 & 0 \\ 0 & -1 & 0 \\ 0 & 0 & -2 \end{bmatrix}\begin{bmatrix} 1 & 0 & 0 \\ 2 & 2 & 0 \\ 2 & 1 & 1 \end{bmatrix}$$

From (2.4),

$$e^{\Lambda_1 t}=\begin{bmatrix} e^{-t} & te^{-t} & 0 \\ 0 & e^{-t} & 0 \\ 0 & 0 & e^{-2t} \end{bmatrix} \text{ and } e^{\Lambda_2 t}=\begin{bmatrix} e^{-t} & 0 & 0 \\ 0 & e^{-t} & 0 \\ 0 & 0 & e^{-2t} \end{bmatrix}$$

From (2.1)–(2.2), for a common initial state $\mathbf{x}(0)=[1 \quad 2 \quad 3]'$ for both systems, the zero-input response of the state $\mathbf{x}(t)$ is

$$e^{\Lambda_1 t}\mathbf{x}(0)=\begin{bmatrix} e^{-t}+6\,te^{-t} \\ 2e^{-t}-6\,te^{-t} \\ -4e^{-t}-6te^{-t}+7e^{-2t} \end{bmatrix} \text{ and } e^{\Lambda_2 t}\mathbf{x}(0)=\begin{bmatrix} e^{-t} \\ 2\,e^{-t} \\ -4e^{-t}+7e^{-2t} \end{bmatrix}$$

The two waveforms of $\mathbf{x}(t)$ for the two systems are shown in Figure 2.3. The second waveform is remarkedly faster and smoother than the first. The difference is caused by the only difference between these two systems: the first system has a generalized eigenvector while the second system does not. This internal difference cannot be revealed by the transfer function model of these two systems.

From (2.2), the difference in Λ and in $e^{\Lambda t}$ between these two systems will also make a considerable difference between the two system's zero-state responses. Hence, the state space model can reveal zero-state response more explicitly than the transfer function model, even though the latter is defined by the system's zero-state response (1.6)–(1.7) only.

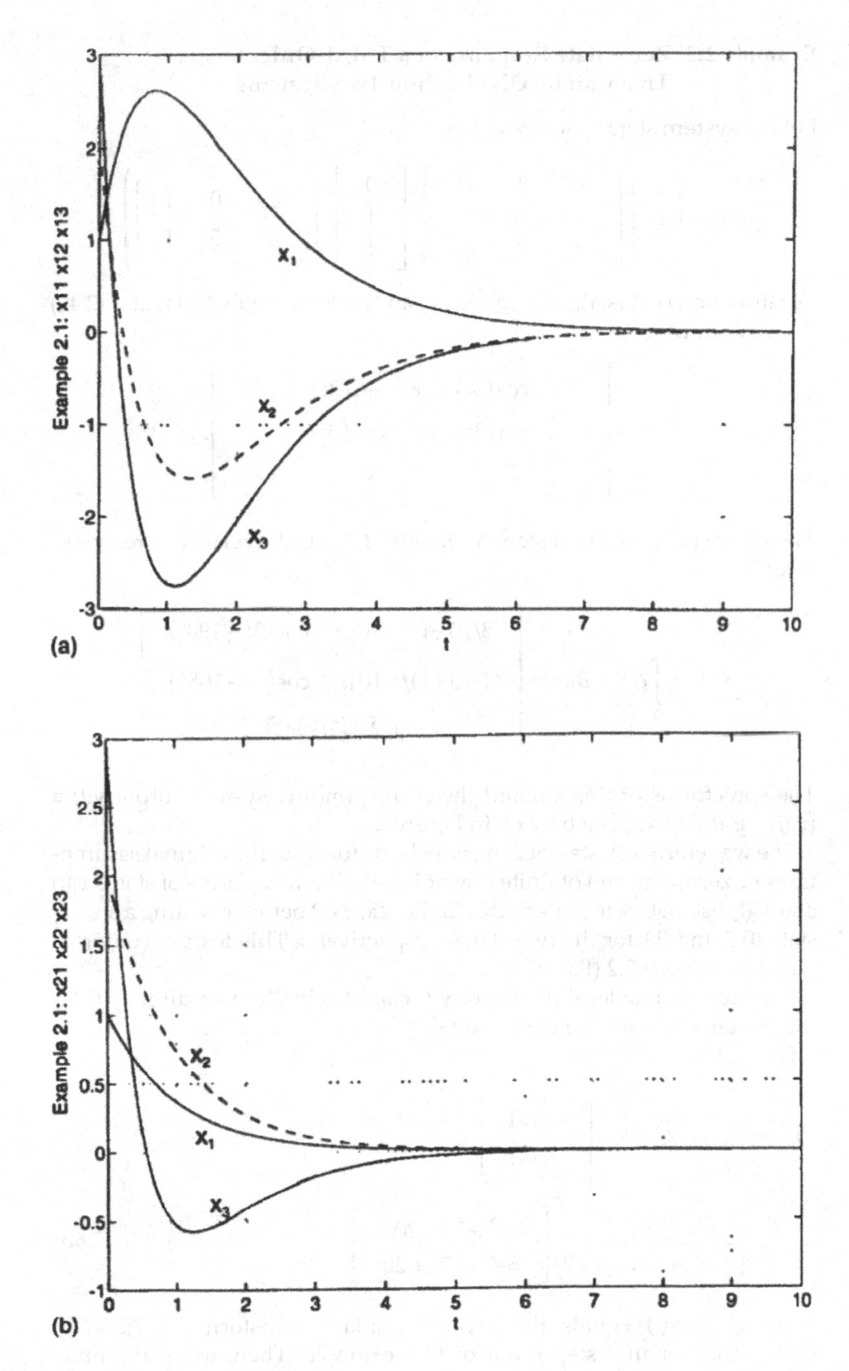

FIGURE 2.3
State zero-input response waveforms of two systems sharing the same poles.

Example 2.2: Zero-State Response of a Third-Order System That Can Be Divided into Two Systems

Let the system state space model be

$$(A,B,C)=\left(\begin{bmatrix}-1 & 3 & 0\\ -3 & -1 & 0\\ 0 & 0 & -2\end{bmatrix},\begin{bmatrix}0\\1\\1\end{bmatrix},\begin{bmatrix}5/3 & 0 & 1\\ 0 & 5 & 1\end{bmatrix}\right)$$

Because matrix A is already in its Jordan form, we apply (2.4a) and (2.4b) directly, then

$$e^{At}=\begin{bmatrix}e^{-t}\cos(3t) & e^{-t}\sin(3t) & 0\\ -e^{-t}\sin(3t) & e^{-t}\cos(3t) & 0\\ 0 & 0 & e^{-2t}\end{bmatrix}.$$

Then from (2.1), for a unit step input ($\mathbf{u}(t)=1, t\geq 0$), the zero-state response of $\mathbf{x}(t)$ is

$$\mathbf{x}(t)=\int_0^t e^{A(t-\tau)}B\,d\tau=\begin{bmatrix}3/10+\left(1/\sqrt{10}\right)e^{-t}\cos(3t-198°)\\ 1/10+\left(1/\sqrt{10}\right)e^{-t}\cos(3t-108°)\\ 1/2-(1/2)e^{-2t}\end{bmatrix}$$

The waveforms of this $\mathbf{x}(t)$ and the corresponding system output $\mathbf{y}(t)\equiv[y_1(t)\quad y_2(t)]'=C\mathbf{x}(t)$ can be seen in Figure 2.4.

The waveforms all start at zero, which conforms to the original assumptions of zero state and of finite power input. The waveforms of states $x_1(t)$ and $x_2(t)$ oscillate with a period of $2\pi/\omega=2\pi/3\approx 2$ before reaching a steady state (0.3 and 0.1 for the two states, respectively). This feature conforms with Conclusion 2.2 (Part B).

The steady-state level of output $\mathbf{y}(t)$ can also be derived directly from the system's transfer function model.

From (1.9),

$$G(s)=C(sI-A)^{-1}B\equiv\begin{bmatrix}g_1(s)\\ g_2(s)\end{bmatrix}$$

$$=\frac{1}{\left(s^2+2s+10\right)(s+2)}\begin{bmatrix}s^2+7s+20\\ 6s^2+17s+20\end{bmatrix}\qquad(2.6b)$$

From (1.7), $\mathbf{y}(t)$ equals the inverse Laplace transform of $Y(s)=G(s)U(s)=G(s)/s$ for unit step input of this example. Then, using the finite value theorem of Laplace transform, the constant steady-state value of $\mathbf{y}(t)$ can be derived directly as $\mathbf{y}(t\rightarrow\infty)=\lim_{s\rightarrow 0}sY(s)=[1\quad 1]'$.

This value is in accordance with Figure 2.4. This derivation shows that the classical control theory can find the steady state of output $\mathbf{y}(t)$ more easily than the modern control theory.

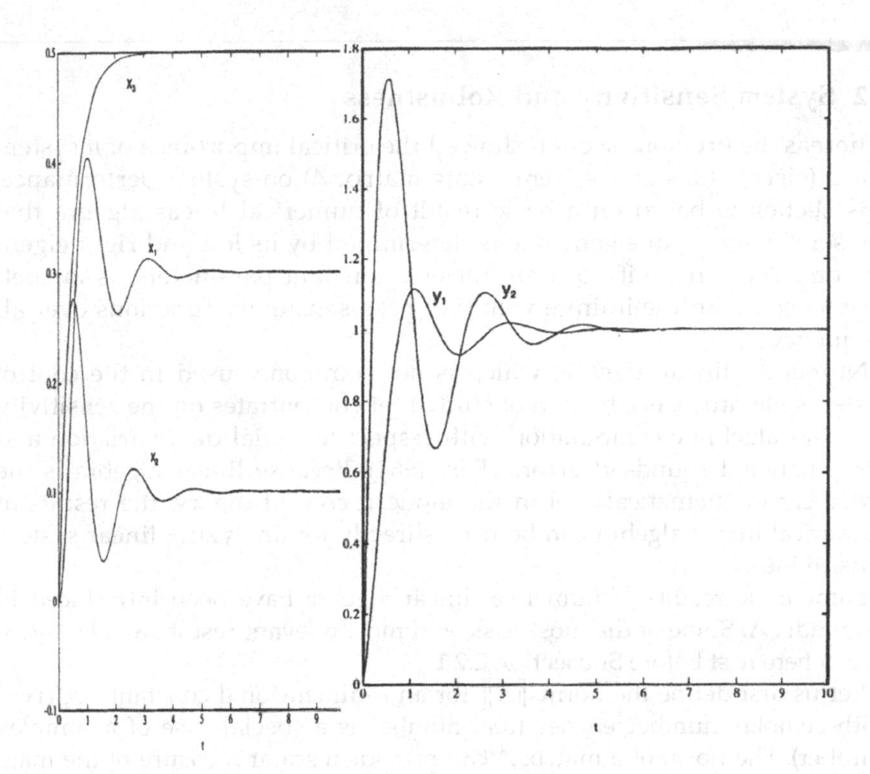

FIGURE 2.4
Zero-state response of system state $\mathbf{x}(t)$ and of system output $\mathbf{y}(t)$.

However, the BW of the classical control theory cannot reflect accurately and reflects far less accurately than the system poles, the transient response, or performance.

From the definition just before (2.6a), we can compute the BW of the two systems $g_1(s)$ and $g_2(s)$ of (2.6b), as $BW_1=3.815$ and $BW_2=9.21$. The huge difference between BW_1 and BW_2 reflects very wrongly the speed of reaching the steady state of $y_1(t)$ and $y_2(t)$, because Figure 2.4 shows that these two responses reach their steady state at about the same time! This similar quickness of the transient response is accurately revealed by the system poles because systems $g_1(s)$ and $g_2(s)$ share the same poles (same state matrix A).

Furthermore, the much higher BW_2 is supposed to indicate the much higher performance of system $g_2(s)$ than that of system $g_1(s)$; yet, Figure 2.4 shows just the sharp opposite! Figure 2.4 shows that the response $y_2(t)$ is actually much less desirable than $y_1(t)$, mainly because it is much less smooth than $y_1(t)$. This difference in smoothness is correctly reflected by the only difference (matrix C) of the state space models of these two systems ([5/3 0 1] vs. [0 5 1]).

Although Example 2.2 is very simple and very common, its sharp conclusion shows convincingly that BW cannot reflect a system's performance reliably and correctly. Therefore, bandwidth cannot be used to guide reliably the high performance and high robustness control system design.

2.2 System Sensitivity and Robustness

Whereas the previous section showed the critical importance of n system poles (eigenvalues of a system's state matrix A) on system performance, this section is based on a basic result of numerical linear algebra that the sensitivity of an eigenvalue is determined by its left and right eigenvectors. The sensitivity of only these n constant parameters, is far better focused than the infinite values of $p \times p$ sensitivity functions over all frequency.

Numerical linear algebra, which is not commonly used in the control systems literature, is a branch of study that concentrates on the sensitivity of linear algebraic computation with respect to initial data variation and computational round-off errors (Fox, 1964). Because linear algebra is the prevalent mathematical tool in the modern control theory, the results of numerical linear algebra can be used directly for analyzing linear system sensitivities.

Some basic results of numerical linear algebra have been introduced in Appendix A. Some of the most basic and most relevant results will be introduced here first before Subsection 2.2.1.

Let us first define the norm $\|A\|$ for an n-dimensional constant matrix A with complex number entries (real number is a special case of a complex number). The norm of a matrix A can provide a scalar measure of the magnitude of matrix A.

Before establishing the matrix norm, let us establish the vector norm $\|\mathbf{x}\|$, for an n-dimensional constant vector $\mathbf{x}=[x_1, \ldots, x_n]'$ with complex number entries x_i $(i=1, \ldots, n)$.

Like the absolute value of a scalar number, the vector norm $\|\mathbf{x}\|$ must satisfy the following properties (Chen, 1984):

1. $\|\mathbf{x}\| \geq 0$ and $\|\mathbf{x}\| = 0$ if and only if $\mathbf{x}=0$ (or all entries of $\mathbf{x}=0$).
2. $\|a\mathbf{x}\| \leq |a|\,\|\mathbf{x}\|$, where a is a scalar quantity.
3. $\|\mathbf{x}+\mathbf{y}\| \leq \|\mathbf{x}\|+\|\mathbf{y}\|$, where $\mathbf{y}$ is also an n-dimensional vector.

This third property is also called the "triangular inequality".

Vector norms have the following important property called "Cauchy-Schwartz inequality" (Chen, 1984).

$$|\mathbf{x}^*\mathbf{y}| = |\mathbf{y}^*\mathbf{x}| = \|\mathbf{x}\|\,\|\mathbf{y}\|\cos(\theta) \leq \|\mathbf{x}\|\,\|\mathbf{y}\| \tag{2.7}$$

Where θ is the angle between the two vectors $\mathbf{x}$ and $\mathbf{y}$.

Definition 2.3

There are three vector norms $\|\mathbf{x}\|$ defined as follows:

1. $\|\mathbf{x}\|_1 = |x_1| + \cdots + |x_n|$
2. $\|\mathbf{x}\|_2 = \left(|x_1|^2 + \cdots + |x_n|^2\right)^{1/2} = \left(\mathbf{x}^*\mathbf{x}\right)^{1/2}$ ("*" stands for the transpose and complex conjugate operation).
3. $\|\mathbf{x}\|_\infty = \max\{|x_i|,\, i = 1,\ldots,n\}$

In most cases only the norm $\|\mathbf{x}\|_2$ is used unless specified otherwise in this book. Therefore, $\|\mathbf{x}\|$ is the default of $\|\mathbf{x}\|_2$ in this book.

The matrix norm $\|A\|$ must also have the following four properties:

1. $\|A\| \geq 0$ and $\|A\| = 0$ if and only if $A = 0$ (all entries a_{ij} of matrix A equal 0)
2. $\|aA\| = |a|\,\|A\|$, where a is a scalar
3. $\|A + B\| \leq \|A\| + \|B\|$, where B is a matrix of same dimension
4. $\|Ax\| \leq \|A\|\|x\|$ (2.8)

Based on these properties, particularly (2.8), there can be three definitions for matrix norm $\|A\|$ corresponding to the three vector norms $\|\mathbf{x}\|$ of Definition 2.3.

Definition 2.4

1. $\|A\|_1 = \max\left\{|a_{1j}| + \cdots + |a_{nj}|,\, j = 1,\ldots,n\right\}$ (greatest column norm 1)
2. $\|A\|_2 = \max\left\{\left[\lambda_i\left(A^*A\right)\right]^{1/2} = \text{singular value } \sigma_i(A),\, i = 1,\ldots,n\right\}$ (2.9)
3. $\|A\|_\infty = \max\left\{|a_{i1}| + \cdots + |a_{in}|,\, i = 1,\ldots,n\right\}$ (greatest row norm 1)

Unless specified otherwise, $\|A\|$ is the default of $\|A\|_2$, which is also called the "spectrum norm".

There is a fourth commonly used matrix norm called "Frobenius norm", which is easiest to compute:

4. $\|A\|_F = \left(\sum_{\text{all } i \text{ and } j} |a_{ij}|^2\right)^{1/2} = \left[\text{Trace}\left(A^*A\right)\right]^{1/2}$ (2.10)

   where the operation "Trace" stands for the sum of all diagonal elements.

The norm $\|A\|_F$ can be used to predict the range of spectrum norm $\|A\|$. Based on the singular value decomposition of matrix $A = U\Sigma V^*$ where Σ is a diagonal matrix with singular values $\sigma_1 \geq \sigma_2 \geq \ldots \geq \sigma_n \geq 0$ (see Section A.3), from (2.9 and 2.10),

$$\|A\|_F = \left[\text{Trace}\left(A^* A\right)\right]^{1/2} = \left[\text{Trace}\left(\Sigma^* \Sigma\right)\right]^{1/2} = \left[\sigma_1^2 + \cdots + \sigma_n^2\right]^{1/2}$$

$$\leq \sqrt{n}\sigma_1 = \sqrt{n}\|A\| \tag{2.11a}$$

$$\text{Or} \quad \geq \sigma_1 = \|A\| \tag{2.11b}$$

In other words, $\|A\|$ is between $(\|A\|_F/\sqrt{n})$ and $\|A\|_F$.

Definition 2.5: Condition Number of a Computational Problem

Let A be data and $f(A)$ be the result of computational problem $f(A)$. Let ΔA be the variation of data A, and let Δf be the corresponding variation of result $f(A)$ due to ΔA such that

$$f(A + \Delta A) = f(A) + \Delta f$$

Then the condition number $k(f)$ of the computational problem $f(A)$ is defined by the following inequality as

$$\|\Delta f\|/\|f\| \leq k(f)\|\Delta A\|/\|A\| \tag{2.12}$$

Therefore, $k(f)$ is the relative sensitivity of problem f with respect to the relative variation of data A. A small $k(f)$ implies low sensitivity of problem f, which is then called a "well-conditioned problem". On the other hand, a large $k(f)$ implies high sensitivity of problem f, which is then called an "ill-conditioned problem" (Wilkinson, 1965).

Example 2.3: Wilkinson (1965), Tsui (1983b)

Let the computational problem be the computation of solution $\mathbf{x}$ of a set of linear equations $A\mathbf{x}=\mathbf{b}$, where A and $\mathbf{b}$ are the given data.

Let $\Delta\mathbf{b}$ be the variation of $\mathbf{b}$ (no variation of data A). Then $A(\mathbf{x}+\Delta\mathbf{x})=\mathbf{b}+\Delta\mathbf{b}$ implies that

$\|\Delta\mathbf{x}\| = \|A^{-1}\Delta\mathbf{b}\| \leq \|A^{-1}\|\,\|\Delta\mathbf{b}\|$.

Thus, from $\|\mathbf{b}\| \leq \|A\|\,\|\mathbf{x}\|$,

$$\|\Delta\mathbf{x}\|/\|\mathbf{x}\| \leq \|A^{-1}\|\|A\|\|\Delta\mathbf{b}\|/\|\mathbf{b}\|$$

From Definition 2.5, this inequality implies that the condition number of this computational problem is $\|A^{-1}\|\,\|A\|$.

Suppose in the same computational problem, ΔA is the variation of data A (no variation of data $\mathbf{b}$), then $(A+\Delta A)\ (\mathbf{x}+\Delta\mathbf{x})=\mathbf{b}$ implies that (assuming $\|\Delta A\ \Delta\mathbf{x}\|$ is negligible)

$$\|\Delta\mathbf{x}\| = \|A^{-1}(-\Delta A\mathbf{x})\| \leq \|A^{-1}\|\|\Delta A\|\|\mathbf{x}\|$$

Thus, $\|\Delta \mathbf{x}\|/\|\mathbf{x}\| \le \|A^{-1}\|\|A\|\|\Delta A\|/\|A\|$,

which implies again that the condition number of this computational problem is $\|A^{-1}\|\,\|A\|$.

Because of the result of Example 2.3, we define the condition number of a matrix A as

$$k(A) = \|A^{-1}\|\|A\| \tag{2.13}$$

In the following two subsections, we will first analyze the sensitivity of the eigenvalues of system state matrix A and then use this result to analyze the sensitivity of the system stability property.

2.2.1 The Sensitivity of Eigenvalues (Robust Performance)

Robust performance is defined as the low sensitivity of system performance with respect to the system model uncertainty and terminal disturbance. Because Section 2.1 indicated that the eigenvalues of the system state matrix (or system poles) most directly and explicitly determine the system performance, it is only obvious that the sensitivities of these eigenvalues most directly and explicitly determine the system's robust performance.

From (1.10), $V^{-1}AV = \Lambda$, where matrix Λ is in Jordan form with all eigenvalues of A. Therefore, if A becomes $A + \Delta A$, then

$$V^{-1}(A + \Delta A)V = \Lambda + V^{-1}\Delta A\ V \equiv \Lambda + \Delta\Lambda \tag{2.14}$$

$$\text{Or}\quad \|\Delta\Lambda\| = \|V^{-1}\Delta A V\| \le \|V^{-1}\|\|V\|\|\Delta A\| \equiv k(V)\|\Delta A\| \tag{2.15a}$$

Inequality (2.15a) indicates that the condition number of the eigenvector matrix V, $k(V)$, can decide the magnitude of $\Delta\Lambda$. However, because $\Delta\Lambda$ is usually not in Jordan form, $\Delta\Lambda$ may not accurately indicate the actual variation of the individual eigenvalues.

Based on (2.14), a bound of the minimum variation of the individual eigenvalues is derived (Wilkinson, 1965):

$$\min\{\Delta\lambda_i \equiv |\lambda_i - \lambda|, i = 1,\ldots,n\} \le k(V)\|\Delta A\| \tag{2.15b}$$

where λ_i and λ are individual eigenvalues of matrices A and $A + \Delta A$, respectively.

However, only the bound of the maximum of $\Delta\lambda_i$ is of our interest.

Nonetheless, an indication of the sensitivity of all eigenvalues (or of Λ), $s(\Lambda)$, can still be derived from (2.15) as

$$s(\Lambda) \equiv k(V) = \|V\|\|V^{-1}\| \tag{2.16}$$

although s(Λ) is not an accurate indication of the variations (or sensitivities) of the individual eigenvalues. An advantage of this bound though is that it is still valid for large $\|\Delta A\|$ (Wilkinson, 1965).

To derive a more accurate measure of the sensitivity of individual eigenvalues, first-order perturbation analysis is applied, and the following result is obtained under the assumption of small enough $\|\Delta A\|$ (Wilkinson, 1965).

Theorem 2.1

Let λ_i, $\mathbf{v}_i$, and $\mathbf{t}_i$ be the i-th eigenvalue, right eigenvector, and left eigenvector of matrix A, respectively (i=1, …, n). Let $\lambda_i+\Delta\lambda_i$ be the i-th eigenvalue of matrix $A+\Delta A$ (i=1, …, n). Then for small enough $\|\Delta A\|$,

$$|\Delta\lambda_i| \le \|\mathbf{t}_i\|\|\mathbf{v}_i\|\|\Delta A\| \equiv s(\lambda_i)\|\Delta A\|, \; i=1,\ldots,n \tag{2.17}$$

Proof

Let $\Delta A = dB$, where d is a positive yet small enough scalar quantity, and B is a matrix of the same dimension as A. Let $\lambda_i(d)$ and $\mathbf{v}_i(d)$ be the i-th eigenvalue and i-th right eigenvector of matrix $A+dB$, respectively, i=1, …, n. Then,

$$(A+dB)\mathbf{v}_i(d) = \lambda_i(d)\mathbf{v}_i(d) \tag{2.18}$$

Without loss of generality, we assume i=1. From the perturbation theory,

$$\lambda_1(d) = \lambda_1 + k_1 d + k_2 d^2 + \cdots \tag{2.19a}$$

and

$$\mathbf{v}_1(d) = \mathbf{v}_1 + (c_{21}\mathbf{v}_2 + \cdots + c_{n1}\mathbf{v}_n)d + (c_{22}\mathbf{v}_2 + \cdots + c_{n2}\mathbf{v}_n)d^2 + \cdots \tag{2.19b}$$

where k_j and c_{ij} (i=2, …, n, j=1, 2, …) are constants. For small enough ΔA or small enough d, (2.19) can be simplified by first-order perturbation as

$$\lambda_1(d) \approx \lambda_1 + k_1 d \equiv \lambda_1 + \Delta\lambda_1 \tag{2.20a}$$

and

$$\mathbf{v}_1(d) \approx \mathbf{v}_1 + (c_{21}\mathbf{v}_2 + \cdots + c_{n1}\mathbf{v}_n)d \tag{2.20b}$$

By substituting (2.20) into (2.18) and from $A\mathbf{v}_i=\lambda_i\mathbf{v}_i$ and $d^2 \ll 1$, we have

$$\left[(\lambda_2 - \lambda)c_{21}\mathbf{v}_2 + \cdots + (\lambda_n - \lambda)c_{n1}\mathbf{v}_n + B\mathbf{v}_1\right]d = k_1\mathbf{v}_1 d \tag{2.21}$$

Upon multiplying the first left eigenvector $\mathbf{t}_1$ ($\mathbf{t}_1\ \mathbf{v}_j = \delta_{1j}$) on the left side of both sides of (2.21), we have

$$\mathbf{t}_1 B \mathbf{v}_1 = k_1.$$

Therefore, from (2.20a) and (2.8)

$$|\Delta\lambda_1| = |\mathbf{t}_1 B \mathbf{v}_1 d| \le \|\mathbf{t}_1\|\|\mathbf{v}_1\|\|dB\| = \|\mathbf{t}_1\|\|\mathbf{v}_1\|\|\Delta A\|$$

The derivation after (2.18) is valid for other eigenvalues and eigenvectors. Hence the proof.

This theorem shows clearly that, similar to the sensitivity of all eigenvalue matrix Λ of (2.16), the sensitivity of an individual eigenvalue λ_i, $s(\lambda_i)$, equals $\|\mathbf{v}_i\|\|\mathbf{t}_i\|$, $i = 1,\ldots,n$.

Example 2.4

Consider the following two matrices

$$A_1 = \begin{bmatrix} n & 1 & 0 & & \cdots & 0 \\ 0 & n-1 & 1 & 0 & & \vdots \\ \vdots & 0 & n-2 & 1 & & \vdots \\ \vdots & & & \ddots & \ddots & 0 \\ \vdots & & & & 2 & 1 \\ 0 & 0 & 0 & \cdots & 0 & 1 \end{bmatrix},$$

$$A_2 = \begin{bmatrix} n & n & 0 & & \cdots & 0 \\ 0 & n-1 & n & 0 & & \vdots \\ \vdots & 0 & n-2 & n & & \vdots \\ \vdots & & & \ddots & \ddots & 0 \\ \vdots & & & & 2 & n \\ 0 & 0 & 0 & \cdots & 0 & 1 \end{bmatrix}$$

The two matrices share the same eigenvalues $n, n-1, n-2, \ldots, 2, 1$. The right and left eigenvector matrices of A_1 and A_2 are derived as follows:

$$V = \begin{bmatrix} 1 & -x_2 & x_3 & -x_4 & \cdots & (-1)^{n-1} x_n \\ 0 & 1 & -x_2 & x_3 & \cdots & (-1)^{n-2} x_{n-1} \\ 0 & 0 & 1 & -x_2 & \cdots & (-1)^{n-3} x_{n-2} \\ \vdots & & & & & \vdots \\ \vdots & & & \ddots & \ddots & \vdots \\ \vdots & & & & 1 & -x_2 \\ 0 & 0 & 0 & \cdots & 0 & 1 \end{bmatrix},$$

$$T = \begin{bmatrix} 1 & x_2 & x_3 & \cdots & \cdots & x_n \\ 0 & 1 & x_2 & \cdots & \cdots & x_{n-1} \\ 0 & 0 & 1 & & \ddots & \vdots \\ \vdots & & \ddots & \ddots & \ddots & \vdots \\ \vdots & & & \ddots & 1 & x_2 \\ 0 & 0 & \cdots & \cdots & 0 & 1 \end{bmatrix}$$

where

For $A_1: \ x_i = x_{i-1}/(i-1) = 1/(i-1)!, i = 2,\ldots,n,$ and $x_1 = 1$.

Or $x_2 = 1, x_3 = 1/2!, x_4 = 1/3!,\ldots,x_n = 1/(n-1)!$

For $A_2: \ x_i = nx_{i-1}/(i-1) = n^{i-1}/(i-1)!, i = 2,\ldots,n,$ and $x_1 = 1$.

Or $x_2 = n, x_3 = n^2/2!,\ldots,x_n = n^{n-1}/(n-1)!$.

The eigenvector matrix parameters x_i are much greater for A_2 than for A_1. Therefore, from (2.17), the sensitivity of eigenvalues of matrix A_2 is much higher than that of matrix A_1.

For example,

$$s(\lambda_1) = \|\mathbf{t}_1\|\|\mathbf{v}_1\| = \|\mathbf{v}_n\|\|\mathbf{t}_n\| = s(\lambda_n) = \left(1 + x_2^2 + \cdots + x_n^2\right)^{1/2}(1) \quad (2.22a)$$

For $A_1, s(\lambda_1) = \left[1 + (1/2)^2 + \cdots + 1/(n-1)!^2\right]^{1/2} \approx 1$

For $A_2, s(\lambda_1) = \left[1 + (n/2)^2 + \cdots + \left(n^{n-1}/(n-1)!\right)^2\right]^{1/2} \approx n^{n-1}/(n-1)!$

$$s(\lambda_{n/2}) = \|\mathbf{t}_{n/2}\|\|\mathbf{v}_{n/2}\| = \|\mathbf{v}_{(n/2)+1}\|\|\mathbf{t}_{(n/2)+1}\| = s\left(\lambda_{(n/2)+1}\right)$$

$$= \left[1 + \cdots + x_{(n/2)+1}^2\right]^{1/2}\left[1 + \cdots + x_{n/2}^2\right]^{1/2} \quad (2.22b)$$

For $A_1, s(\lambda_{n/2}) \approx (1)(1) = 1$

For $A_2, s(\lambda_{n/2}) \approx \left[n^{n/2}/(n/2)!\right]^2$

The following table (Table 2.1) shows the approximate values of $s(\lambda_1)$ and $s(\lambda_{n/2})$, at different matrix sizes n, and for the two matrices A_1 and A_2 of Example 2.4:

TABLE 2.1

Approximate Eigenvalue Sensitivities of Matrices A_1 and A_2 of Example 2.4

	$s(\lambda_1)$	$s(\lambda_1)$	$s(\lambda_1)$	$s(\lambda_{n/2})$	$s(\lambda_{n/2})$
	$n=5$	$n=10$	$n=20$	$n=10$	$n=20$
For A_1	1	1	1	1	1
For A_2	5	275	2×10^6	7×10^6	8×10^{12}

Table 2.1 shows that eigenvalue sensitivities are the lowest possible (= 1) for matrix A_1. Hence the computation of eigenvalues of matrix A_1 is well conditioned. For matrix A_2, the eigenvalue sensitivities are very high and are about 10^6 times higher for eigenvalue $\lambda_{n/2}$ than for eigenvalue λ_1. Hence the computation of eigenvalues of Matrix A_2 is ill-conditioned.

The difference between matrices A_1 and A_2 is only at the upper diagonal line. From Example 1.5 and Eq. (2.4c), the upper diagonal line elements of A_1 and A_2 are the coupling links between the adjacent eigenvalues of A_1 and A_2. The weaker these coupling links, the lower the sensitivity of each eigenvalue such as matrix A_1, as well as for A_2. In addition, the middle eigenvalue $\lambda_{n/2}$ accumulates more link effects on both sides and therefore has a higher sensitivity than the side eigenvalue λ_1 or λ_n.

From another perspective, the smaller the off-diagonal coupling elements of the matrix, the weaker the effect of matrix variation on the eigenvalues of that matrix [Gershgorin Theorem, see Wilkinson (1965)], and the more similar that matrix compares to the Jordan form.

The theoretical analysis on the eigenvalue sensitivity of matrix A_2 was also carried out by Wilkinson (1965) and Chen (1984) using a special ΔA, in which all other elements are zero except $\Delta A_{n1}=d$. The characteristic polynomial of matrix $A_2+\Delta A$ is $(s-n) \ldots (s-2)(s-1)+d(-n)^{n-1}$. A root locus plot of this polynomial with respect to d demonstrates the root variations with respect to the increase of d (Chen, 1984).

Using the same special ΔA, it is of interest to compare the characteristic polynomial of matrix $A_1+\Delta A$: $(s-n) \ldots (s-2)(s-1)+d(-1)^{n-1}$. The constant term of the characteristic polynomial of matrix A_1 is almost $n^{n-1}/n!$ times smaller than that of matrix A_2.

To summarize, only the comparisons of matrices A_1 and A_2 of Example 2.4 of this book show very clearly the extreme importance of the coupling effects on the eigenvalue sensitivity. This understanding will be used in Chapter 9 for analytical eigenvector assignment. In reality, it is also common sense that if a system relies more heavily on more system components to work together, then that system is more sensitive to each of its components, although such systems usually have higher performance – another evidence of the contradiction between low sensitivity/high robustness and high performance.

2.2.2 The Sensitivity of System Stability (Robust Stability)

Stability is the foremost system property. Therefore, the sensitivity of this property called "robust stability" with respect to system model uncertainty, is also critically important.

Consequently, a generally accurate measure of robust stability is also critically important to guide control system design in real practice, as shown by such design experience of the past 70+ years (Doyle and Stein 1981). This measure has also been the main design criterion of the classical control theory such as gain and phase margins (GM and PM).

In addition, for a really successful practical design, the robust stability measure not only must be generally accurate but also must be relatively easily optimized by design computation.

From Conclusion 2.1, the most direct and most basic stability criterion is that all system poles have negative real parts. Hence the real part of eigenvalues of state matrix A and their sensitivities are the most direct and most important factors in measuring robust stability. That is why robust stability is determined by state matrix A *exclusively* (see the Lyapunov stability criterion).

Let us compare with the Routh–Hurwitz stability criterion, where the system's characteristic polynomial coefficients must be computed first. This computation is proven to be almost as unreliable as the computation of system poles, see Examples 1.6 and 2.4, and Wilkinson (1965). The Routh–Hurwitz test then works on these coefficients based on Conclusion 2.1 and is therefore indirect as compared to Conclusion 2.1 and system poles. The Routh–Hurwitz test does not reveal at all the sensitivity of the test itself either.

Let us compare the gain and phase margins (GM and PM) which are based on Nyquist plot and which uses all $p \times p$ functions of the loop transfer function and at all frequency (Rosenbrock, 1974; Safonov et al., 1977; Postlethwaite et al., 1979; Huang et al., 1982; Doyle et al., 1992). Hence, GM and PM are based on too much raw data, are determined based on the Cauchy integral theorem/Nyquist test, which is indirect compared to Conclusion 2.1, and are known to be not generally accurate (Vidyasagar, 1984). In addition, it is almost impossible to maximize both GM and PM generally and effectively. This should be the main reason that control systems research switched to the modern control theory in the 1960s and 70s in the first place.

In general, stability property is essentially a system's internal convergence property over the time domain (see Definition 2.1 and Conclusion 2.1), while loop transfer function is a gross relationship between the system's terminals over frequency. This is the reason that in the classical control theory, robust stability is based on too much raw data and is far less focused than the modern control theory, in which robust stability is determined fully by only $n \times n$ elements of the system's internal state matrix A. These are the fundamental reasons due to which the classical control theory is inferior to the modern control theory.

This subsection will compare three existing robust stability measures of the modern control theory, named M_1, M_2, and M_3. All three measures use system poles and their sensitivities, which are determined exclusively by the system's state matrix A. All three measures are defined based on the assumption that matrix A is already stable (all eigenvalues have negative real parts or $\text{Re}(\lambda_i)<0$ for all i, with $\text{Re}(\lambda_n)$ being the least negative), and all three measures are defined such that the higher the measure (or the margin), the less likely the system will become unstable due to the variation of matrix A, and the more robust stable the system.

Among these three measures, M_1 and M_2 were developed in the 1980s (Kautsky et al., 1985; Qiu and Davison, 1986; Juang et al., 1986; Dickman, 1987; Lewkowicz and Sivan, 1988), while M_3 was proposed by this author in the early 1990s (Tsui, 1990, 1994a).

Let us first introduce these three measures.

$$M_1 = \min\{\underline{\sigma}(A - j\omega I), \omega = 0 \text{ to} \infty\} \; (\underline{\sigma} \text{ is the smallest singular value}), \quad (2.23)$$

$$M_2 = s(\Lambda)^{-1}|Re(\lambda_n)|, \; (s(\Lambda) \text{ is defined in } (2.16)) \quad (2.24)$$

$$M_3 = \min\{s(\lambda_i)^{-1}|Re(\lambda_i)|, i = 1,2,\ldots,n\}, s(\lambda_i) \text{ is defined in } (2.17) \quad (2.25)$$

Now, we will compare these three measures using two standards, the general accuracy and the easiness of maximization by design computation while accommodating other important design objectives such as pole assignment (best guarantee of performance or stabilization).

M₁: Because $\underline{\sigma}$ is the smallest possible matrix variation norm for the matrix to become singular (Theorem A.8), M_1 equals the smallest possible ΔA for matrix A to have an unstable eigenvalue $j\omega$ for all ω. Therefore, M_1 should be a generally accurate robust stability measure, although Example 2.5 shows that it is still less accurate than M_3. A reason may be that it is based on all $n \times n$ elements of matrix A, while M_3 is based only on the n eigenvalues (of matrix A) relevant to stability.

The main drawback of M_1 is at design. It is very difficult to design a feedback gain matrix K (see Section 3.1) such that the new state matrix $A-BK$ is stable or has arbitrarily assigned eigenvalues, while maximizing the M_1 of matrix $A-BK$. The only existing theoretical result related to M_1 is that M_1 will be at its maximum possible value (= $|\text{Re}(\lambda_n)|$) when $s(\lambda_n)$ is at its minimum (=1) (Lewkowicz and Sivan, 1988). Unfortunately, this is unachievable in almost all cases.

M₂: The term $|\text{Re}(\lambda_n)|$ is the smallest variation for any eigenvalue to reach unstable region (see Figure 2.1). M_2 equals this variation divided by the sensitivity of all eigenvalues $s(\Lambda)$, and is therefore the likelihood margin for λ_n to become unstable. The lower the sensitivity $s(\Lambda)$, the higher the M_2. However, it has been emphasized in the whole Subsection 2.2.1 that $s(\Lambda)$ is *not* an accurate measure of the sensitivity of λ_n as compared to the sensitivity of λ_n itself, $s(\lambda_n)$.

The advantage of M_2 is at design. There exist several general and effective algorithms that can compute matrix K such that arbitrarily given (stable) eigenvalues of matrix $A-BK$ are placed while $s(\Lambda)^{-1}$ or M_2 is maximized (Kautsky et al., 1985; MATLAB, 1990). Some of these algorithms will be introduced in Chapter 9.

M₃: It equals the smallest stability likelihood margin among *all* eigenvalues (not just λ_n)! Because stability requires *every* eigenvalue (not just λ_n) to be stable, M_3 is much more comprehensive than M_2, which concentrates only on λ_n. Furthermore, unlike M_2, the stability likelihood margin for each eigenvalue λ_i is measured by the sensitivity $s(\lambda_i)$ of that λ_i itself, and is therefore far more accurate than the all-eigenvalue sensitivity $s(\Lambda)$ of M_2. This advantage can be proven by the following inequality, which shows that M_3 has much better resolution than M_2:

$$\text{Because} \quad s(\Lambda) \equiv \|V\|\|V^{-1}\| > \|\mathbf{v}_i\|\|\mathbf{t}_i\| \equiv s(\lambda_i), \tag{2.26}$$

$$\text{Therefore} \quad M_2 = s(\Lambda)^{-1}|\text{Re}(\lambda_n)| \le M_3 \le |\text{Re}(\lambda_n)| = \text{maximal possible } M_2 \tag{2.27}$$

Because M_1 shares the same bounds of (2.27) as M_3 (Kautsky et al., 1985; Lewkowicz and Sivan, 1988), it is also more accurate than M_2.

From (2.26 and 2.27), all three robust stability measures/margins will reach the same maximal possible value $|\text{Re}(\lambda_n)|$, if $s(\Lambda)=k(V)=1$ (or $s(\lambda_i)=1$ for all i), even though this ideal case is unachievable in almost all cases.

On the design prospect, the existing algorithms of maximizing M_2 can be used directly to maximize M_3, by simply adding a weighting factor $|\text{Re}(\lambda_i)|^{-1}$ to each eigenvector $\mathbf{v}_i$.

To summarize, M_3 is more generally accurate than M_1 (see Example 2.5) and far more generally accurate than M_2. M_3 and M_2 are far better than M_1 in the actual design computation.

Example 2.5: Lewkowicz and Sivan (1988), Tsui (1994a)

Let the two system state matrices be

$$A_1 = \begin{bmatrix} -3 & 0 & 0 \\ 4.5 & -2 & 0 \\ 0 & 0 & -1 \end{bmatrix} \text{ and } A_2 = \begin{bmatrix} -3 & 0 & 0 \\ 1.5 & -2 & 0 \\ 3 & 0 & -1 \end{bmatrix}$$

The two matrices share the same eigenvalues but different eigenvectors. Hence, the eigenvalue sensitivity and the robust stability are different for these two state matrices.

The eigenstructure decompositions of these two matrices are

$A_1 = V_1 \Lambda_1 V_1^{-1}$

$$= \begin{bmatrix} -0.217 & 0 & 0 \\ -0.976 & 1 & 0 \\ 0 & 0 & 1 \end{bmatrix} \begin{bmatrix} -3 & 0 & 0 \\ 0 & -2 & 0 \\ 0 & 0 & -1 \end{bmatrix} \begin{bmatrix} 4.61 & 0 & 0 \\ 4.5 & 1 & 0 \\ 0 & 0 & 1 \end{bmatrix} \begin{matrix} (\|t_1\| = 4.61) \\ (\|t_2\| = 4.61) \\ (\|t_3\| = 1) \end{matrix}$$

and

$A_2 = V_2 \Lambda_2 V_2^{-1}$

$$= \begin{bmatrix} 0.4264 & 0 & 0 \\ -0.6396 & 1 & 0 \\ -0.6396 & 0 & 1 \end{bmatrix} \begin{bmatrix} -3 & 0 & 0 \\ 0 & -2 & 0 \\ 0 & 0 & -1 \end{bmatrix} \begin{bmatrix} 2.345 & 0 & 0 \\ 1.5 & 1 & 0 \\ 1.5 & 0 & 1 \end{bmatrix} \begin{matrix} (\|t_1\| = 2.345) \\ (\|t_2\| = 1.803) \\ (\|t_3\| = 1.803) \end{matrix}$$

In the above result, all $\|\mathbf{v}_i\| = 1$. Therefore, all $s(\lambda_i) = \|\mathbf{t}_i\|$, which are shown on the right side.

Based on this result, the three robust stability measures M_i ($i = 1, 2$, and 3) for the two system state matrices are listed in Table 2.2.

TABLE 2.2

Robust Stability Measures of Two State Matrices

	A_1	A_2
$M_1 =$	1	0.691
$M_2 = s(\Lambda)^{-1}\|-1)\| = s(\Lambda)^{-1}$	0.1097	0.2014
$s(-3)^{-1}$	0.2169	0.4264
$s(-2)^{-1}$	0.2169	0.5546
$s(-1)^{-1}$	1	0.5546
$M_3 =$	$s(-2)^{-1}\|-2\| = 0.4338$	$s(-1)^{-1}\|-1\| = 0.5546$

Inspection of matrices A_1 and A_2 shows that, unlike A_2, the λ_n (= −1) of A_1 is completely decoupled and thus has sensitivity $s(-1) = 1$. This feature is reflected in M_1 (which reached the maximal possible value $|-1|$) and also in M_3 (which has considered the fact that $s(-1) = 1$).

Inspection of matrices A_1 and A_2 also shows that, unlike A_2, the large element 4.5 of A_1 caused high sensitivity of other two adjacent eigenvalues −3 and −2 and Λ. This feature is reflected in M_2 (which reached the minimal possible value $s(\Lambda)^{-1} = 0.1097$) and also in M_3 (which has considered the fact that both $s(-3)^{-1}$ and $s(-2)^{-1}$ have a low value of 0.2169).

Therefore, only M_3 considered comprehensively and accurately, these two conflicting features of matrix A_1. Based on M_3 of (2.25), among all three eigenvalues of matrix A_1, eigenvalue −2 has a smaller stability likelihood margin (0.4338) than that (1) of eigenvalue −1 and is therefore more likely to become unstable due to the variation of A_1, even though eigenvalue −1 is closer to the unstable region than eigenvalue −2.

These two conflicting features of matrix A_1 can be averaged out so that the robust stability of A_1 and A_2 should be about the same. M_3 reflected this reality more correctly in the sense that it has similar values 0.4338 and 0.5546 for A_1 and A_2, respectively. In contrast, M_1 does not reflect the high sensitivity of other eigenvalues −2 and −3 of matrix A_1 so that the value of M_1 for matrix A_1 (1) is much higher than that for matrix A_2 (0.691), while M_2 does not reflect the low sensitivity of λ_n of matrix A_1 so that the value of M_2 for matrix A_1 (0.1097) is only half of that for matrix A_2 (0.2014).

To summarize, this example shows quite convincingly that M_3 is far more generally accurate than other robust stability measures of the modern control theory such as M_1 and M_2.

This example showed again the importance of decoupling on low eigenvalue sensitivity and high robust stability: because of complete decoupling, the likelihood margin for eigenvalue −1 of matrix A_1 to become unstable is more than double of (1 vs. 0.4338) that of eigenvalue −2 which is coupled by the matrix element 4.5, even though eigenvalue −2 has double the distance to the unstable region than eigenvalue −1.

2.3 Conclusion

The state space control theory provides distinctly general, accurate, and clear analysis on linear time-invariant systems, especially their performance and sensitivity properties. *Only* this kind of analysis and understanding can be used to guide generally and effectively the design of complex control systems. For example, the transfer function and loop transfer function-based performance measure (bandwidth) and robust stability measure (gain and phase margins) are far less generally accurate, and hence cannot be used to guide generally and effectively the design for improving the system performance and robustness.

This is also the reason that linear time-invariant system control results form the basis of the study of other more complicated systems such as non-linear, distributive, and time-varying systems, even though most practical systems belong to those categories. From the next chapter, we will see some basic, practical, critical, yet unsolved design problems being solved simply and satisfactorily, and being solved using state space techniques.

Exercises

2.1 a) Based on the eigenstructure decomposition result of Exercise Problem 1.12 of Chapter 1 and based on (2.1–2.4), derive e^{At} for state matrix A of both systems of Exercise 1.12.

b) Derive the same e^{At} using inverse Laplace transform of $(sI-A)^{-1}$ for both systems.

c) Derive zero-input response $e^{At}\,\mathbf{x}(0)$ with $\mathbf{x}(0)=[1\ \ 2\ \ 3]'$ for both systems. Plot and compare the waveforms of these two responses.

2.2 Repeat Exercise 2.1 for the two systems of Example 2.5.

2.3 Consider the system

$A=A_1$ of Exercise Problem 1.12, $B=[0\ 1\ -1]'$, and $C=\begin{bmatrix} 1 & 0 & -1 \\ 0 & 1 & -1 \end{bmatrix}$.

a) Derive zero-state response $\mathbf{y}(t)$ of this system using (2.3), for a unit step input.

b) Derive the same $\mathbf{y}(t)$ using the inverse Laplace transform of $Y(s)=G(s)/s$. Derive the steady-state value of $\mathbf{y}(t)$ using the final value theorem of $Y(s)$.

c) Compute the bandwidth of the two single-output systems of $G(s)$.

d) Plot and compare the waveforms of the two signals of output $\mathbf{y}(t)$.

2.4 Consider the system

$$(A,B,C)=\left(\begin{bmatrix}-1 & 3 & 0\\ -3 & -1 & 0\\ 0 & 0 & -2\end{bmatrix},\begin{bmatrix}0\\0\\1\end{bmatrix},\begin{bmatrix}5/3 & 5 & 0\\ 0 & 5 & 1\end{bmatrix}\right)$$

Repeat Exercise 2.3 for this system.

Hint: $G(s)=\begin{bmatrix}5(s+2)^2\\ 6s^2+17s+20\end{bmatrix}\Big/\left[(s^2+2s+10)(s+2)\right].$

2.5 Verify the expression (2.22) of Example 2.4.

2.6 Analyze robust stability of the two state matrices of Example 2.1, using the three measures of (2.23–2.25). For Jordan blocks of size greater than 1×1, the eigenvalue sensitivity Definition (2.17) should be modified (Golub and Wilkinson, 1976). A simple method is to add all eigenvalue sensitivities in the same Jordan block together.

For example, the eigenvalue λ_1 (= –1) of matrix A_1 has a 2×2 Jordan block with two pairs of corresponding eigenvectors $\mathbf{v}_1$, $\mathbf{t}_1$ and $\mathbf{v}_2$, $\mathbf{t}_2$. Then,

$$s(\lambda_1)=\|\mathbf{v}_1\|\|\mathbf{t}_1\|+\|\mathbf{v}_2\|\|\mathbf{t}_2\|=(1)(\sqrt{3})+(\sqrt{8})(\sqrt{(1/2)})$$

2.7 Repeat Exercise 2.6 for the two state matrices of Exercise 1.12.

2.8 Using (2.25), calculate robust stability measure M_3 for the following four state matrices

a) $\begin{bmatrix}-1 & 1 & 0\\ 0 & -2 & 0\\ 0 & 0 & -3\end{bmatrix}$

Answer: $M_3=\min\{0.707\times|-1|, 0.707\times|-2|, 1\times|-3|\}=0.707\times|-1|$

Eigenvalue –1 is most likely to become unstable among the three.

b) $\begin{bmatrix}-1 & 0 & 0\\ 0 & -2 & 1\\ 0 & 0 & -3\end{bmatrix}$

Answer: $M_3=\min\{1\times|-1|, 0.707\times|-2|, 0.707\times|-3|\}=1\times|-1|$

Eigenvalue –1 is still most likely to become unstable among the three, though the likelihood is reduced

c) $\begin{bmatrix}-1 & 0 & 0\\ 0 & -2 & 2\\ 0 & 0 & -3\end{bmatrix}$

Answer: $M_3=\min\{1\times|-1|, 0.447\times|-2|, 0.447\times|-3|\}=0.447\times|-2|$

Eigenvalue –2 is most likely to become unstable among the three.

d) $\begin{bmatrix} -1 & 0.5 & 0 \\ 0 & -2 & 0 \\ 0 & 0 & -3 \end{bmatrix}$

Answer: M_3=min{0.89× |–1|, 0.89× |–2|, 1× |–3|}=0.89× |–1|
Comparing part (a), weaker coupling makes higher M_3.

2.9 Repeat Exercise 2.6 for the following six state matrices:

$$\begin{bmatrix} -3 & 0 & 0 \\ 3 & -2 & 0 \\ 1.5 & 0 & -1 \end{bmatrix}, \begin{bmatrix} -3 & 0 & 0 \\ -3 & -2 & 0 \\ 1.5 & 0 & -1 \end{bmatrix}, \begin{bmatrix} -3 & 0 & 0 \\ 2 & -2 & 0 \\ 2.5 & 0 & -1 \end{bmatrix},$$

$$\begin{bmatrix} -3 & 0 & 0 \\ 2.5 & -2 & 0 \\ 2 & 0 & -1 \end{bmatrix}, \begin{bmatrix} -3 & 0 & 0 \\ 4 & -2 & 0 \\ 0.5 & 0 & -1 \end{bmatrix}, \begin{bmatrix} -3 & 0 & 0 \\ 0.5 & -2 & 0 \\ 4 & 0 & -1 \end{bmatrix}$$

2.10 Modifying M_3 of (2.25) to $\min\{s(\lambda_i)^{-1}\ |\text{Re}(\lambda_i)|\ \cos^{-1}(\theta_i)\}$, where θ_i is (if known) the direction of variation of complex λ_i (called "departure angle" in root locus, instead of the pre-assumed 0°), would make the M_3 even more accurate (Tsui, 2015). Derive this new M_3 for the complex eigenvalues of Example 2.2, and try θ_i=0°, 30°, 45°, and 60°. To find the sensitivity of these eigenvalue $s(\lambda_i)$ in 2×2 Jordan blocks, use the method of Exercise 2.6.

3

Feedback System Sensitivity

The feedback system discussed in this book consists of two basic subsystem components – an "open-loop system", which contains the given "plant system", and a feedback controller system called a "compensator". Hence, the analysis of such feedback systems is different from that of single systems.

Of the two critical properties of performance and low sensitivity (robustness) of feedback systems, sensitivity has been less clearly analyzed in the state space control theory. It is analyzed in this chapter, which is divided into two sections.

Section 3.1 highlights a concept of the classical control theory about feedback system sensitivity – the decisive role of the loop transfer function. This concept will guide the design throughout this book, even though the focus remains on state space models of the systems.

Section 3.2 analyzes the sensitivity properties of three basic existing feedback control structures of the state space control theory – direct state feedback, static output feedback, and observer feedback. The emphasis is on the observer feedback structure, which is far more commonly studied and whose compensator is the only dynamic system among the three feedback structures. A key design requirement on the robustness property of this feedback structure, called "loop transfer recovery" (LTR), is introduced.

3.1 Sensitivity and Loop Transfer Function of the Feedback Systems

The basic feedback control structure of the control systems theory is shown in Figure 3.1

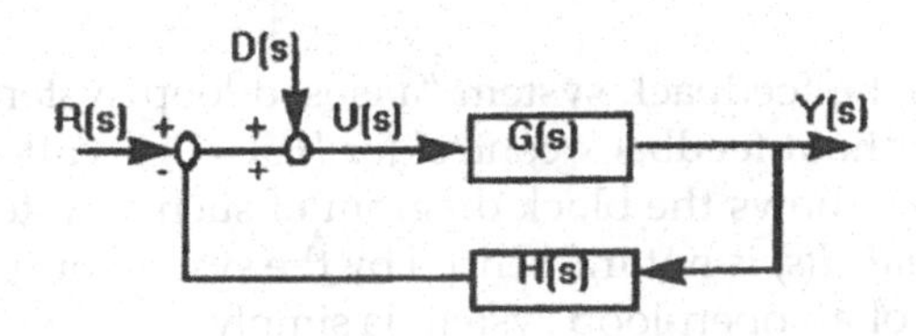

FIGURE 3.1
The basic structure of feedback (closed-loop) system.

DOI: 10.1201/9781003259572-3

In this system structure, there is a feedback path from the plant system output $Y(s)$ to plant system input $U(s)$, through a general feedback control system called "compensator" $H(s)$. The signals $R(s)$ and $D(s)$ are the Laplace transforms of an external reference signal $\mathbf{r}(t)$ and an input disturbance signal $\mathbf{d}(t)$, respectively.

The plant system whose output $Y(s)$ is the subject of control, is either $G(s)$ or a component of $G(s)$ and is treated in this book as $G(s)$. Therefore, the controller to be designed is $H(s)$.

The structure of Figure 3.1 is very basic but is the most common among control structures. For more complicated control system configurations, the analysis and design are usually carried out block by block and module by module, with each block or module structured as shown in Figure 3.1.

Because input $U(s)$ can control the behavior of the plant system's output $Y(s)$, we call $U(s)$ the "control input signal". The generation and actuation of a high-power control input will cause disturbance. Such disturbance is treated as an additional signal $D(s)$ in Figure 3.1.

The purpose or requirement of control systems is generally the control of the system output (or response) $Y(s)$ so that it can quickly reach and stabilize at its desired state, such as the desired vehicle and engine speed, the desired radar and airborne system angle, the desired robot arm position, and the desired container pressure and temperature, etc. The desired system output state is specified by the reference signal $R(s)$. Hence, how well the system output reaches its desired state determines the performance of the system.

The final steady state of $Y(s)$ is relatively easy to analyze (using the final value theorem such as Example 2.2) and relatively easy to satisfy. Hence, the transient response properties such as convergence speed are critical factors to system performance and are the main challenges of feedback control design.

The most basic feature of the feedback system in Figure 3.1 is that the control signal $U(s)$, which controls the signal $Y(s)$, is itself controlled based on $Y(s)$. This feedback of $Y(s)$ to $U(s)$ creates a loop that starts and ends at $U(s)$ in Figure 3.1 and whose transfer function is called the "loop transfer function", represented as

$$L(s) = -H(s)G(s) \tag{3.1}$$

We therefore call the feedback system "a closed-loop system". On the other hand, a system without feedback control (or $H(s) = 0$) is called an "open-loop system". Figure 3.2 shows the block diagram of such a system, in which the control input signal $U(s)$ is not influenced by the system output $Y(s)$. The loop transfer function of an open-loop system is simply

$$L(s) = 0 \tag{3.2}$$

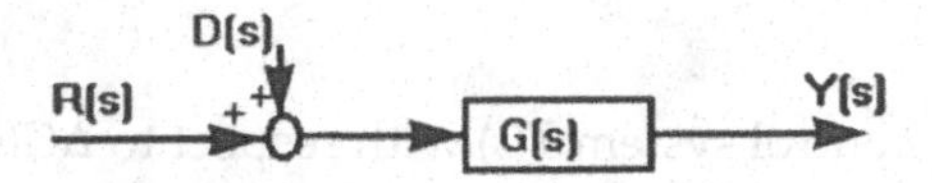

FIGURE 3.2
The block diagram of an open-loop system.

The main difference between feedback control and control without feedback, is the sensitivity to the plant system mathematical model uncertainty defined as $\Delta G(s)$ and to the control input disturbance $D(s)$. This section shows that this difference is determined almost solely by the difference of the loop transfer function $L(s)$. These two sensitivities are treated in Subsections 3.1.1 and 3.1.2, respectively.

To simplify the description of this concept, only SISO systems are studied in this section. However, this basic concept is general to MIMO systems as well.

3.1.1 Sensitivity to System Model Uncertainty

In most practical situations, the given mathematical model (either state space or transfer function) of the plant system is inaccurate. This is because the practical physical system is usually nonlinear, and its parameters are usually distributive and are difficult to measure accurately. Even for an initially accurate model, the actual plant system will inevitably experience wear-out and accidental damage.

To summarize, there is an inevitable and unknown difference between the actual plant system and its mathematical model $G(s)$. This difference is called "model uncertainty" and is defined as $\Delta G(s)$. Therefore, it is essential that the control systems, which are designed based on the given and available mathematical model $G(s)$, have low sensitivity to $\Delta G(s)$.

In SISO systems, the overall transfer function from $R(s)$ to $Y(s)$, of the closed-loop and open-loop control systems of Figures 3.1 and 3.2 are, respectively,

$$T_c(s) = \frac{G(s)}{1 + H(s)G(s)} \tag{3.3a}$$

and

$$T_o(s) = G(s) \tag{3.3b}$$

Let $\Delta T(s)$ be the uncertainty (or deviation) of the overall control system $T(s)$ caused by $\Delta G(s)$. We will use the relative plant system model uncertainty $\Delta G(s)/G(s)$ and the relative control system uncertainty $\Delta T(s)/T(s)$, to define the sensitivity of $T(s)$ with respect to $\Delta G(s)$.

Definition 3.1

The sensitivity of a control system $T(s)$ with respect to $\Delta G(s)$ is defined as

$$s(T)|_G = \frac{|\Delta T(s)/T(s)|}{|\Delta G(s)/G(s)|} \tag{3.4a}$$

For small enough $\Delta G(s)$ and $\Delta T(s)$,

$$s(T)|_G \approx \left|\frac{\partial T(s)G(s)}{\partial G(s)T(s)}\right| \tag{3.4b}$$

where $\partial T(s)/\partial G(s)$ stands for the partial derivative of $T(s)$ with respect to $G(s)$. Equation (3.4b) is the general formula for deriving $s(T)|_G$.

By substituting $T_c(s)$ of (3.3a) and $T_o(s)$ of (3.3b) into (3.4b), we have

$$s(T_c)|_G = \left|\frac{1}{1+H(s)G(s)}\right| = \left|\frac{1}{1-L(s)}\right| \tag{3.5a}$$

and

$$s(T_o(s))|_G = 1 \tag{3.5b}$$

A comparison of (3.5a) and (3.5b) shows clearly that the sensitivity to the plant system model uncertainty of a closed-loop system, $s(T_c(s))|_G$, can be much lower than that of an open-loop system $s(T_o(s))|_G$. The difference is determined solely by the loop transfer function $L(s)$.

Example 3.1: Sensitivity to the Uncertainty of Individual Plant System Parameters

Let $G(s) = \dfrac{K}{s+\lambda}$ and $H(s) = 1$.

Then from (3.3a) and (3.3b),

$$T_c(s) = \frac{K}{s+\lambda+K}$$

and

$$T_o(s) = \frac{K}{s+\lambda}.$$

Thus from (3.4b),

$$s(T_c)\big|_K = \left|\frac{\partial T_c(s)}{\partial K}\frac{K}{T_c(s)}\right| = \left|\frac{s+\lambda+K-K}{(s+\lambda+K)^2}\frac{K}{K/(s+\lambda+K)}\right|$$

$$= \left|\frac{1}{1-L(s)}\right|$$

$$s(T_o)\big|_K = \left|\frac{\partial T_o(s)}{\partial K}\frac{K}{T_o(s)}\right| = \left|\frac{1}{(s+\lambda)}\frac{K}{K/(s+\lambda)}\right| = 1$$

$$s(T_c)\big|_\lambda = \left|\frac{\partial T_c(s)}{\partial \lambda}\frac{\lambda}{T_c(s)}\right| = \left|\frac{-K}{(s+\lambda+K)^2}\frac{\lambda}{K/(s+\lambda+K)}\right|$$

$$= \left|\frac{-\lambda/(s+\lambda)}{1-L(s)}\right|$$

and

$$s(T_o)\big|_\lambda = \left|\frac{\partial T_o(s)}{\partial \lambda}\frac{\lambda}{T_o(s)}\right| = \left|\frac{-K}{(s+\lambda)^2}\frac{\lambda}{K/(s+\lambda)}\right|$$

$$= \left|\frac{-\lambda}{(s+\lambda)}\right|$$

Therefore, the sensitivity to either parameter K or parameter λ of a closed-loop system $T_c(s)$ equals that of an open-loop system divided by $1 - L(s)$. For open-loop control systems $T_o(s)$, at $s = 0$, this sensitivity equals $1 = 100\%$, which is quite high.

3.1.2 Sensitivity to Control Input Disturbance

As introduced at the beginning of this section, disturbance $D(s)$ associated with the generation of high-power control input $U(s)$ is serious and inevitable. Hence, a practical control system must have low sensitivity to $D(s)$.

In practice, the controller component which actually generates and asserts the control, is called the "actuator".

From the superposition principle of linear systems, in the presence of $D(s)$, the closed-loop system (Figure 3.1) and open-loop system (Figure 3.2) responses are, respectively,

$$Y_c(s) = \frac{G(s)}{1-L(s)}R(s) + \frac{G(s)}{1-L(s)}D(s) \tag{3.6a}$$

and

$$Y_o(s) = G(s)R(s) + G(s)D(s) \tag{3.6b}$$

If among the two terms of $Y_c(s)$ and $Y_o(s)$, the first term is the desired response when $D(s) = 0$, then the second term is the deviation from the desired first term and is the whole effect of disturbance $D(s)$. Therefore, the gain (magnitude of the transfer function) of this second term represents the sensitivity of the corresponding system toward $D(s)$. The higher the gain, the higher the sensitivity toward $D(s)$.

Definition 3.2

A system's sensitivity to its control input disturbance is represented by its gain from this disturbance to its output.

Similar to the Conclusions of Subsection 3.1.1, a comparison of the second term (or the gain from disturbance to output) of (3.6a) and (3.6b) shows clearly that, the sensitivity to disturbance of a closed-loop system is $(1 - L(s))^{-1}$ times that of an open-loop system.

Because of these conclusions on sensitivities to system model uncertainty and to control input disturbance, function $[I - L(s)]^{-1}$ is called the well-known "sensitivity function".

For MIMO systems, the gain or the norm of the $p \times p$ loop transfer matrix $L(s)$ equals its largest singular value (see (2.9)).

Example 3.2: Sensitivity to Output Measurement Noise

In addition to the sensitivities to system model uncertainty and input disturbance, it is also important to consider the sensitivity to system output measurement noise, which is the measurement error and is also inevitable. The measurement noise is represented by an additional signal $N(s)$ to output measurement $Y(s)$ in the block diagram of Figure 3.3.

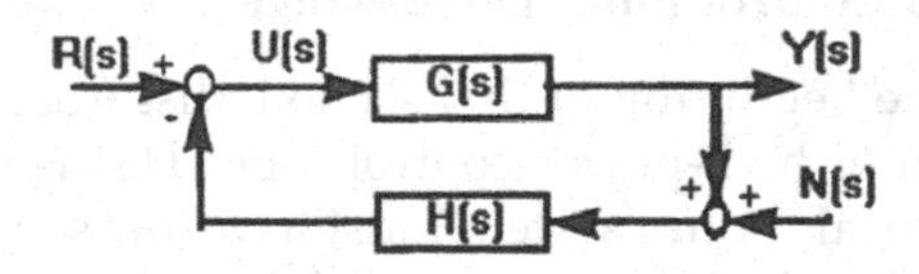

FIGURE 3.3
Feedback control system with output measurement noise.

In many practical analog systems, in particular, non-electrical systems, the output signals $Y(s)$, such as velocity, angle, position, pressure, and temperature, are very difficult to measure accurately. In addition, these nonelectric analog measurements must be converted to electrical signals, to be displayed or be used to compute the corresponding feedback control, and this conversion can also cause error.

The device that makes this conversion of two different physical signals and measurements is called a "transducer" in practice. It is a challenge to make accurate and reliable transducers.

Because the purpose of measuring feedback system output $Y(s)$ is to use it to determine a corresponding feedback control input $U(s)$, the undesirable effect of measurement noise is reflected mainly on signal $U(s)$.

Applying Mason's formula to the system of Figure 3.3 and assuming $R(s) = 0$ as in (3.6), then

$$U(s)=\frac{-H(s)}{1+H(s)G(s)}N(s)=\frac{-H(s)}{1-L(s)}N(s) \quad (3.7)$$

This is the effect of measurement noise $N(s)$ on control input $U(s)$. Similar to Definition 3.2, the lower the magnitude of the transfer function of (3.7), the lower the sensitivity against $N(s)$.

By substituting (3.7) into (1.9) ($Y(s) = G(s)U(s)$), we get

$$Y(s)=\frac{-G(s)H(s)}{1-L(s)}N(s)=\frac{L(s)}{1-L(s)}N(s) \quad (3.8)$$

This is the effect of output measurement noise $N(s)$ on the output $Y(s)$, quite similar to (3.7).

It is very clear from (3.7) that the lower the feedback controller gain $|H(s)|$, the lower the sensitivity and the effect of $N(s)$. For example, in open-loop systems which have no feedback ($H(s) = 0$) and in which output measurement noise does not affect the input $U(s)$ at all, the sensitivity and the effect of output measurement noise $N(s)$ to $U(s)$ are zero.

This conclusion is in sharp contrast to the previous two conclusions on the sensitivity to system model uncertainty and input disturbance, because those two conclusions both imply that increasing the feedback controller gain $|H(s)|$ or the loop gain $|L(s)| = |H(s)G(s)|$ reduces the corresponding two sensitivities $[I-L(s)]^{-1}$!

In fact, indiscriminately increasing the feedback control gain or loop gain, not only is against reducing the sensitivity to output measurement noise, but also has more serious harms to the overall feedback control system, as listed in the following:

1. A high loop gain $|L(s)|$ is likely to cause feedback system instability, from root locus rules. This is especially true for plant systems with unstable zeros or with the number of pole-zero excess exceeding two.
2. A high loop gain $L(s)$ generally reduces the feedback system ($T_c(s)$) performance. From the definitions of $T_c(s)$ of (3.3a) and bandwidth of Section 2.1, a higher $H(s)$ can reduce $T_c(s)$ and hence reduce the bandwidth.
3. A higher loop gain or a higher feedback controller gain $H(s)$ is more difficult to realize and implement in practice. A higher controller gain generally requires a higher control power and is more likely to inflict disturbance, saturation, and component failure (Hu and Lin 2001).

After all, a high feedback controller gain $H(s)$ at high-performance systems, is like making sharp steering turns at high-speed driving. Nothing can be more dangerous than that.

To summarize, the critical factor of feedback system sensitivity is the whole loop transfer function itself, not the high gain of the loop transfer function, nor the gain of the loop transfer function only. In other words, it may be wiser not to design a feedback controller from the prospect of loop gain of the classical control theory.

3.2 Sensitivity of Feedback Systems of the Modern Control Theory

Section 3.1 described the critical importance of the loop transfer function on feedback system sensitivity. The same concept will be used to analyze the sensitivity of three existing and basic control structures of the modern control theory. These three structures are state feedback, static output feedback, and observer feedback. Of the three structures, the dynamic observer feedback system is more general and more challenging to design, than the other two.

As the loop transfer function is determined by the internal system structure and not by an external signal, we will let $R(s) = 0$. We also assume that system $G(s)$ is irreducible.

3.2.1 State Feedback Control Systems

The state feedback control (or direct state feedback control) is defined by the control signal

$$\mathbf{u}(t) = -K\mathbf{x}(t) \tag{3.9}$$

where $\mathbf{x}(t)$ is the system state vector and K is called the "state feedback gain" or "state feedback control law" and is constant.

The block diagram of this feedback system, based on Figure 1.5, is shown in Figure 3.4.

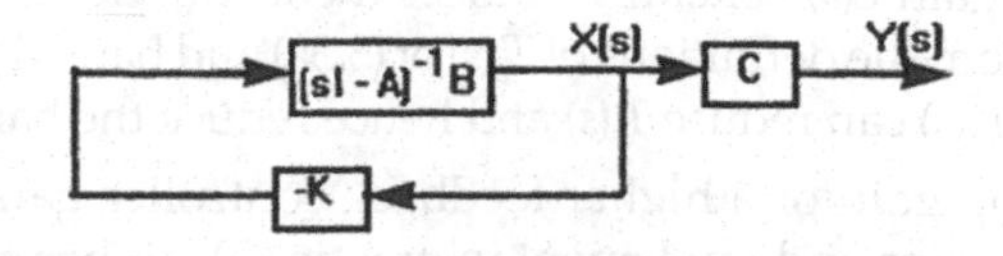

FIGURE 3.4
Block diagram of direct state feedback control system.

It is clear from Figure 3.4 that the loop transfer function of this system is

$$L(s) = -K(sI - A)^{-1}B \equiv L_{KX}(s) \tag{3.10}$$

By substituting (3.8) into (1.1a), the dynamic equation of this feedback system becomes

$$d\mathbf{x}(t)/dt = (A - BK)\mathbf{x}(t) + B\mathbf{u}(t) \tag{3.11}$$

Hence, matrix $A–BK$ is the state matrix of this feedback system, and its eigenvalues are the poles of this feedback system.

From Section 1.1, the system state is the most detailed information of the current state of the system, and the state space model provides the most detailed and explicit information on the system's internal structure. Therefore, the state feedback control design is based on the best possible information. Hence, if this control is designed properly, it should improve system performance and robustness most effectively, even though this design is not aimed directly at the shaping of the loop transfer function $L_{Kx}(s)$.

Theorem 3.1

For any controllable plant system, the direct state feedback control can assign all n arbitrarily given eigenvalues to state matrix $A–BK$, and the feedback system remains controllable. Eigenvalues of state matrix or system poles most directly determine system performance.

Proof

Any controllable system is similar to its block controllable canonical form ($A = A'$, $B = C'$), which is the dual of its corresponding observable canonical form of (A, C) of (1.16).

The form (1.16) implies that there exists a matrix K' such that all unknown parameters of matrix $A–K'C$ can be arbitrarily assigned and that matrix $A–K'C$ remains to be in the observable canonical form for any K'. Hence, arbitrarily given eigenvalues of matrix $A–K'C$ can be assigned, and the system ($A–K'C$, C) remains observable for any K'.

From duality, the above conclusion implies that arbitrarily given eigenvalues can be assigned to matrix $A'–C'K$ and that system ($A–C'K$, C') remains controllable for all K.

However, matrix $A–BK$ cannot in general preserve the same original observable canonical form of matrix A. Hence, direct state feedback control cannot in general preserve the observability of the original system (A, B, C).

In addition, eigenvectors can also be assigned if $p>1$, thus achieving robustness (Section 2.2).

The actual design algorithm for matrix K and for eigenvalue/eigenvector assignment (or eigen-structure assignment) will be presented in Chapters 8 and 9.

Besides the ability to assign eigenstructure, state feedback control can also realize the so-called "linear-quadratic (LQ) optimal control", whose design will be introduced in Chapter 10. It has been proved that the loop transfer function $L_{Kx}(s)$ of this LQ-optimal control satisfies "Kalman inequality" such that

$$\left[I - L_{Kx}(j\omega)\right]^* R\left[I - L_{Kx}(j\omega)\right] \geq R \text{ for all } \omega \tag{3.12a}$$

where R is symmetrical positive definite ($R = R^* > 0$) (Kalman, 1960).

It has been proven from (3.12a) that for $R = rI$ ($r > 0$ is a scalar),

$$\sigma_i\left[I - L_{Kx}(j\omega)\right] \geq 1 \text{ for all } \omega \tag{3.12b}$$

where σ_i ($i = 1, \ldots, p$) is the i-th singular value of the matrix. Equation (3.12b) indicates that the gain margin of this system is between ½ and ∞, while the phase margin of this system is greater than 60° (Lehtomaki et al., 1981). This result can be very clear from the SISO cases.

The plot of all values of a SISO $-L_{Kx}(j\omega)$ for all ω satisfying (3.12b) is shown in the shaded area of Figure 3.5.

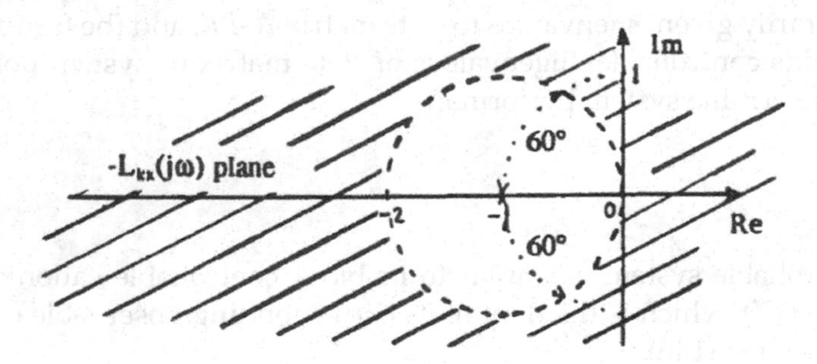

FIGURE 3.5
Loop transfer frequency response of state feedback LQ-optimal control.

It is very clear from Figure 3.5 that the margin between all values in the shaded area to the –1 point is at least ½ in gain magnitude and at least 60° in phase angle. Because the number of encirclements of the –1 point determines feedback system stability (Nyquist test), the plot of Figure 3.5 implies fairly large stability margins and very good robust stability of this feedback system.

Notice that at this good robust stability, no large loop gain or large value of loop transfer function $|L_{Kx}(j\omega)|$ (distance to the origin) is required at all. However, this control or Figure 3.5 does require that at low loop gain $|L_{Kx}(j\omega)|$, the phase angles of $L_{Kx}(j\omega)$ must be between 0° and ±90° (or $L_{Kx}(j\omega)$ must have positive real parts or must be at the right half-plane). This requirement implies the relative importance of the loop phase vs. the loop gain. However, it seems that no other existing designs of $L(j\omega)$ are aimed at satisfying this important requirement.

Although the above good robust stability margins can be guaranteed, it can be shown at the beginning of Chapter 10 that the LQ optimality criterion can be set to neglect robustness in the first place (such as the criterion for minimum-time control). This is another proof that gain/phase margins are not generally accurate measures of the actual system robustness (see Subsection 2.2.2). Even with this critical drawback of inaccuracy, these two margins themselves cannot be calculated accurately if there are $p \times p$ loop transfer functions in matrix $L(j\omega)$.

Furthermore, it is almost unimaginable that an MIMO controller $H(j\omega)$ can be designed generally, simply, and in closed form, such that the corresponding $L(j\omega) = -H(j\omega)G(j\omega)$ can have a really optimized shape in both gain and phase (such as to meet the requirement of Figure 3.5), or such that the norm of the corresponding sensitivity function $[I-L(j\omega)]^{-1}$ (which does not guarantee performance at all) can be minimized, whether this minimized norm is H_2 or H_∞ (see (2.9)).

On the other hand, such an optimized loop transfer function $L_{Kx}(j\omega)$ *can* be generally and effectively designed by state feedback control using state space techniques! For example, this control can guarantee performance by eigenvalue assignment (Theorem 3.1) and simultaneously improve robustness by eigenvector assignment and by optimizing a far more generally accurate robust stability margin M_3 of (2.25) (Subsection 2.2.2).

All of the above means that the transfer function model-based control design is inferior to the state space model-based control design such as state feedback control design. The only kind of information provided by the transfer function model is a lot of raw data of frequency response for $\omega = 0-\infty$, while the information of the state space model can be simplified and pinpointed to only n eigenvalues essential to performance, and only $2n$ parameters of eigenvectors (of each eigenvalue) essential to the sensitivity. This should be the reason why the control research community turned from the classical control theory to the modern control theory in the 1960s and 70s.

The main and critical drawback of direct state feedback control is that it cannot be generally implemented and realized. This is because in almost all systems (1.1), only a few terminal inputs and outputs can be measured, not all n internal state variables. In other words, the measurement information available to the controller cannot be as ideal and complete as the complete set of $\mathbf{x}(t)$. Therefore, direct state feedback control is only a theoretically ideal control based on the ideal information, while almost all real plant systems are not that ideal.

An observer (see Subsection 3.2.3) is needed to implement the $K\mathbf{x}(t)$ control for those non-ideal systems, and thus the new beginning of the modern control theory from the late 1950s, with R. E. Kalman as its founder. We will see from Theorem 3.4 that only a special class of observers called the output feedback compensator *(Definition 3.3)* can fully realize the direct state feedback control including its robustness properties. But we will also see from Section 4.3 that under the existing separation principle, which means to realize a predesigned and arbitrarily given $K\mathbf{x}(t)$-control (or arbitrarily given K), such an output feedback compensator does not exist for the great majority of plant systems.

3.2.2 Static Output Feedback Control Systems

In static output feedback control, the control input is

$$\mathbf{u}(t) = -K_y\mathbf{y}(t) = -K_yC\mathbf{x}(t) \tag{3.13}$$

where $\mathbf{y}(t) = C\mathbf{x}(t)$ is the measured system output and K_y is a constant. The block diagram of this feedback system is shown in Figure 3.6.

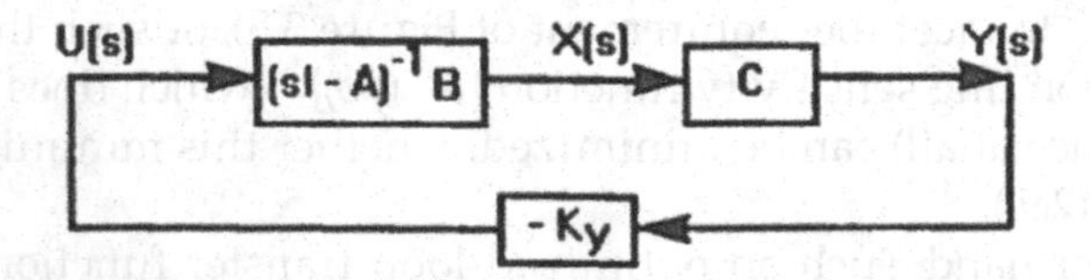

FIGURE 3.6
Block diagram of a static output feedback control system.

The state matrix and the loop transfer function of this feedback system are $A–BK_yC$ and

$$L(s) = -K_yC(sI - A)^{-1}B \tag{3.14}$$

respectively.

The loop transfer function (3.14) indicates that the full realization of this control including its loop transfer function is guaranteed.

This control is very similar to the state feedback control with the only difference that the state feedback control gain is now

$$K = K_yC \tag{3.15}$$

where C is given and K_y is the control gain to be designed. Therefore, the static output feedback control implements a constrained state feedback control, with its state feedback control law K constrained to be linear combinations of the rows of a given matrix C, or K' is constrained to be within the range space of C' (see Subsection A.1.2).

Because the number of rows of C or the dimension of range space of C' is m, which is less than n and which can be much less than n, this constraint exists and can be very severe, and the corresponding $K_yC\mathbf{x}(t)$–control can be weaker and can be much weaker than the direct state feedback control. This is the main disadvantage of the static output feedback control.

Example 3.3

In a second-order SISO system, if C = either [1 0] or [0 1], then from (3.15), the static output feedback control $K_yC\mathbf{x}(t)$ can only implement a constrained state feedback control $K\mathbf{x}(t) = [k_1\ k_2]\mathbf{x}(t) = k_1x_1(t) + k_2x_2(t)$ such that $K\mathbf{x}(t)$ can only be either $k_1x_1(t)$ or $k_2x_2(t)$. This constraint weakens the corresponding control very severely.

As a result of this constraint, unlike the direct state feedback control, which can assign all n eigenvalues and all eigenvectors (if $p>1$), the static output feedback control can assign all n eigenvalues and only $n–m$ or $n–p$ eigenvectors only if $m+p>n$ (Kimura, 1975; Tsui, 1999a), and can assign all n eigenvalues only generically only if $m\times p>n$ (Wang, 1996).

We can grade the overall level of feedback system performance and robustness as follows. It is reasonable to claim that the level of direct state feedback is "ideal", the level of pole assignment and partial (less than n) eigenvector assignment is "excellent", and the level of generic pole assignment is "very good".

The reason for claiming that the combined performance and robustness level of generic pole assignment control is "very good", is as follows. The all-pole assignment is very difficult and requires very strong control to achieve. Thus, the control that is capable of pole assignment can be considered strong enough, so that it can be modified from this original design objective to take care more about robustness.

Therefore, it is reasonable to consider only the plant systems with $m+p\leq n$ nontrivial, because only this kind of systems require a dynamic controller like an observer (static output feedback control is not good enough) (Tsui, 2015).

If $m = n$ and C is nonsingular (it has n linearly independent rows or rank(C) = n or C^{-1} exists), then (3.15) will no longer be a constraint and static output feedback control becomes direct state feedback control in the sense that the full $\mathbf{x}(t)$ is available as $C^{-1}\mathbf{y}(t)$ and that any $K\mathbf{x}(t)$ signal can be realized by static output feedback control as $KC^{-1}\mathbf{y}(t)$ for any K (or $K_y = KC^{-1}$ for any K). Hence, direct state feedback control is a special case of static output feedback, in the special case when $m = n$.

In the same sense, a constrained state feedback control $\underline{KC}\mathbf{x}(t)$ can be called a "generalized state feedback control", where parameter $\underline{C}$ (instead of C) has rank equal to $m+r$, and this rank ranges between n and m. This control can unify direct state feedback control and static output feedback control as its two extreme special cases (no constraint and most constrained).

Because $m+r>m$, this generalized state feedback control overcomes the main drawback of static output feedback control – too few rows of matrix C (only m) and thus too much constraint on the control law K. This is because more independent signals (say r $T\mathbf{x}(t)$ signals) are estimated and added to the output measurements, for generating the generalized state feedback control signal.

The new design principle of this book designs and implements such a generalized state feedback control, because this control has a distinct advantage over the existing observer/state feedback control – the critical loop transfer function and robust property of this control are automatically guaranteed of full realization! This is because the generalized state feedback control can be realized by an output feedback compensator (Theorem 3.4), which is valid for a great majority of plant systems (Chapter 6).

To summarize, generalized state feedback control overcomes the main disadvantage of static output feedback control because $r + m > m$, and overcomes the critical disadvantage of direct state feedback control because loop transfer function and robustness of this control can be far more generally realized.

The actual design methods of this control and static output feedback control for eigenstructure assignment will be presented in Chapters 8 and 9, while the design for LQ-optimal control will be presented in Chapter 10.

Finally, for theoretical integrity, static output feedback control will maintain both the controllability and observability of the original system (A, B, C), because this control will not alter either the controllable canonical form or the observable canonical form of the original plant system (A, B, C).

3.2.3 Observer Feedback System – Loop Transfer Recovery

An observer feedback system does not require the direct measurement of all system states while implementing (at least) a generalized state feedback control, which is much stronger than the static output feedback control. Therefore, the observer feedback control structure overcomes the main disadvantages of both direct state feedback control and static output feedback control. This feedback system is the most important among all three feedback systems of the modern control theory. This should be why, that the time of Kalman filter which is a special observer (see Section 4.1), is considered the start the state space control theory.

An observer is itself a linear time-invariant dynamic system, whose general state space model is defined in this book as

$$d\mathbf{z}(t)/dt = F\mathbf{z}(t) + L\mathbf{y}(t) + TB\mathbf{u}(t) \tag{3.16a}$$

$$-K\mathbf{x}(t) = -K_z\mathbf{z}(t) - K_y\mathbf{y}(t) \tag{3.16b}$$

where $\mathbf{z}(t)$ is the state vector of the observer system and has dimension r, matrix B comes from the plant system state space model (A, B, C), and other observer parameters (F, T, L, K_z, K_y) are of the appropriate dimensions and are free to be designed.

This observer definition is more general than the existing ones. It is a most general and standard model for a system that takes $\mathbf{y}(t)$ and $\mathbf{u}(t)$ as its input and $-K\mathbf{x}(t)$ as its output. The distinct advantages of this general observer model will be made obvious in the rest of this book.

Let us first analyze the condition for an observer of (3.16) to generate a desired state feedback control signal $K\mathbf{x}(t)$.

Because both $\mathbf{x}(t)$ and $\mathbf{y}(t) = C\mathbf{x}(t)$ are time-varying signals, and because both K and C are constants, it is obvious that to generate $K\mathbf{x}(t)$ in (3.16b), the observer state $\mathbf{z}(t)$ must converge to $T\mathbf{x}(t)$ for a constant T. This is the foremost important requirement of observer design (Tsui, 2015).

Theorem 3.2

The necessary and sufficient condition for observer state $\mathbf{z}(t)$ to converge to $T\mathbf{x}(t)$ for a constant T, or for observer output to converge to $K\mathbf{x}(t)$ for a constant K, and for any $\mathbf{x}(0)$ and $\mathbf{z}(0)$, is

$$TA - FT = LC \tag{3.17}$$

where all eigenvalues of matrix F must be stable.

Proof (Luenberger, 1971)

From (1.1a)

$$T\,d\mathbf{x}(t)/dt = T\,A\mathbf{x}(t) + T\,B\mathbf{u}(t) \tag{3.18}$$

By subtracting (3.18) from (3.16a), we have

$$d\left[\mathbf{z}(t) - T\mathbf{x}(t)\right]/dt = F\mathbf{z}(t) + LC\mathbf{x}(t) - TA\mathbf{x}(t) \tag{3.19}$$

$$= F\mathbf{z}(t) - FT\mathbf{x}(t) + FT\mathbf{x}(t) + LC\mathbf{x}(t) - TA\mathbf{x}(t)$$

$$= F\left[\mathbf{z}(t) - T\mathbf{x}(t)\right] \tag{3.20}$$

if and only if (3.17) holds. Because the solution of (3.20) is

$$\left[\mathbf{z}(t) - T\mathbf{x}(t)\right] = e^{Ft}\left[\mathbf{z}(0) - T\mathbf{x}(0)\right]$$

$\mathbf{z}(t)$ converges to $T\mathbf{x}(t)$ for any $\mathbf{z}(0)$ and $\mathbf{x}(0)$, if and only if all eigenvalues of F are stable.

This proof also shows that it is necessary to have the observer gain to $\mathbf{u}(t)$ be defined as TB in (3.16a), for (3.19) to hold.

After $\mathbf{z}(t) \rightarrow T\mathbf{x}(t)$ is guaranteed, replacing $\mathbf{z}(t)$ by $T\mathbf{x}(t)$ in the output part of observer (3.16b) yields

$$K = K_z T + K_y C = \left[K_z : K_y\right]\left[\,T' : C'\,\right]' \equiv \underline{K}\underline{C} \tag{3.21}$$

Therefore, (3.17) (with stable F) and (3.21) together form the necessary and sufficient conditions for an observer (3.16) to generate a desired state feedback signal $K\mathbf{x}(t)$.

The above introduction shows clearly that conditions (3.17) and (3.21) have natural and completely separate physical meanings. More explicitly, (3.17) determines the dynamic part (F, T, L) of the observer (3.16a) exclusively and guarantees that the observer state $\mathbf{z}(t)$ converges to $T\mathbf{x}(t)$ exclusively, while (3.21) determines the output part ($[K_z: K_y]$ or $\underline{K}$) of the observer (3.16b) exclusively under the assumption that (3.17) and $\mathbf{z}(t) \rightarrow T\mathbf{x}(t)$ are satisfied already.

Many design algorithms can compute the solutions of (3.17) and (3.21), for arbitrarily given stable eigenvalues of F and arbitrarily given K. This book will present only one of such algorithms (Algorithm 7.1), which has an additional feature of minimized observer order. This is because (3.17) and (3.21) guarantee only the generation of the $K\mathbf{x}(t)$ signal, but *not* the realization of the loop transfer function $L_{Kx}(s)$ (3.14) or of the critical robustness properties, of the corresponding $K\mathbf{x}$(t)-control. This is the critical problem of the entire state space control theory ever since the 1960s.

Solving this critical problem is the most important technical contribution of this book. We will describe and analyze this problem in detail.

As stated at the beginning of this subsection, observer feedback systems have been the most general control structure of the state space control theory since the 1960s. Because an observer can generate the desired $K\mathbf{x}(t)$ signal and because the poles of this observer feedback system equal the union of observer poles and eigenvalues of matrix A–BK (Theorem 4.1), people assumed that the observer feedback system has the same robustness as the direct state feedback system since the 1960s.

However, ever since the 1960s, bad robustness properties of observer feedback systems have commonly been experienced in practice, even though the observer generates a state feedback control signal that is supposed to guarantee good robustness (see Subsection 3.2.1). In other words, the observer failed to realize the robust properties of direct state feedback control even though the signal of this control is generated (or (3.17) and (3.21) are satisfied). Because robustness is the foremost purpose and property of feedback control, the state space control theory has not found many successful practical applications since the 1960s (Levis 1987).

Meanwhile, the application of rational polynomial matrix has extended the classical control theory into MIMO systems (Rosenbrock, 1974; Wolovich, 1974; Kaileth, 1980; Chen, 1984; Vidyasagar, 1985). Using the concept of loop transfer functions, the classical control theory can consider feedback system robustness relatively easily (see Section 3.1), though far less accurately. In addition, matrix singular values can indicate matrix norm (2.9) (such as loop transfer function norm/loop gain) accurately and can be computed by computer relatively easily. As a result, the classical control theory, especially in the area of robust control, has witnessed significant development since the 1980s (Doyle et al., 1992). For example, minimization of sensitivity function called the H_∞ problem whose brief formulation can be

$$\text{Min}\left\{\max_\omega\left\{\left\|\left[I-L(j\omega)\right]^{-1}\right\|_\infty \text{ at all } \omega = 0 \text{ to } \infty\right\}\right\}$$

has received much attention (Zames, 1981; Francis, 1987; Doyle et al., 1989; Kwakernaak, 1993; Zhou et al., 1995).

Until the end of the 1970s, the cause of the failed robust realization by the observer was understood – it is caused by the failed realization of the loop transfer function $L_{Kx}(s)$ of direct state feedback control (Doyle, 1978). Let us describe this cause in the following.

The overall feedback system of the general observer (3.16) can be depicted in Figure 3.7.

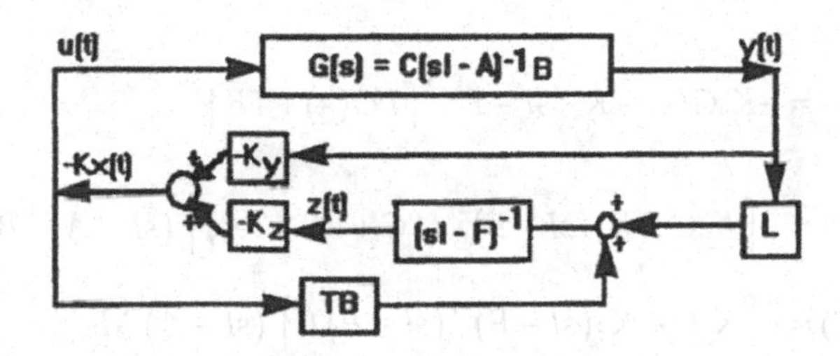

FIGURE 3.7
Block diagram of general observer feedback system.

Figure 3.7 shows that an observer can be considered a feedback controller $H(s)$ whose input is $\mathbf{y}(t)$ and output is $\mathbf{u}(t)$, where

$$\begin{aligned} U(s) &= -H(s)Y(s) \\ &= -\left[I + K_z(sI-F)^{-1}TB\right]^{-1}\left[K_y + K_z(sI-F)^{-1}L\right]Y(s). \end{aligned} \tag{3.22}$$

It should be noted from (3.16) that the transfer function from signal $\mathbf{y}(t)$ only (not signal $\mathbf{u}(t)$) to signal $K\mathbf{x}(t)$ is, from (3.16) and (1.8),

$$H_y(s) = -\left[K_y + K_z(sI-F)^{-1}L\right] \tag{3.23}$$

which is different from $-H(s)$ of (3.22). This difference is caused only by the additional feedback of $\mathbf{u}(t)$ into the observer. If this feedback, which is defined by its path gain TB and its loop gain $K_z(sI-F)^{-1}TB$, equals zero, then and only then $-H(s) = H_y(s)$.

Theorem 3.3

The loop transfer function $L_{Kx}(s)$ at the breaking point $-K\mathbf{x}(t)$ of Figure 3.7 equals the loop transfer function of the corresponding direct state feedback control system (3.10)

$$L_{Kx}(s) = -K(sI - A)^{-1}B \tag{3.24}$$

Proof (Tsui, 1988a)

From Figure 3.7,

$$L_{Kx}(s) = -H_y(s)G(s) - K_z(sI - F)^{-1}TB \tag{3.25a}$$

by (3.23)

$$= -K_yG(s) - K_z(sI - F)^{-1}\left[LG(s) + TB\right] \tag{3.25b}$$

$$\text{by (1.9)} = -\left[K_yC + K_x(sI - F)^{-1}(LC + sT - TA)\right](sI - A)^{-1}B$$

$$\text{by (3.17)} = -\left[K_yC + K_z(sI - F)^{-1}(sI - F)T\right](sI - A)^{-1}B$$

$$\text{by (3.21)} = -K(sI - A)^{-1}B.$$

Figure 3.7 also shows that $-K\mathbf{x}(t)$ is only an internal signal of the observer (3.16), while $\mathbf{u}(t)$ is the real analog signal that is attributed to the plant system $G(s)$ and that is where the disturbance is introduced (see Subsection 3.1.2). Therefore, the loop transfer function $L(s)$ of the observer feedback system which determines the feedback system sensitivity, should be defined at the breaking point $\mathbf{u}(t)$ (Doyle, 1978).

Now by (3.22),

$$L(s) = -H(s)G(s)$$

$$= -\left[I + K_z(sI - F)^{-1}TB\right]^{-1}\left[K_y + K_z(sI - F)^{-1}L\right]G(s) \tag{3.26}$$

Because $L(s) \neq L_{Kx}(s)$, which is the loop transfer function of the direct state feedback system (Theorem 3.3), and because loop transfer function determines the sensitivity and robustness of the corresponding feedback system, the observer feedback system of (3.16) and Figure 3.7 has different sensitivity and robustness properties of the corresponding direct state feedback system (Doyle, 1978).

Example 3.4

To understand further the difference of the two-loop transfer functions $L(s)$ and $L_{Kx}(s)$, defined at two different nodes $\mathbf{u}(t)$ and $K\mathbf{x}(t)$ of Figure 3.7, respectively, we will analyze the following two more equivalent system diagrams of this overall observer feedback system.

The first diagram is called signal flow diagram (Figure 3.8).

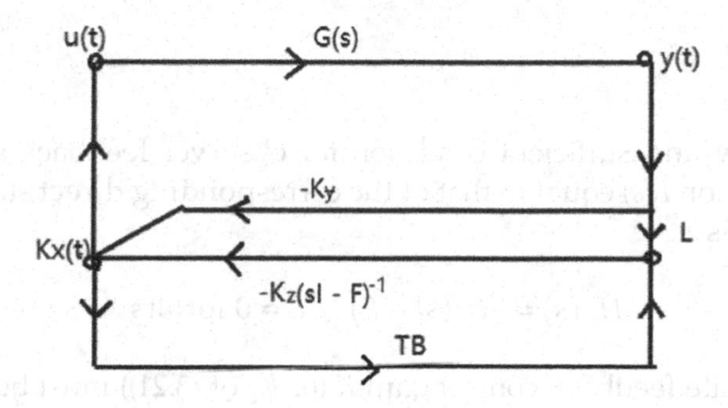

FIGURE 3.8
Signal flow diagram of observer feedback system.

Figure 3.8 shows that at node $\mathbf{u}(t)$, there is only one loop path. The bottom loop with gain $-K_z(sI-F)^{-1}TB$ is only attached to this main single loop path. In contrast, at node $K\mathbf{x}(t)$, there are two independent loop paths. The loop path with gain $-K_z(sI-F)^{-1}TB$ is an independent loop path among the two.

The second diagram Figure 3.9 is a simplified but equivalent block diagram of Figure 3.7.

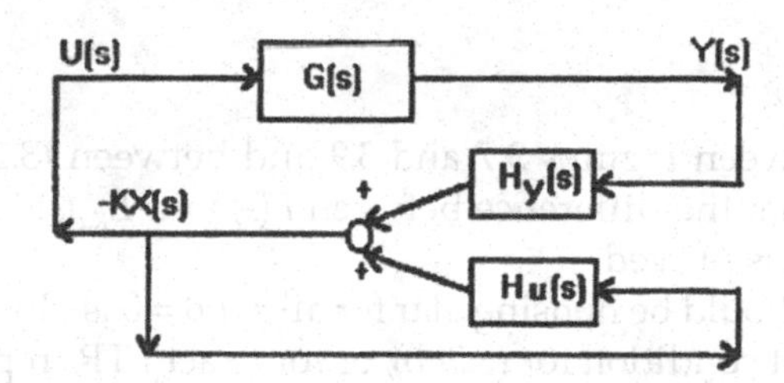

FIGURE 3.9
A simplified and equivalent block diagram of observer feedback system.

Of the two transfer functions, $H_y(s)$ and $H_u(s)$ of Figure 3.9, $H_y(s)$ is already defined in (3.23) while

$$H_u(s) = -K_z(sI - F)^{-1}TB \tag{3.27}$$

We should draw the same conclusions of (3.26) and (3.24 and 3.25a), from Figures 3.8 and 3.9 that

$$L(s) = [I - H_u(s)]^{-1} H_y(s)G(s)$$

$$= -\left[I + K_z(sI - F)^{-1}TB\right]^{-1}\left[K_y + K_z(sI - F)^{-1}L\right]G(s),$$

and $L_{Kx}(s) = -H_y(s)G(s) + H_u(s)$

$$= -K_yG(s) - K_z(sI - F)^{-1}\left[LG(s) + TB\right] = -K(sI - A)^{-1}B$$

Theorem 3.4

The necessary and sufficient condition for observer feedback system loop transfer function $L(s)$ equal to that of the corresponding direct state feedback control $L_{Kx}(s)$ is

$$H_u(s) = -K_z(sI - F)^{-1}TB = 0 \text{ for all } s \tag{3.28a}$$

In practice, state feedback control gain K (or K_z of (3.21)) must be completely free to design for a good feedback system state matrix $A–BK = A–B[K_z\text{: } K_y]\underline{C}$ (see Sections 3.2.1 and 3.2.2). Therefore, in practice, (3.28a) becomes

$$H_u(s) = -K_z(sI - F)^{-1}TB = 0 \text{ for all } s \text{ and all } K_z \tag{3.28b}$$

Moreover, the necessary and sufficient condition for (3.28b) is

$$TB = 0 \tag{3.29}$$

Proof

Comparisons between Figures 3.7 and 3.9 and between (3.26) and (3.25) all indicate clearly that the difference between $L(s)$ and $L_{Kx}(s)$ is caused only by $H_u(s)$. Thus (3.28a) is proved.

Because $(sI–F)$ should be nonsingular for all s, $TB = 0$ is obviously the necessary and sufficient condition for (3.28b), or for exact LTR in practical design.

If the freedom of K_z is not completely used for a good feedback system matrix $A–B[K_z : K_y]\underline{C}$, then some individual values of K_z and other system parameters may be cooked so that (3.28a), (3.17), and (3.21) can be satisfied without satisfying (3.29) (Chen et al. 1991). In other words, (3.29) is not a necessary condition for $L(s) = L_{Kx}(s)$, if K_z is cherry-picked only to satisfy (3.28a).

However, if the state feedback control law K or K_z is designed not for a good performance and robust control, but instead only to satisfy (3.28a) (or just to be realized), then such a control design is not for a good control but for the sake of design only (or is purposeless)! Nonetheless, this bizarre argument and criticism have prevented the public appearance of Theorem 3.4 and condition (3.29) ($TB = 0$) for years.

Definition 3.3

Theorem 3.4 and (3.29) imply that Figure 3.9 becomes Figure 3.1 (when $TB = 0$), which is the basic feedback structure of the classical control theory, and there is no input feedback in this structure. We call an observer (3.16) that does not take input feedback or ($TB = 0$), an "output feedback compensator". Theorem 3.4 indicates that only an output feedback compensator can fully realize the state feedback $K\mathbf{x}(t)$-control including its robustness properties.

In the studies of (Doyle and Stein, 1979, 1982) subsequent to (Doyle, 1978), the problem of making $L(s) = L_{Kx}(s)$, called "Loop transfer recovery (LTR)", is imposed. This is an additional requirement of observer design – the observer is not only to generate the state feedback control signal $K\mathbf{x}(t)$ but also to fully realize the loop transfer function and robustness of that control. In other words, the observer must satisfy not only (3.17) and (3.21) but also (3.29).

Because robustness is the foremost purpose and property of feedback control systems, the LTR problem is critically important to all state space control design theories. It has attracted great attention since its proposition (Sogaard Anderson, 1986; Tsui, 1987b; Stein and Athans, 1987; Dorato, 1987; Moore and Tay, 1989; Saberi and Sannuti, 1990; Liu and Anderson, 1990; Niemann et al., 1991; Saberi et al. 1991; Saeki, 1992; Tsui, 1992, 1993b; Saberi et al., 1993; Tsui, 1996a, b, 1998b, 1999b, c, 2002, 2006, 2015).

It should be noted that the proposition of a design problem does not mean the problem is solved. In practice, the solving of a problem is usually much more difficult than the proposition of that problem, and usually determines the practical use and significance of that problem. In addition, the derivation of a preliminary solution of a problem does not mean that the problem is solved satisfactorily. In practice, deriving a really satisfactory solution is usually much more difficult than deriving an unsatisfactory solution, and also usually determines decisively the practical use and significance of that theoretical problem.

Unfortunately, for a great majority of given plant systems, it is impossible to design an observer that can satisfy (3.17), (3.21), and (3.29), if the state feedback control gain K is to be previously and independently designed or is to be arbitrarily given (see Section 4.3). In other words, the robust properties of an independently designed state feedback control cannot be fully realized for the great majority of plant systems.

Because the independent and separate design of state feedback control law K is the only design practice of the 1950s at the very start of state space control design, and is the defining feature of the "separation (design) principle" since the 1960s at the very start of the state space control theory (see Section 4.3), the conclusion of the above paragraph is the reason that the modern control theory has not found many successful practical applications since the 1960s.

Because of this reason, this book proposes a fundamentally novel design principle, in which the state feedback control law K is not previously and separately designed before the observer design. Instead, the observer parameter $[K_z: K_y]$, not K, is designed. Only after $[K_z:K_y]$ is designed, then K is determined as $K = [K_z: K_y]\ [T':C']'$ (Eq. 3.21). In other words, K is not predesigned and arbitrarily given before the observer design, and thus matrix $[T':C']'$ is not required to be nonsingular so that (3.21) can hold for arbitrarily given K. In matrix $[T':C']'$, the observer parameter T is designed even earlier with a flexible number of rows, to make sure that the much more important conditions (3.17) and (3.29) are satisfied. The feature of flexible and reduced number of rows of matrix T makes the exact solution of (3.17) and (3.29) exists for a great majority of plant systems. In addition, approximate and far more approximate solutions of (3.17) and (3.29) can be easily computed for all plant system conditions.

Therefore, completely different from the separation principle, the design of $[K_z: K_y]$ (or K) of this book is now based on the key observer parameter T and parameter C, which is key to system output measurement information. This new design principle is therefore called a "synthesized design principle".

Therefore, the critical and unsolved design problem of realizing the robustness of state feedback control (or LTR) is now decisively solved by this novel synthesized design principle of this book!

Chapter 4 introduces the concept and outlines the advantages of this new design principle. Chapters 5 and 6 will describe in detail the design computation of the solution of (3.17) and (3.29). The design of $[K_z: K_y]$ will be introduced in Chapters 8–10.

3.3 Summary

The main purpose or the critical requirement of feedback/automatic control is improving and guaranteeing feedback system robustness or low sensitivity against model uncertainty and input disturbance. This is achieved by automatic feedback control correction and adjustment, using system output measurement. The loop transfer function, created by the feedback control itself, determines the sensitivity and robustness of the corresponding feedback system.

State feedback control is defined as a control signal $K\mathbf{x}(t)$ where K is a constant (may be constrained by (3.21) or not), and is the general and basic form of the modern control theory. This control is based on far more direct and detailed information of system structure and state, and is therefore far better in improving feedback system performance and robustness, than any other basic forms of feedback control, if this K is not designed for other useless purposes.

The only remaining design problem is to fully realize the $K\mathbf{x}(t)$ control. An observer is needed to realize this control when the entire set of $\mathbf{x}(t)$ cannot be measured (true in almost all practical plant systems). The necessary and sufficient condition to generate the signal $T\mathbf{x}(t)$ for a constant T (K is then equal to $\underline{K}[T'\!:\!C']'$ of (3.21) for whatever $\underline{K}$) is (3.17). The necessary and sufficient condition to realize the loop transfer function and robust properties of this control is simply $TB = 0$ (3.29).

4

A New Feedback Control Design Principle/Approach

Chapter 3 analyzed the observer design requirements, which can be re-outlined as follows.

To guarantee observer state $\mathbf{z}(t) \to T\mathbf{x}(t)$ for a constant T requires (Theorem 3.2)

$$TA - FT = LC\left(F \text{ is stable}\right) \tag{4.1}$$

To have observer output $\to K\mathbf{x}(t)$ requires (assuming $\mathbf{z}(t) \to T\mathbf{x}(t)$)

$$K = K_z T + K_y C = \left[K_z : K_y\right]\left[T' : C'\right]' \equiv \underline{K}\underline{C} \tag{4.2}$$

To realize the loop transfer function and robust property of $K\mathbf{x}(t)$-control requires (Theorem 3.4)

$$TB = 0 \tag{4.3}$$

The real challenge is how to satisfy and best satisfy these three requirements generally and systematically. For this purpose, a fundamentally novel design principle/approach is proposed in this chapter, which is divided into four sections.

Section 4.1 points out a basic and general observer design procedure that (4.1) is satisfied separately and before (4.2). In most existing observer designs and all existing loop transfer recovery (LTR) observer designs, only state observers are designed which satisfy (4.1) and (4.2) simultaneously. If only (4.1) is satisfied and signal $T\mathbf{x}(t)$ is estimated, then the desired state feedback control signal $K\mathbf{x}(t)$ can be generated directly from this $T\mathbf{x}(t)$ signal and system output signal $C\mathbf{x}(t)$, without estimating the system state $\mathbf{x}(t)$ (with too many ($= n$) elements) first.

Section 4.2 analyzes the poles and performance of the observer feedback system. A generalized version of "separation property" is proven that the observer feedback system poles are formed by the eigenvalues of A–BK (for whatever K=$\underline{K}\underline{C}$) and F, if (4.1) is satisfied.

Section 4.3 lists eight severe drawbacks of the most basic principle of modern control design – the "separation principle" (Willems, 1995). The most severe among which is the failed realization of the loop transfer function and robust property of state feedback control for a great majority of plant systems.

DOI: 10.1201/9781003259572-4

Section 4.4 summarizes the previous sections, in particular Section 4.3, and proposes a design principle that is completely different from the separation principle. This new design principle satisfies (4.1) and (4.3) first and only then designs $\underline{K}$ instead of K (= $\underline{K}$ $\underline{C}$), or designs $\underline{K}$ based on $\underline{C}$ instead of separated from $\underline{C}$. The decisive advantage of this new design principle is the full realization of the loop transfer function and robust properties of this $\underline{KC}\mathbf{x}(t)$ control for a great majority of plant systems and the complete overcoming of the eight drawbacks of Section 4.3.

4.1 Basic Observer Design Concept – Generating State Feedback Signal Directly Without Generating Explicit System States

We will use three observer design examples to explain this concept, that is, generating $K\mathbf{x}(t)$ signal directly from observer state $\mathbf{z}(t)$ ($\rightarrow T\mathbf{x}(t)$) and $C\mathbf{x}(t)$, instead of from $\mathbf{x}(t)$. As explained following Theorem 3.2, this design concept implies satisfying (4.1) first and separate from satisfying (4.2). We will show that this design concept not only fits the physical meanings of (4.1) and (4.2) but also follows the existing design procedures.

Example 4.1: Full-Order State Observer Design (1960s)

A state observer is an observer (3.16) that estimates system state $\mathbf{x}(t)$, or parameter K of (3.16b) is given as an identity matrix I. A full-order state observer has order equal to n, the system order. Because observer order r equals the number of rows of parameter T, matrix T of this observer has n linearly independent rows.

For example, let $T=I$ and $F=A-LC$, then (4.1) is satisfied and the observer state $\mathbf{z}(t)\rightarrow I\mathbf{x}(t)$ if matrix F is stable. The corresponding observer dynamic part (3.16a) becomes

$$d\mathbf{z}(t)/dt = (A-LC)\mathbf{z}(t)+L\mathbf{y}(t)+B\mathbf{u}(t) \quad (4.4)$$

$$= A\mathbf{z}(t)+B\mathbf{u}(t)+L\left[\mathbf{y}(t)-C\mathbf{z}(t)\right] \quad (4.5)$$

Observer (4.5) is the structure of the Kalman filter, from which the modern control theory was started in 1960 (Anderson, 1979; Balakrishnan, 1984), where L is the Kalman filter gain.

If $T\neq I$, then to generate $K\mathbf{x}(t)=I\mathbf{x}(t)=K_z\,T\mathbf{x}(t)$ in (3.16b), $K_z=T^{-1}K$ in order to satisfy (4.2). Hence, a full-order state observer design satisfies (4.1) first and then uses result T of (4.1) to satisfy (4.2).

A full-order state observer feedback control system is represented in Figure. 4.1.

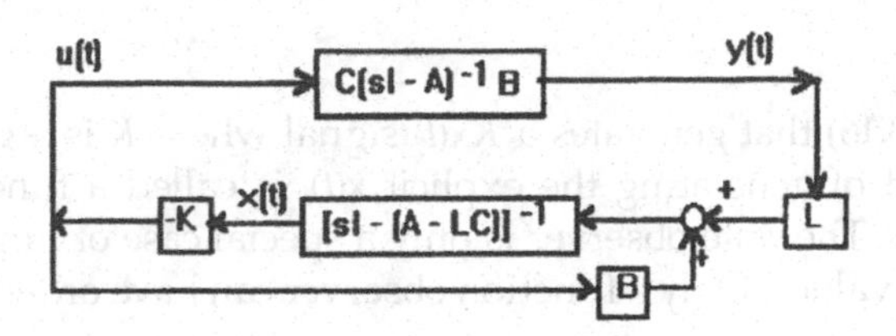

FIGURE 4.1
Block diagram of full-order state observer feedback control system.

Example 4.2: Reduced-Order State Observer Design (1970s)

The order of a reduced-order state observer is $n-m$, or matrix T of (4.1) has $n-m$ linearly independent rows.

Thus, to satisfy (4.2), $K=I=\underline{K}\underline{C}$ or $\underline{K}=\underline{C}^{-1}$, matrix $\underline{C}\equiv[T':C']'$ must have m rows of matrix C added to matrix T so that $\underline{C}$ can have n linearly independent rows. Again, the reduced-order state observer design satisfies (4.1) first and then uses result T of (4.1) and matrix C to satisfy (4.2).

The block diagram of a reduced-order state observer feedback control system is shown in Figure 4.2.

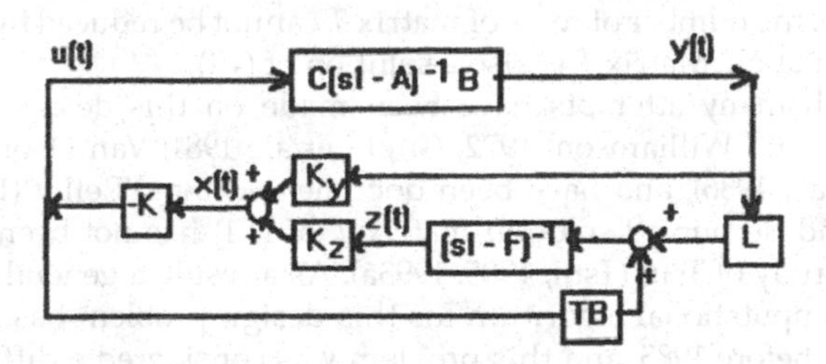

FIGURE 4.2
Block diagram of reduced-order state observer feedback control system.

Comparing Figures 4.1 and 4.2, the main difference is that the reduced-order state observers have an additional path from **y**(t) to **x**(t) with gain K_y or the data of m output measurements C**x**(t) are used to estimate **x**(t). We define the observer of Figure 4.1 ($K_y=0$) as strictly proper, and the observer of Figure 4.2 ($K_y\neq 0$) proper (Chen, 1984).

In modern control systems, the ultimate purpose of an observer is to realize K**x**(t)-control. This is why as soon as **x**(t) is estimated by a state observer of either Figure 4.1 or Figure 4.2, this signal **x**(t) will be multiplied immediately by K.

However, satisfactory K**x**(t) signals can be generated directly from $\underline{C}$**x**(t) (or many satisfactory parameter K can satisfy $K=\underline{K}\,\underline{C}$ of (4.2)) without necessarily n signals in $\underline{C}$**x**(t) (or without necessarily rank($\underline{C}$)$=n$), especially if K**x**(t) has only p control input signals (or rank(K)$=p$) and p is much less than n. This is a major advantage of generating the K**x**(t) signal from signal $\underline{C}$**x**(t) directly instead of from the n signals of I**x**(t).

Definition 4.1

The observer (3.16) that generates a $K\mathbf{x}(t)$ signal where K is a given constant, directly instead of generating the explicit $\mathbf{x}(t)$, is called a function (or functional) observer. The state observer is only a special case of function observer for $K=a$ special value I. Only a function observer can have order less than $n-m$.

Example 4.3: Overview of Minimal Order Function Observer Design (Tsui, 2015)

This design problem has attracted continued attention for 50 years since 1970, and system order reduction has also been an important problem of the control systems theory for even longer time (Kung and Lin, 1981). A lower-order controller is also arguably much easier to simulate. In addition, this design problem is the only existing application of the basic concept of this section – generating the $K\mathbf{x}(t)$ signal directly from $\underline{C}\mathbf{x}(t)$ (= $[T': C']'\mathbf{x}(t)$) instead of from $\mathbf{x}(t)$.

It is clear that minimizing the function observer order r is minimizing the number (r) of rows of matrix T while satisfying Eq. (4.2), $K=\underline{K}\,[T': C']'$.

It is also obvious that to make this minimization really general and systematic, every row of matrix T must be decoupled from each other. Otherwise, the number of rows of matrix T cannot be reduced by systematic computation. Matrix T is also a solution of (4.1).

Although many attempts have been made on this design problem (Fortmann and Williamson, 1972; Gupta et al., 1981; Van Dooren, 1984; Fowell et al., 1986), and have been documented by O'Reilly (1983), the general and decoupled solution matrix T of (4.1) has not been derived until the study of Tsui (Tsui, 1985, 1993a). As a result, a general and systematic computational algorithm for this design problem has not been developed before 1985, and this problem was considered a difficult and unsolved problem (Kaileth, 1980; Chen, 1984).

Only with the general and decoupled solution of (4.1), and applying this solution T to the remaining design requirement (4.2), does (4.2) become truly a set of linear equations – the simplest possible theoretical formulation of this design problem (Tsui, 1998a, 2003a). Only then a really general and systematic design algorithm for that set of linear equations was developed by Tsui (1985, 1986b). This design algorithm for (4.2) will be introduced in Chapter 7, after this decoupled solution T of (4.1) is introduced in Chapter 5. This design problem is proven to be essentially solved by Tsui (1985, 1986b). See Tsui (2015).

Examples 4.1–4.3 demonstrate the basic concept of observer design – design observer dynamic part (3.16a) first and separately (or satisfy (4.1) first) before the design of (3.16b) or (4.2). An obvious advantage of this design concept is observer order reduction and adjustment, and this advantage cannot be realized until a general and decoupled solution T of (4.1) is derived, as demonstrated in Example 4.3.

This should be the reason why this basic design concept and its advantage have not really been well understood by control researchers and not well introduced in the control literature, even more than three decades after its derivation in 1985.

Besides function observer order reduction, an even much higher significance of this basic design concept and its observer order adjustment is at the new design principle of this book, which is defined procedure wise as satisfying (4.1) and (4.3) first. From Theorem 3.4, (4.3) (or $TB=0$) is the key condition to realize the robust properties of state feedback control, and feedback system robustness is much more important than observer order reduction. Not bothered by satisfying Eq. (4.2) or not fixing the observer order at $n-m$, the exact solution to (4.1) and (4.3) exists in a great majority of plant systems, as will be described in Section 4.4.

4.2 Performance of the Observer Feedback System – Separation Property

In Section 4.1, we introduced the basic design concept of satisfying (4.1) first and separately without considering (4.2). In this section, we will prove that (4.1) alone is sufficient for the separation property – the observer feedback system poles are formed by the eigenvalues of F and $A-B\underline{KC}$. This implies that (4.1) alone guarantees that the performance of the observer feedback system is on par with the performance of the observer and the state feedback system.

This is an essential validation of the new design principle of this book – satisfying (4.1) and (4.3) first, because prioritizing (4.3) implies prioritizing the robustness of the observer feedback system. Both performance and robustness are essential to the feedback control system.

Theorem 4.1: Separation Property

If (4.1) is satisfied, then the poles of the feedback system that is formed by plant system (1.1) and observer (3.16) equal the eigenvalues of matrices F and $A-B\underline{KC}$ of (3.16) and (4.2).

Proof (Tsui, 1993b)

By substituting (3.16b) into the plant system input $\mathbf{u}(t)$ and then by substituting this $\mathbf{u}(t)$ and $\mathbf{y}(t)=C\mathbf{x}(t)$ into the dynamic parts of plant system (1.1a) and observer (3.16a), the combined observer feedback system state matrix is

$$\begin{bmatrix} d\mathbf{x}(t)/dt \\ d\mathbf{z}(t)/dt \end{bmatrix} = \begin{bmatrix} A-BK_yC & -BKz \\ LC-TBK_yC & F-TBK_z \end{bmatrix} \begin{bmatrix} \mathbf{x}(t) \\ \mathbf{z}(t) \end{bmatrix} \equiv A_c \begin{bmatrix} \mathbf{x}(t) \\ \mathbf{z}(t) \end{bmatrix} \tag{4.6}$$

By multiplying $Q^{-1} = \begin{bmatrix} I & 0 \\ -T & I \end{bmatrix}$ and $Q = \begin{bmatrix} I & 0 \\ T & I \end{bmatrix}$

on the left and right side of A_c of (4.6) respectively, we have

$$\underline{A}_{\underline{c}} = Q^{-1} A_c Q = \begin{bmatrix} A - BK_y C - BK_z T & -BK_z \\ -TA + FT + LC & F \end{bmatrix} \tag{4.7}$$

$$= \begin{bmatrix} A - B\underline{KC} & -BK_z \\ 0 & F \end{bmatrix} \tag{4.8}$$

if and only if (4.1) holds. The eigenvalues of matrix $\underline{A}_c$ of (4.8) equal the eigenvalues of $A–B\underline{KC}$ and F, while matrices A_c and $\underline{A}_c$ are similar and have the same eigenvalues.

In all existing designs, matrices F and $A–B\underline{KC}$ are designed separately without considering the overall observer feedback system poles. The separation property guarantees the same poles of the overall observer feedback system and these two subsystems, and thus guarantees the similar performance of the overall observer feedback system. For example, this property guarantees the stability of the observer feedback system if F and $A–B\underline{KC}$ are both stable.

The novel synthesized design principle of this book inherited the existing separated design results of (4.1) and of $A–B\underline{KC}$. The differences between this new principle and the existing separation principle are that $\underline{K}$ of (4.7) is designed instead of K (= $\underline{KC}$) and that the dimension of matrix F is now flexible. Therefore, Theorem 4.1 shows that this novel design principle also inherited the existing separation property of (4.1), and has the corresponding guarantee of performance of its final overall design results.

However, the new design principle of this book satisfies (4.1) first without (4.2), and uses a more general observer formulation (3.16). Thus, the proof for this new version of separation property that reflects this new design procedure and new observer formulation is needed. This new proof first appeared in the work of Tsui (1993b).

Finally, for the sake of theoretical integrity, we shall point out that although (4.1) is a sufficient condition of separation property, it is not a necessary condition for this property. This will be proved by the following counter-example.

Example 4.4

Let a matrix $\underline{A}_c$ of (4.7) and its characteristic polynomial be

$$|sI - \underline{A}_c| = \begin{vmatrix} sI - (A - B\underline{KC}) & BK_z \\ TA - FT - LC & sI - F \end{vmatrix} = \begin{vmatrix} s-a & -b & : & 1 \\ -b & s-a & : & 1 \\ - & - & - & - \\ c & c & : & s-f \end{vmatrix}$$

where a, b, c, and f are scalars. Then,

$$|sI-\underline{A}_c| = (s-f)|sI-(A-B\underline{KC})| + \begin{vmatrix} s-a & -b \\ c & c \end{vmatrix} + \begin{vmatrix} -b & s-a \\ c & c \end{vmatrix}$$

$$= (s-f)\left|sI-(A-B\underline{KC})\right| \quad (4.9)$$

The equality of (4.9) holds, or the separation property holds, even if $c \neq 0$, or even if $(TA-FT-LC) \neq 0$, or even if condition (4.1) does not hold.

In any practical design, parameters of the matrix of (4.7) must be designed to satisfy (4.1), (4.3), and a satisfactory state matrix $A-B\underline{KC}$, but not to fit the special case of Example 4.4. Thus, the argument that (4.1) is not a necessary condition of separation property is totally meaningless in practice.

This argument is very similar to the argument described after Theorem 3.4 that (4.3) is not a necessary condition to (3.28a) and $L(s) = L_{Kx}(s)$. Because parameters (K, F, T) must be designed to satisfy (4.1) and a satisfactory state matrix $A-B\underline{KC}$, they simply cannot be designed to fit the special case of satisfying (3.28a), (4.1) and (4.2) without satisfying (4.3). Thus, this argument is totally senseless in practice. Nonetheless, that senseless argument has in fact prevented the publication of Theorem 3.4 for years.

4.3 Eight Drawbacks and Irrationalities of the Modern Control Design and Separation Principle

Separation principle first assumes that the information $\mathbf{x}(t)$ is already available, designs a state feedback control $K\mathbf{x}(t)$ based on this assumption, and only then designs an observer to realize this control (Tsui, 2006, 2012). Hence, the designs of K and its realizing observer are completely separate. Because K is separately designed, it is arbitrarily given when the observer is designed.

The modern control design has always followed separation principle for over six decades since its start in 1960 (Willems, 1995). This is natural since state feedback control (when $\mathbf{x}(t)$ is fully measurable) was first developed in the 1950s even before the observers and the modern control theory were developed in 1960. Nonetheless, this design principle has eight severe drawbacks and irrationalities, as will be listed in this section.

4.3.1 Drawback 1 of Separation Principle: Invalid Basic Assumption

The most basic assumption that the ideal information of complete set of $\mathbf{x}(t)$ is available before the design of state feedback control law K, is invalid. All systems have the number of output measurements less than n, the number of state variables in $\mathbf{x}(t)$ $(m<n)$, when an observer is needed. These systems are

not as ideal as the systems with $m=n$. Even the not ideal yet still trivial systems ($m+p>n$, see Subsection 3.2.2) are very rare. Furthermore, the great majority of real plant systems are very non-trivial, like $m+p<n$ and non-minimum phase. Hence, it is simply irrational to assume that all of these systems are ideal. It is simply irrational to expect that a control law K designed based on this very invalid assumption can be fully or even approximately realized (including its robust properties) in reality.

This irrationality also implies the complete ignorance of the realization of the so-designed $K\mathbf{x}(t)$ control, even though this realization is very difficult – it requires an entire observer or feedback compensator to realize the $K\mathbf{x}(t)$ control.

4.3.2 Drawback 2 of Separation Principle: Ignor Key Parameters

Under separation principle, the design of K considered only the system parameter (A, B) in the feedback system state matrix $A–BK$, but ignored completely parameters T and C in matrix $\underline{C}=[T'|C']'$. Here, matrix T is the key parameter of an observer relied upon to realize the $K\mathbf{x}(t)$ control, and C is the key parameter of output measurement information $C\mathbf{x}(t)$ relied upon to realize the $K\mathbf{x}(t)$ control. Hence, ignoring these two key parameters implies complete ignorance of the realization of the so-designed $K\mathbf{x}(t)$ control.

For example, ignoring parameter C means ignoring completely whether the number of output measurements is 100%, or 50%, or 10%, or even 1%, of the number of plant system states $\mathbf{x}(t)$, when designing the state feedback control law K and assuming the measurements of 100% of $\mathbf{x}(t)$ are available. How can such an ignorant design be rational?

4.3.3 Drawback 3 of Separation Principle: Wrong Design Priority

This design principle also implies irrational design priority of the observer/controller – it designs the observer output part parameter (K, or $[K_z: K_y]$) of (3.16b) first and then designs the dynamic part of observer (3.16a). The dynamic part of any system is always far more important than the static output part of that system, such as an observer (3.16).

Because we rely entirely on the observer to realize the $K\mathbf{x}(t)$-control, this wrong priority of designing the observer's main dynamic part and minor output part also implies the ignorance of the realization of the so-designed $K\mathbf{x}(t)$-control.

If the first three irrationalities of separation principle all focus on the ignorance of the realization of the so-designed $K\mathbf{x}(t)$-control, then the next three drawbacks of separation principle demonstrate the consequence of the first three irrationalities – failed realization, even the failed approximate realization, of the critical robust properties of the so-designed $K\mathbf{x}(t)$control, for a great majority of plant systems, which are far from ideal.

4.3.4 Drawback 4 of Separation Principle: Unnecessary Design Requirement

Because separation principle means K is arbitrarily given when an observer is designed, Eq. (4.2) ($K=\underline{K}\ \underline{C}$) must be satisfied for an arbitrarily given K. This means that rank($\underline{C}$) must be n or the observer must be a state observer that can estimate system state $\mathbf{x}(t)$ (= $\underline{C}^{-1}$ [$\mathbf{z}(t)'$: $\mathbf{y}(t)'$]′, see Examples 4.1 and 4.2). This requirement of having the information of the entire set of $\mathbf{x}(t)$ is also pre-assumed by the separation principle, see Subsection 4.3.1. This is why under the separation principle, almost all existing observers (except function observers) are state observers and all LTR observers are state observers.

However, we actually need an observer to estimate only p control input signals of $K\mathbf{x}(t)$, and p is generally much less than n (see Chapter 1 and Example 4.3). Hence, the requirement of the separation principle to generate all n signals of $\mathbf{x}(t)$ is unnecessary and too excessive, and is therefore irrational.

4.3.5 Drawback 5 of Separation Principle: Abandon Existing Control Structure

Because of this unnecessary and too excessive observer design requirement of the separation principle, the corresponding observer must be a state observer and thus *cannot* satisfy (4.3) ($TB=0$) for a great majority of plant systems.

A state observer satisfying $TB=0$ is defined as an "unknown input observer" if matrix B is the gain of the unknown input (Wang et al., 1975). Based on the study of Kudva et al. (1980), an unknown input observer exists if and only if the plant system satisfies all of the following three conditions: (a) minimum-phase (all transmission zeros are stable); (b) $m \geq p$ (number of output measurements is no less than the number of control inputs); and (c) rank (CB)$=p$ (matrix product CB is full column rank). The proof of these three conditions can be seen in Conclusions 6.3 and 6.4.

A great majority of plant systems cannot satisfy these three conditions. First, it is very hard to require that among all transmission zeros, every zero be stable. Exercise 4.2 proves that almost all plant systems that have the same number of outputs and inputs (and have n–m transmission zeros) cannot satisfy condition (a). Second, condition (c) cannot be satisfied by many real plant systems such as airborne systems.

For example, if we assume 50% of the plants are $m=p$ and 30% are $m>p$, and if we assume condition (c) cannot be satisfied by half of the systems with $m>p$, then about 85% of plants cannot have unknown input observers. To summarize, under the separation principle, the state observers of a great majority of the plants cannot satisfy (4.3) or $TB=0$ or must have input feedback. Theorem 3.4 indicates that for these plants, the robust properties of $K\mathbf{x}(t)$-control *cannot be realized.*

In addition, almost all well-established feedback compensators in the classical control theory do not take input feedback, because their basic purpose is to uphold robustness against plant model uncertainty *and* input disturbance (see Section 2.1). Now under the separation principle, this well-established and well-rationalized feedback structure *must be abandoned*, for a great majority of the plant systems!

4.3.6 Drawback 6 of Separation Principle: Failed Robust Realization

According to Theorem 3.4, unable to satisfy (4.3) ($TB=0$) means unable to realize robust properties of state feedback control that an observer is supposed to realize! This is obviously the gravest and the most critical drawback of separation principle.

In fact, under separation principle, even the approximate realization of loop transfer function $L_{Kx}(s)$, called "loop transfer recovery LTR" because recovery means gradual and approximate but not full and exact, is very unsatisfactory, as shown in the following.

The main result of approximate LTR is called asymptotic LTR. It is either to design a Kalman filter (see Example 4.1) while asymptotically increasing the plant input noise level q (Doyle and Stein, 1979; Doyle, 1981), or to design a state observer while asymptotically increasing the time scale of observer poles (Saberi and Sannuti, 1990; Saberi et al., 1991). Both approaches are aimed at increasing the observer gain L to the plant output, so that the observer gain TB to the plant input is overwhelmed.

However, as clearly warned by conclusions following Example 3.2, the indiscriminate increase of observer gain L is unpractical and prohibitive especially in robust control design (the large gain L will cause instability for non-minimum phase plants, for example), even though this asymptotic LTR result has received the most attention (Fu, 1990: Shaked et al. 1985; Tahk and Speyer 1987).

Another result of LTR is seeking an upper bound on the difference $\|L(j\omega)-L_{Kx}(j\omega)\|_\infty$ over all ω (Moore and Tay, 1989). However, this bound can be unacceptably large – it will ever increase until a boundary-valued Riccati equation is solvable (Weng and Shi, 1998). In addition, the important phase angle of $L(j\omega)-L_{Kx}(j\omega)$ is not considered at all. Hence, this result is unsatisfactory either mainly because $L_{Kx}(j\omega)$ or K is arbitrarily given or mainly because of separation principle.

The following numerical example, raised by Doyle and Stein (1979), can demonstrate that the result of asymptotic LTR is far from being satisfactory.

Example 4.5

Let the plant system and its state feedback control law be

$$(A, B, C, K)=\left(\begin{bmatrix} 0 & -3 \\ 1 & -4 \end{bmatrix}, \begin{bmatrix} 2 \\ 1 \end{bmatrix}, \begin{bmatrix} 0 & 1 \end{bmatrix}, \begin{bmatrix} 30 & -50 \end{bmatrix}\right)$$

Its transfer function $G(s)=\dfrac{s+2}{(s+1)(s+3)}$,

which has a stable transmission zero –2.

The loop transfer function to be recovered is

$$L_{Kx}(s) = -K(sI - A)^{-1}B = \frac{-(10s+50)}{(s+1)(s+3)}$$

Doyle and Stein (1979) listed two asymptotic LTR results (two Kalman filters with increased input noise level q):

(1) A Kalman filter with poles at $-7 \pm j2$:

$$(F = A - LC, L, T, K_z, K_y) = \left(\begin{bmatrix} 0 & -53 \\ 1 & -14 \end{bmatrix}, \begin{bmatrix} 50 \\ 10 \end{bmatrix}, I, K, 0 \right).$$

Its loop transfer function $L(s)$ according to (3.26) is

$$L(s) = \frac{-100(10s+26)G(s)}{s^2+24s-797}$$

and is very different from the target loop transfer function $L_{Kx}(s)$.

(2) A Kalman filter with a higher input noise level $q = 100$:

$$(F = A - LC, L, T, K_z, K_y) = \left(\begin{bmatrix} 0 & -206.7 \\ 1 & -102.4 \end{bmatrix}, \begin{bmatrix} 203.7 \\ 98.4 \end{bmatrix}, I, K, 0 \right)$$

Its loop transfer function $L(s)$ according to (3.26) is

$$L(s) = \frac{-(1191s+5403)G(s)}{s^2+112.4s+49.7}$$

The second result is the best of Doyle and Stein (1979). The high input noise level q caused a very high filter/observer gain $\|L\| = 226.2$. Yet the corresponding loop transfer function $L(s)$ is still very different from the target $L_{Kx}(s)$, especially at $\omega < 10$ (see Figure 3 of Doyle and Stein (1979)).

Using design methods other than asymptotic LTR, the best result on the same example was derived directly by this author (Tsui, 1988b)!

$$(F, L, T, K_z, K_y) = (-2, 1, [1 - 2], 30, 10)$$

It can be verified that (4.1), (4.2), and (4.3) are all satisfied. This best yet simple result shows clearly that the asymptotic LTR design method is unsatisfactory.

The state space model of Example 4.5 is actually dual (similar) to that of the example of Doyle and Stein (1979). However, all transfer functions and loop transfer functions of Example 4.5 and Doyle and Stein (1979) are the same.

The plant system of Example 4.5 satisfies the three conditions of unknown input observers of Subsection 4.3.5. The real advantage of the new design principle of this book, as will be proposed in Section 4.4, will be evident on much more general plant systems and the great majority of plant systems, which do not satisfy those three conditions.

To summarize, the ultimate purpose of an observer is to realize the $K\mathbf{x}(t)$-control. Can the existing design of observers really be rational if the critical robust property of $K\mathbf{x}(t)$-control *cannot* be realized? And if the critical robust properties of a $K\mathbf{x}(t)$-control cannot be actually realized, then what is any actual use of the optimality of that $K\mathbf{x}(t)$-control itself? This is the key reason for the irrationality and the failed practical applications of separation principle of the existing modern control system design theory.

4.3.7 Drawback 7 of Separation Principle: Two Extreme Controls

The last two obvious irrationalities of the existing modern control design/separation principle and its two exiting controls introduced in Subsections 3.2.1 and 3.2.2, are that the structures, strength and constraints of these two controls are at *two extremes*. These two existing controls are state feedback control $K\mathbf{x}(t)$ and static output feedback control $K_yC\mathbf{x}(t)$.

The K of $K\mathbf{x}(t)$-control is not constrained by (4.2) ($K=\underline{K}\underline{C}$) at all and is ideally powerful because matrix $\underline{C}$ is supposed to have rank n; on the other hand, the K of $K_yC\mathbf{x}(t)$-control is most constrained by (4.2) ($K=\underline{K}\underline{C}=K_yC$) and is therefore the weakest, because rank ($\underline{C}=C$) has the minimal rank m.

4.3.8 Drawback 8 of Separation Principle: Two Extreme Control Structures

On the feedback structures that can fully realize the above two existing controls, including their robust properties. Such a feedback observer structure for state feedback control is an unknown input observer, which has the highest possible order ($=n-m$) and which does not exist at a great majority of plant systems, while such a feedback observer structure for static output feedback control is not needed at all – order is the lowest possible ($=0$) and this control is fully realizable for all plant systems.

Order is well-known to be the most important parameter of any controller/observer system. How can a design methodology that fixes this most important parameter (order) at either maximum $n-m$ or minimum 0 be rational?

All of these extremes of Subsections 4.3.7 and 4.3.8 are summarized in Table 6.2.

People cannot help but ask, why under the existing modern control design theory and separation principle, must the forms and structures of these existing two controls be locked at such unsatisfactory extremes? Is it much more rational to have a control that can unify these two extremes?

All of the above eight irrationalities and drawbacks of the existing modern control design and separation principle are very obvious, serious, and unacceptable, even though people have always followed separation principle for more than six decades since 1960. This is the reason that the modern control theory has not found many successful practical applications in the past six decades since its start.

It is the author's belief that, if it is natural that people followed the existing separate designs of the state feedback control design of the 1950s and the state observers design since the 1960s at the beginning of the state space control theory, then it is time and overdue to overcome the eight critical and fatal drawbacks of this existing design principle.

It is the author's belief also that there is a simple technical reason for people not breaking away from separation principle. That is the lack of knowledge of a general and decoupled solution of Eq. (4.1) or (3.17), as shown in Example 4.3 (Tsui, 1993a). Without a general and decoupled solution T of (4.1), the number of rows of T or $\underline{C}$ (= $[T' \vdots C']'$) is fixed so that the observer order cannot be freely adjustable and reducible. That fixed observer order made the existing observers fixed at state observers like the ones in Example 4.5, and made it technically impossible to break away from the separation principle (Tsui, 2015).

4.4 A New Design Principle That Guarantees the General and Full Realization of Robustness of the Generalized State Feedback Control

Armed with a general and decoupled solution of (4.1) and the ability to freely adjust and reduce the observer order (Tsui, 1985), armed with the clear understanding that (4.3) ($TB=0$) is the condition to realize the loop transfer function of $Kx(t)$ control (Tsui, 1987b), armed with the realization of the superiority of the $\underline{K}\underline{C}x(t)$-control even if it is constrained by $\underline{C}$ when rank($\underline{C}$) is less than n (Subsection 3.2.2), and alarmed by the drawbacks of separation principle that rank ($\underline{C}$)$=n$ and $TB=0$ cannot be satisfied for a great majority of plant systems (see Subsections 4.3.5 and 4.3.6), a new design principle that simply breaks away from the existing separation principle is envisaged in 1990.

The following is an excerpt from Tsui (2015): "In 1990 and in an 86°F room one summer afternoon, while stuck by the dilemma that making $TB=0$ would make rank($\underline{C}$)$<n$, it suddenly occurred to this author that rank($\underline{C}$) does not need to be n (or observer order needs not to be fixed at n–m), that state feedback control law K needs not to be pre-designed (it can be designed afterwards and indirectly by designing $\underline{K}$ in the form of $K=\underline{K}\underline{C}$) using the existing static output feedback design techniques, and separation principle needs not to be adhered after all!".

Thus, a fundamentally new design principle can be outlined in the following two major steps:

Step 1: Design the dynamic part of observer (3.16a) satisfying (4.1) and (4.3).

Equation (4.1) guarantees (a) the convergence of observer state $\mathbf{z}(t)$ to $T\mathbf{x}(t)$; (b) the observer output convergences to $K\mathbf{x}(t)$ ($K=\underline{K}\underline{C}$ of (4.2)) (Theorem 3.2); and (c) the observer feedback system poles equal eigenvalues of F and $A–BK$ (Theorem 4.1). Equation (4.3) guarantees $L(s)=-K(sI–A)^{-1}B$ and the full realization of robustness of $K\mathbf{x}(t)$-control for whatever constant $K=\underline{K}\underline{C}$ of (4.2) (Theorem 3.4).

The design computations of the solution of (4.1) and (4.3) will be described in Chapters 5 and 6, respectively. The remaining freedom of (4.1) is fully used to satisfy (4.3), and the remaining design freedom of (4.1) and (4.3) will be fully used to maximize rank($\underline{C}$). Conclusion 6.1 proves that the exact solution of (4.1) and (4.3) exists if the plant system either has more outputs than inputs ($m>p$) or has at least one stable transmission zero. Examples 4.6 and 4.7 show that the great majority of plant systems satisfy either of these two conditions. Thus, the critical drawbacks 4.3.5 and 4.3.6 of the separation principle are decisively overcome.

Section 6.4 further describes the adjustment of rank($\underline{C}$) ($=m+r$) or observer order r (= row rank of T) based on the design requirements – a higher design requirement usually requires a more powerful $\underline{K}\underline{C}\mathbf{x}(t)$-control and a higher rank($\underline{C}$), while a lower rank($\underline{C}$) makes (4.3) and realization of control easier to satisfy. In case that exact solution of (4.1) and (4.3) does not exist for a required high rank ($\underline{C}=[T'{:}C']'$) or a high row rank(T), then an exact solution of (4.1) and an approximate solution of (4.3) ($TB=0$) can always be easily found.

Step 2: Design the output part of observer (3.16b) or parameter $\underline{K}$ ($K=\underline{K}\underline{C}$).

A very highly recommended goal of this design is to maximize robust stability measure M_3 of (2.25) by eigenvalue and eigenvector assignment, because M_3 is by far the most accurate measure of system performance and robustness. This design indirectly yet much more effectively designs the loop transfer function $L_{Kx}(s)=-K(sI–A)^{-1}B$. The actual design procedures of eigenvalue and eigenvector assignments will be described in Chapters 8 and 9, respectively.

The $K\mathbf{x}(t)$-control ($K=\underline{K}\underline{C}$ as of (4.2)) of this new design principle unifies completely and in all technical aspects, the existing direct state feedback control and static output feedback control. This unification is under the overall condition of $L(s)=L_{Kx}(s)$ and is in the simplest possible term of rank($\underline{C}$): maximum n (or maximum $r=n–m$) $\geq$ rank($\underline{C}$) $\geq$ minimum m (or minimum $r=0$). Thus, the last two drawbacks 4.3.7 and 4.3.8 of the separation principle are also completely overcome by the new design principle of this book.

Definition 4.2

The control defined as $\mathbf{u}(t)=-\underline{K}\underline{C}\mathbf{x}(t)$, where rank (known matrix $\underline{C}$) ranges between n and m, is called the "generalized state feedback control". This control unifies completely the existing direct state feedback control (rank($\underline{C}$)=maximum n) and static output feedback control (rank ($\underline{C}$=C)=minimum m).

To summarize, the new design principle of this book is superior to the existing separation principle, in the sense that it overcomes all eight drawbacks and irrationalities of separation principle listed in Section 4.3. Besides the overcoming of the last four drawbacks 4.3.5–4.3.8, the first four irrationalities are also simply avoided: Irrationalities 4.3.1 and 4.3.4 are avoided because the invalid assumption and the excessive requirement of having all n signals of $\mathbf{x}(t)$ is now replaced by the reliably (because TB=0) estimated r signals of $T\mathbf{x}(t)$. Irrationalities 4.3.2 and 4.3.3 are avoided because the design of the $\underline{K}\underline{C}\mathbf{x}(t)$ (= $\underline{K}[T'$: $C']'\mathbf{x}(t)$) control is now fully based on parameters T and C, and is now designed after the dynamic part (with key parameter T) of the observer is already designed in Step 1.

This new design principle fully used the decoupled property of solution T of (4.1) and the successful elimination of the existing design requirement of rank(T)=n–m and rank($\underline{C}$)=n, and thus made the exact solution of (4.1) and (4.3) much more general and general to a great majority of plant systems, for the first time! In other words, the *first general* output feedback compensator (Definition 3.3) that can fully realize a generalized state feedback control (Definition 4.2) including its signal and its loop transfer function $L_{Kx}(s)$, is claimed fully developed by the new design principle of this book (Tsui, 1998b).

The block diagram of the new output feedback compensator of this book is shown in Figure 4.3.

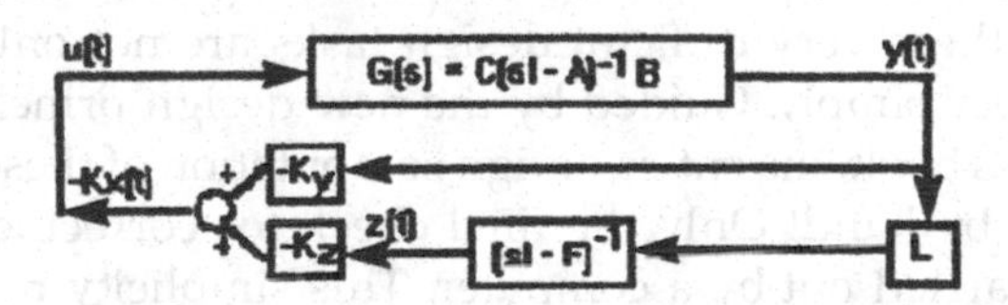

FIGURE 4.3
Block diagram of output feedback compensator that fully realizes and implements generalized state feedback control (including its loop transfer function and robustness).

In Figure 4.3, if there are n–m linearly independent signals in $\mathbf{z}(t)$, then the total number of signals of $[T'$:$C']'\mathbf{x}(t)$ available before the $\underline{K}$ block is n – the complete set of signals of $\mathbf{x}(t)$. If the branch for the observer dynamic part (F, L, T) and its signal $\mathbf{z}(t)$ are eliminated, then Figure 4.3 becomes the block diagram of static output feedback control of Figure 3.7. Overall, Figure 4.3 resembles and recovers the basic feedback structure of the classical control system (Figure 3.1), which is well established with successful practical applications. However, Figure 4.3 implements a $K\mathbf{x}(t)$ control which is far superior to any control designed by the classical control theory (see Subsections 3.2.1 and 3.2.2).

Let us use Figure 4.3 to illustrate one more time, the main features and main advantages of this design result of the new design principle of this book.

1. Unlike the feedback system of the general observer (3.16) (Figure 3.8), the feedback of input $\mathbf{u}(t)$ is eliminated in Figure 4.3. Only this feature

guarantees that the loop transfer function and robustness properties of the $\underline{KC}\mathbf{x}(t)$ control of Figure 4.3 are fully realized (Theorem 3.4)!
2. Unlike the feedback system of reduced-order state observer (Figure 4.2), only r (instead of n–m) signals of $T\mathbf{x}(t)$ are estimated. The allowance of lower row rank (T) greatly improved the generality of eliminating the input feedback (TB=0), from Figure 4.2 (invalid for a great majority of plants) to Figure 4.3 (valid for the great majority of the plants)!
3. Unlike the general feedback system of the classical control theory (Figure 3.1), the control signal of Figure 4.3 is $K\mathbf{x}(t)$, which is far superior to any control of the classical control theory.

In this new edition, two run-through design examples (8.3 and 8.4) that clearly demonstrate these advantages are added in Section 8.3. The given plant systems of these two examples are third order and two inputs and one output. The classical control theory cannot even analyze accurately the 2×2 loop transfer function of this example, while the existing modern control theory and separation principle cannot realize the loop transfer function $L_{Kx}(s)$ of the $K\mathbf{x}(t)$ control or cannot guarantee feedback system robustness at all, because condition (b) ($m \geq p$) of Subsection 4.3.5 is not met. Examples 8.3 and 8.4 fully realized (including robust properties) a $K\mathbf{x}(t)$ control that can assign all three poles and one of the eigenvectors! Neither of these two design tasks – exact LTR and full pole assignment with partial (one out of three) eigenvector assignment, can be achieved by other existing design theories and design methodologies.

Furthermore, these very difficult design tasks are not only achieved but also achieved very simply. Guided by the new design principle and design algorithms of this book, the entire design computation of these two examples was carried out by hand! Only the final check for correctness of the three eigenvalues is carried out by a computer. This simplicity reveals very convincingly the power and effectiveness of the new design principle and new design methods of this book, and it also enabled the relatively easy creation of ten more exercise problems similar to these two examples. We fully expect many more successful practical applications of the new design principle and new design algorithms of this book.

Finally, we will rebuke the only substantial criticism that, when the existing direct state feedback control's robust properties cannot be realized by an observer, then our generalized state feedback control (Definition 4.2) will be constrained and weaker than the direct state feedback control (see Subsection 3.2.2).

However, if the critical robust properties of a direct state feedback control cannot be actually realized by an observer, then the actual level of overall observer feedback system performance and robustness is like zero – no guarantee of robustness at all! The analysis at the end of Subsection 4.3.5 shows that this situation is true for about 85% of plant systems conditions based on some reasonable assumptions. Thus, the existing separation design principle must be abandoned, and in fact, it has been abandoned in practice since 1960 for over six decades.

On the other hand, for a great majority of those plant systems that direct state feedback control "must be abandoned", our new design principle can design an output feedback compensator that can realize fully (including the robust property) a generalized state feedback control! See Step 1 at the beginning of this section. We will compare the what overall observer feedback system performance and robustness level achieved for what percentage of plant systems, by these two design principles, in the following Examples 4.6 and 4.7.

Example 4.6

Conclusion 6.3 of Section 6.2 listed the three plant system conditions to have exact LTR under separation principle: (a) minimum-phase (all transmission zeros are stable); (b) $m \geq p$ (number of output measurements is no less than the number of control inputs); and (c) rank $(CB)=p$ (matrix product CB is full column rank).

Exercise 4.2 proves that because almost all plant systems with $(m=p)$ have $n-m$ transmission zeros, almost all of these systems have at least one unstable transmission zero (non-minimum phase). Condition (c) cannot be satisfied by many (say half of) real plant systems such as airborne systems even with $m>p$.

Let us assume 50% of the plants are $m=p$ and 30% are $m>p$. Then, under separation principle, about 85% of plants cannot have exact LTR. Even if we assume 50% of the plants are $m=p$ and 40% are $m>p$, then about 80% of plants still cannot have exact LTR. Because the existing approximate LTR design is unsatisfactory (Subsection 4.3.6), the overall level of feedback system performance and robustness is almost zero for these systems.

On the other hand, Conclusion 6.1 proves that our new design principle can achieve exact LTR if the plant system either has more outputs than inputs $(m>p)$ or has at least one stable transmission zero. Exercises 4.3 and 4.7 show that almost all plant systems with $m=p$ have at least one stable transmission zero. In addition, Exercises 4.4 and 4.8 show that in the practical world, at least half of systems with $m=p$ can satisfy $(m+r)+p>n$, and almost all of these systems can satisfy $(m+r)\times p>n$, where r is the number of stable transmission zeros. The results of systems with $m>p$ should be at least better than that of systems with $m=p$, because of more output measurement information to utilize and less control inputs to realize.

From Subsection 3.2.2, the level of overall feedback system performance can be considered ideal, excellent, and very good for output feedback compensator system conditions $q=n$, $q+p>n$, and $q\times p>n$, respectively, where $q=\text{rank}(\underline{C})=r+m$.

Hence, using the new design principle, if 50% of the plants are $m=p$ and 30% are $m>p$, then the percentages of plant systems for these three levels are 15%, 40%, and 80%, respectively. If 50% of the plants are $m=p$ and 40% are $m>p$, then the percentages of plant systems for these three levels are 20%, 45%, and 90%, respectively.

Let us summarize these conclusions in Figures 4.4 and 4.5.

Figure 4.4 is based on the assumption that 50% of the plants are $m=p$ and 30% are $m>p$. The levels of the overall feedback system performance and robustness are represented by the broken line and the solid line, for the design results of separation principle and the design results of the new design principle of this book, respectively.

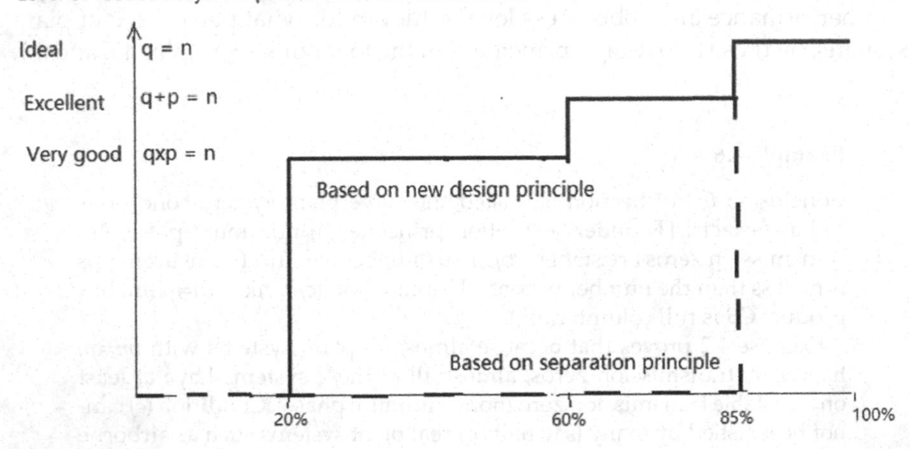

FIGURE 4.4
Levels of feedback system performance and robustness of two design principles.

The solid line is higher than the broken line for 65% (from 20% to 85% or 15% to 80% from right to left) of all plants! The raise of the level of feedback system performance and robustness is from very low (no guarantee of robustness at all) to at least very good, for these systems, even though this level is lower than the ideal level of direct state feedback control! This means that the new design principle can improve feedback system performance and robustness much better than separation principle, for 65% of the plant systems!

Figure 4.5 is based on the assumption that 50% of the plants are $m=p$ and 40% are $m>p$. The levels of the overall feedback system performance and robustness are represented by the broken line and the solid line, for the design results of separation principle and the design results of the new design principle of this book, respectively.

The solid line is much higher than the broken line for 70% (10%–80%, or 20%–90% from right to left) of all plants! This raise of the level of feedback system performance and robustness from very low (no guarantee of robustness at all) to at least very good, for these systems, even though this level is lower than the ideal level of direct state feedback control! This means that the new design principle can improve feedback system performance and robustness much better than separation principle, for 70% of the plant systems!

The actual level lines, whether solid or broken, should have a curved shape instead of straight lines. The curves should turn sharply at the passing of each of the turning points of these two figures. For example, the waveforms at the rising turning points should look like the current versus voltage curve of a diode at 0.7 V.

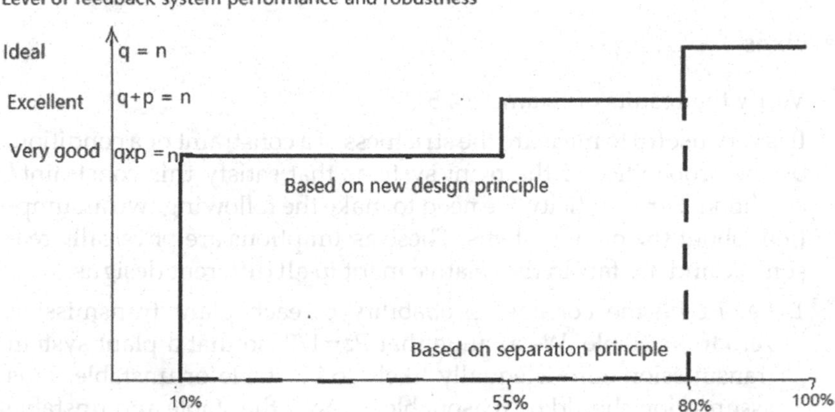

FIGURE 4.5
Levels of feedback system performance and robustness of two design principles under another assumption.

Example 4.7 (Tsui, 2015)

This example is for all fifth ($n=5$) order systems with two outputs and two inputs ($m=p=2$). Although this example is not as general as Example 4.6 that is for all plant systems with all situations of parameters n, m, and p, it should still be much more general than the examples in other control system literature.

Based on Exercise 8.13 and Tsui (2015), based on the fact that systems with $m=p$ have generically $n-m$ transmission zeros (Davison and Wang, 1974), and based on the reasonable assumption of Exercise 4.2 that the probability of each transmission zero is stable is 50%, the probability for this system to be minimum-phase is only 12.5%. Thus, only 12.5% of these systems can really realize state feedback control under the separation principle. In addition, the existing static output feedback control cannot assign all poles and cannot guarantee the stability of the feedback system since $m \times p < n$. Thus, all systems of this example are hopeless from the point of view of the existing modern control design theory.

However, based on the new design principle of this book, an output feedback compensator of order r can be designed that can fully realize a generalized state feedback control $\underline{K}\underline{C}\mathbf{x}(t)$ with rank$(\underline{C})=r+m$, where r is the number of stable transmission zeros of the plant. Hence, according to Exercise 4.2, the probability to achieve excellent feedback system performance and robustness (or $(r+m)+p>n$) is 50%, while the probability to achieve very good feedback system performance and robustness (or $(r+m)\times p>n$) is 87.5%!

To summarize, the new design principle of this book always achieves the overall feedback system performance and robustness, either at the same level of the existing separation principle when exact LTR can be achieved, or at a much higher level when exact LTR cannot be achieved (the majority of the plant system conditions).

Exercises

4.1 Verify the results of Example 4.5.

4.2 It is very useful to measure the strictness of a constraint or a condition, by the probability of the plant systems that satisfy this constraint/condition. For simplicity, we need to make the following two assumptions about the plant systems. These assumptions are physically reasonable and are fair in the relative merit to all different designs.

1. Let Pz be the constant probability of each plant transmission zero to be stable. We assume that $Pz=1/2$, so that a plant system transmission zero is equally likely to be stable or unstable. This assumption should be reasonable because the stable and unstable regions are about half and half.
2. We assume $m=p$ so that the number of system transmission zeros is simply $n-m$ (Davision and Wang, 1974), and so that rank $[T':C']'$ of our new design principle is $r+m$, where r is the number of stable transmission zeros out of the $n-m$ transmission zeros of the plant.

(a) Based on these two assumptions, compute the Pr as the probability of r stable transmission zeros out of $n-m$ transmission zeros. Formula: $Pr=[n-m: r]Pz^r(1-Pz)^{n-r}$, where $[n-m: r]$ is the combination of r elements out of $n-m$ elements.

Answer:

$n-m=$	1	2	3	4	5	6	7	8	9
$r=$									
0	1/2	$1/2^2$	$1/2^3$	$1/2^4$	$1/2^5$	$1/2^6$	$1/2^7$	$1/2^8$	$1/2^9$
1	1/2	$2/2^2$	$3/2^3$	$4/2^4$	$5/2^5$	$6/2^6$	$7/2^7$	$8/2^8$	$9/2^9$
2		$1/2^2$	$3/2^3$	$6/2^4$	$10/2^5$	$15/2^6$	$21/2^7$	$28/2^8$	$36/2^9$
3			$1/2^3$	$4/2^4$	$10/2^5$	$20/2^6$	$35/2^7$	$56/2^8$	$84/2^9$
4				$1/2^4$	$5/2^5$	$15/2^6$	$35/2^7$	$70/2^8$	$126/2^9$
5					$1/2^5$	$6/2^6$	$21/2^7$	$56/2^8$	$126/2^9$
6						$1/2^6$	$7/2^7$	$28/2^8$	$84/2^9$
7							$1/2^7$	$8/2^8$	$36/2^9$
8								$1/2^8$	$9/2^9$

(b) Based on the result of Part (a), find the probability of minimum-phase $P(r=n-m)$ for $n-m=1$–8.

Answer: $Pn-m=0.5$, 0.25, 0.125, 0.0625, 0.03125, 0.0156, 0.0078, and 0.0036. This probability is very low after $n-m>2$ and is rapidly approaching 0.

4.3 One of the sufficient conditions for the new design principle of this book is to have at least one stable transmission zero ($r>0$) (see Conclusion 6.1). Calculate the probability of $r>0$ based on the assumption of Exercise 4.2, for $n–m=1–8$.

Answer: $P(r>0)=1–P(r=0)=0.5$, 0.75, 0.875, 0.9375, 0.9688, 0.9844, 0.9922, and 0.9964. This probability is almost 100% when $n–m>3$. This means that almost all systems with $m=p$ satisfy this sufficient condition of the new design principle of this book.

4.4 The sufficient condition for the generalized state feedback control of this book to assign arbitrarily given poles and some eigenvectors is $r+m+p>n$ or $r>n–m–p$ [see Step 2 of Algorithm 8.1]. Based on the result of Exercise 4.2, calculate the probability of $r>n–m–p$ (= 100% if $m+p>n$).

Answer:

$n=$	3	4	5	6	7	8	9	10
$m=p=$	%	%	%	%	%	%	%	%
2	100	75	50	31.25	18.75	10.94	6.25	3.52
3	100	100	100	87.5	68.75	50	34.38	22.66
4	100	100	100	100	100	93.75	81.25	65.63

The probability of the existing static output feedback control to assign all poles and some eigenvectors is 0% in the above table if the entry is not 100%. Thus, the improvement of our generalized state feedback control from the existing static output feedback control is very significant.

4.5 The condition for the generalized state feedback control of this book to assign generically arbitrarily given poles and to guarantee stability is $(r+m)p>n$ or $r>n/p–m$ (Wang, 1996). Based on the result of Exercise 4.2, calculate the probability of $r>n/p–m$ (= 100% if $n/p–m<0$).

Answer:

$n=$	3	4	5	6	7	8	9	10	11	12
$m=p=$	%	%	%	%	%	%	%	%	%	%
2	100	75	88	69	81	66	77	63	75	
3	100	100	100	100	100	100	98	99	99+	98

The probability is very high as soon as m is higher than 2 and decreases very slowly so that no substantial decrease can be seen in the above table. For example, it can be calculated that when $n=16$ and $m=3$, the probability is still at 98%; when $n=26$ and $m=4$, the probability is still at 99.7%. This result means that our generalized state feedback control can assign all poles and can guarantee stability for almost all plant systems (with $m=p$).

The probability of the existing static output feedback control to assign all poles is 0% in the above table if the entry is not 100%. Thus, the improvement of our generalized state feedback control from the existing static output feedback control is very significant.

4.6 Repeat Exercise 4.2. Change Pz from (1/2) to (3/4). This new Pz implies that each plant system transmission zero is three times more likely to be stable than to be unstable. This new Pz is considerably higher than 1/2 and is a considerably more favorable assumption. Therefore, we expect the real practical values of Pz would be between these two values.

(a) Answer: $Pr=[n-m: r](3/4)^r(1-¾)^{n-r}$

$n-m=$	1	2	3	4	5	6	7	8
$r=$								
0	1/4	$1/4^2$	$1/4^3$	$1/4^4$	$1/4^5$	$1/4^6$	$1/4^7$	$1/4^8$
1	3/4	$6/4^2$	$9/4^3$	$12/4^4$	$15/4^5$	$18/4^6$	$21/4^7$	$24/4^8$
2		$9/4^2$	$27/4^3$	$54/4^4$	$90/4^5$	$135/4^6$	$189/4^7$	$252/4^8$
3			$27/4^3$	$108/4^4$	$270/4^5$	$540/4^6$	$945/4^7$	$1,512/4^8$
4				$81/4^4$	$405/4^5$	$1215/4^6$	$2835/4^7$	$5,670/4^8$
5					$243/4^5$	$1458/4^6$	$5103/4^7$	$13608/4^8$
6						$729/4^6$	$5103/4^7$	$20412/4^8$
7							$2187/4^7$	$17496/4^8$
8								$6,561/4^8$

(b) Based on the result of Part (a), find the probability of minimum-phase $P(r=n-m)$ for $n-m=1$–8.

Answer: $Pn-m=0.75$, 0.56, 0.42, 0.32, 0.24, 0.18, 0.13, and 0.1. This probability is lower than one-half after $n-m>2$ and decreases as $n-m$ increases. This implies that the existing exact LTR can rarely be achieved for most non-trivial plant systems.

4.7 One of the sufficient conditions for the new design principle of this book is to have at least one stable transmission zero ($r>0$) (see Conclusion 6.1). Calculate the probability of $r>0$ based on the assumption of Exercise 4.6, for $n-m=1$–8.

Answer: $P(r>0)=1-P(r=0)=0.75$, 0.9375, 0.9844, 0.9964….

This probability is almost 100% when $n-m>1$. This means that almost all systems with $m=p$ satisfy this sufficient condition of the new design principle of this book.

Someone published a numerical example of $r=0$, to show that the new design principle of this book does not work on that example, and was referenced quite a few times (Tsui, 1996b), even though the result of this Exercise shows that the probability of $r=0$ is almost zero.

Using a highly unlikely individual example to deny a whole general design principle, is to cover and to mix up the difference between "general to a great majority of systems" and "general to all systems", and to ignore and to cover deliberately the basic fact that a perfect design method for every plant system (no matter how bad conditioned) cannot exist in the real world.

4.8 The sufficient condition for the generalized state feedback control of this book to assign arbitrarily given poles and some eigenvectors is $r+m+p>n$ or $r>n-m-p$ [see Step 2 of Algorithm 8.1]. Based on the result of Exercise 4.6, calculate the probability of $r>n-m-p$ (= 100% if $m+p>n$).

Answer:

$n=$	3	4	5	6	7	8	9	10
$m=p=$	%	%	%	%	%	%	%	%
2	100	94	84	74	63	53	44	37
3	100	100	100	98	97	90	83	76
4	100	100	100	100	100	99.6	98	96

The probability of the existing static output feedback control to assign all poles and some eigenvectors is 0% in the above table if the entry is not 100%. Thus, the improvement of our generalized state feedback control from the existing static output feedback control is very significant.

4.9 The condition for the generalized state feedback control of this book to assign generically arbitrarily given poles and to guarantee stability is $(r+m)p>n$ or $r>n/p-m$ (Wang, 1996). Based on the result of Exercise 4.6, calculate the probability of $r>n/p-m$ (= 100% if $n/p-m<0$).

Answer:

$n=$	3	4	5	6	7	8	9	10	11	12
$m=p=$	%	%	%	%	%	%	%	%	%	%
2	100	94	98	95	98	96	99	97		
3	100	100	100	100	100	100	99+	99+	99+	99+

The probabilities are almost all 100%. That means our generalized state feedback control can assign all poles and guarantee feedback system stability for all of these plant systems. The probability of the existing static output feedback control to assign generically all poles is 0% in the above table if the entry is not 100%. Thus, the improvement of our generalized state feedback control from the existing static output feedback control is very significant.

5

Solution of Matrix Equation TA–FT=LC

Chapter 4 proposed the new design principle of this book. Procedure wise, that is to satisfy (4.1) and (4.3) first and in priority. The necessity and advantages of this new design principle are also explained. The problem of computing the solutions of (4.1) and (4.3) was first raised directly by Tsui (1987b). The satisfactory solution of (4.1) and (4.3) first appeared in the work of Tsui (1992), much delayed from its verbal presentation at the 1990 American Control Conference. This solution has made this new design principle possible (Tsui, 2000, 2015).

The design computation algorithm for (4.1) is presented in this chapter. The design computation algorithm for (4.3) using the remaining freedom of (4.1) will be presented in Chapter 6. This chapter has two sections.

Section 5.1 presents the algorithm for computing the block-observable Hessenberg form of the plant system's state space model. Although this computation is unnecessary for the analytical solution of (4.1), it significantly improves the numerical computation of this solution, reveals important analytical parameters such as observability indices, and separates the observable part of the system from the unobservable part.

Section 5.2 presents the solution of (4.1) and its computational algorithm. It demonstrates also the analytical and numerical advantages of this solution and its computation. Equation (4.1) is by far the most important matrix equation in the state space control design (Tsui, 1987a, 1993a, 2015).

5.1 Computation of System's Observable Hessenberg Form

5.1.1 Single-Output Systems

The Hessenberg form matrix is defined as follows

$$A = \begin{bmatrix} x & * & 0 & \dots & \dots & 0 \\ x & x & * & 0 & \dots & 0 \\ \vdots & \vdots & \ddots & \ddots & & \vdots \\ \vdots & \vdots & & \ddots & \ddots & 0 \\ \vdots & \vdots & & & x & * \\ x & x & \dots & \dots & \dots & x \end{bmatrix} \tag{5.1}$$

DOI: 10.1201/9781003259572-5

where the "x" elements are arbitrary and the "*" elements are nonzero. This matrix is also called the "lower Hessenberg form". Its transpose is called the "upper Hessenberg form".

The Hessenberg form is the simplest matrix form, which can be computed from a general state matrix A by orthogonal matrix operations without iteration for convergence. For example, the next simplest form – Schur triangular form, which differs from the Hessenberg form by making all * elements in (5.1) equal to 0, is computed by iteration for convergence (QR method).

In the established computational algorithms of some basic numerical linear algebra problems, whether in the QR method of computing the eigenvalues (Wilkinson, 1965) and singular value decomposition (Golub and Reinsch, 1970), or in the computation of solutions of the Sylvester equation (Golub et al., 1979) and the Riccati equation (Laub, 1979; Section 10.1), the computation of the Hessenberg form has always been the first step (Laub and Linnemann, 1986).

As the first step of the design algorithm for the solution of Eq. (4.1), a special form of system matrix pair (A, C) called the "observable Hessenberg form", in which matrix A is in the lower Hessenberg form, is also computed first (Van Dooren et al., 1978; Van Dooren, 1981).

The single-output case of this form is:

$$\begin{bmatrix} CH \\ \cdots\cdots\cdots \\ H'AH \end{bmatrix} = \begin{bmatrix} * & 0 & \cdots & \cdots & \cdots & 0 \\ \cdots & \cdots & \cdots & \cdots & \cdots & \cdots \\ x & * & 0 & \cdots & \cdots & 0 \\ x & x & * & 0 & \cdots & 0 \\ \vdots & \vdots & & \ddots & & \vdots \\ \vdots & \vdots & & & \ddots & 0 \\ \vdots & \vdots & & & x & * \\ x & \cdots & \cdots & \cdots & \cdots & x \end{bmatrix} \tag{5.2}$$

where matrix H is a unitary similarity transformation matrix ($H'H=I$), which transforms system matrix pair (A, C) into observable Hessenberg form (5.2), by the following algorithm:

Algorithm 5.1: Computation of a Single-Output Observable Hessenberg Form System Matrix

Step 1: Let $j=1$, $H=I$, $\mathbf{c}_1=C$, and $A_1=A$.

Step 2: Compute the unitary matrix H_j such that $\mathbf{c}_jH_j=[c_j, 0 \ \dots \ \dots \ 0]$ (see Section A.2).

Step 3: Compute

$$H_j'A_jH_j = \begin{bmatrix} a_{jj} & : & \mathbf{c}_{j+1} \\ \cdots & \cdots & \cdots \\ \mathbf{x} & : & A_{j+1} \end{bmatrix} \}n-j \tag{5.3}$$

Step 4: Update matrix

$$H = H\begin{bmatrix} I_{j-1} & : & 0 \\ \cdots & \cdots & \cdots \\ 0 & : & H_j \end{bmatrix}$$

where I_{j-1} is a $(j-1)$-dimensional identity matrix.

Step 5: If $\mathbf{c}_{j+1}$ of (5.3) equals 0, then go to Step 7.

Step 6: Let $j=j+1$ so that $(A_{j+1}, \mathbf{c}_{j+1})$ of (5.3) becomes $(A_j, \mathbf{c}_j)$. If $j=n$, then go to Step 7; otherwise, go back to Step 2.

Step 7: The final result is

$$\begin{bmatrix} CH \\ \cdots\cdots \\ H'AH \end{bmatrix} = \begin{bmatrix} c_1 & 0 & \cdots & \cdots & \cdots & 0 & : & 0 & \cdots & 0 \\ \cdots & \cdots & \cdots & \cdots & \cdots & \cdots & : & & & \\ a_{11} & c_2 & 0 & \cdots & \cdots & 0 & : & 0 & \cdots & 0 \\ : & a_{22} & c_3 & 0 & \cdots & 0 & : & & & \\ : & : & & \ddots & & : & : & & & \\ : & : & & & \ddots & 0 & : & & & \\ : & : & & & & c_j & : & & & \\ a_{j1} & \cdots & \cdots & \cdots & \cdots & a_{jj} & : & 0 & \cdots & 0 \\ \cdots & \cdots & \cdots & \cdots & \cdots & \cdots & \cdots & \cdots & \cdots & \cdots \\ & & & X & & & : & & \bar{A}_o & \end{bmatrix}$$

$$\equiv \begin{bmatrix} C_o & : & 0 \\ \cdots & : & \\ A_o & : & 0 \\ \cdots & \cdots & \cdots \\ X & : & \bar{A}_o \end{bmatrix} \begin{matrix} \\ \\ \}\, j \\ \\ \}\, n-j \end{matrix} \tag{5.4}$$

The matrix pair (A_o, C_o) of (5.4) is the observable part of the system, with dimension j. This result will be used in the later analysis and design as (A, C) and with dimension n.

5.1.2 Multiple-Output Systems

In multiple-output systems of m outputs, C is now a matrix of m rows. The corresponding observable Hessenberg form in this case is extended to the so-called "block-observable Hessenberg form" as follows:

$$\begin{bmatrix} CH \\ \cdots\cdots \\ H'AH \end{bmatrix} = \begin{bmatrix} C_1 & 0 & \cdots & \cdots & \cdots & 0 \\ \cdots & \cdots & \cdots & \cdots & \cdots & \cdots \\ A_{11} & C_2 & 0 & \cdots & \cdots & 0 \\ \vdots & A_{22} & C_3 & 0 & \cdots & 0 \\ \vdots & \vdots & & \ddots & & \vdots \\ \vdots & \vdots & & & \ddots & 0 \\ \vdots & \vdots & & & & C_v \\ A_{v1} & \cdots & \cdots & \cdots & \cdots & A_{vv} \end{bmatrix} \begin{matrix} \} m_0 \\ \\ \} m_1 \\ \} m_2 \\ \\ \\ \} m_{v-1} \\ \} m_v \end{matrix} \tag{5.5}$$

$$\quad m_1 \quad m_2 \quad \cdots \quad m_v$$

where C_j $(j=1, \ldots, \nu)$ are $m_{j-1} \times m_j$ dimensional matrix blocks in lower echelon form as shown below, A_{ij} are $m_i \times m_j$ matrix blocks, and $m_0 = m \geq m_1 \geq m_2 \geq \ldots \geq m_\nu > 0$.

Example 5.1: All Lower Echelon Form Matrices with Three Rows Are

$$\begin{bmatrix} * & 0 & 0 \\ x & * & 0 \\ x & x & * \end{bmatrix}, \begin{bmatrix} * & 0 \\ x & * \\ x & x \end{bmatrix}, \begin{bmatrix} * & 0 \\ x & 0 \\ x & * \end{bmatrix}, \begin{bmatrix} 0 & 0 \\ * & 0 \\ x & * \end{bmatrix}, \begin{bmatrix} * \\ x \\ x \end{bmatrix}, \begin{bmatrix} 0 \\ * \\ x \end{bmatrix}, \begin{bmatrix} 0 \\ 0 \\ * \end{bmatrix}$$

From Section A.2 of Appendix A, there exists a unitary matrix H_j such that $\underline{C}_j H_j = [C_j : 0 \ldots 0]$ where C_j is in its lower echelon form.

From Example 5.1, all m_j columns of the echelon form matrix C_j are linearly independent. There are also m_j rows of C_j marked with the symbol * that are linearly independent of each other. Each of the remaining $m_{j-1} - m_j$ rows of C_j and the corresponding linearly dependent rows in matrix $H'AH'$ of (5.5) can be expressed as a linear combination of the linearly independent rows above that linearly dependent row (see Section A.2).

Example 5.2: Linearly Dependent Rows of Lower Echelon Form Matrices of Example 5.1

Of the three rows of seven matrices C_i $(i=1\text{–}7)$ of Example 5.1, the linearly dependent rows (the rows without the mark *) are

Matrix Blocks:	C_1	C_2	C_3	C_4	C_5	C_6	C_7
Linearly dependent rows	none	Third	Second	First	Second and third	First and third	First and second

In matrix blocks C_i $(i=2\text{–}6)$ with linearly dependent rows, there exist at least one $\mathbf{d}_i$ vector such that $\mathbf{d}_i C_i = 0$. Each of these $\mathbf{d}_i$ vectors has a "1" element whose position at the vector is corresponding to the position of a linearly dependent row of C_i, and the x elements are linear combination coefficients of that linearly dependent row:

Matrix Blocks	C_2	C_3	C_4	C_5	C_6	C_7
$\mathbf{d}_i$ vector such that $\mathbf{d}_i C_i = 0$	$[x\ x\ 1]$	$[x\ 1\ 0]$	$[1\ 0\ 0]$	$[x\ 0\ 1]$ or $[x\ 1\ 0]$	$[0\ x\ 1]$ or $[1\ 0\ 0]$	$[0\ 1\ 0]$ or $[1\ 0\ 0]$

Without loss of generality, we assume $m_1 = m$, or all m rows of system matrix C are linearly independent. During the computation of block-observable Hessenberg form (5.5), these m linearly independent rows will become dependent. If a row of block C_j becomes linearly dependent, then the corresponding rows/columns will disappear at the subsequent C_i blocks ($i > j$), or the corresponding output will no longer be connected to any more of the system states (see Figure 1.9). With this adaptation, Algorithm 5.1 can be generalized to the following Algorithm 5.2 for multi-output cases:

Algorithm 5.2: Computation of Block-Observable Hessenberg Form

Step 1: Let $j = 1$, $H = I$, $\underline{C}_1 = C$, and $A_1 = A$, $m_0 = m$, and $n_0 = 0$.

Step 2: Compute the unitary matrix H_j such that $\underline{C}_j H_j = [C_j : 0 \ldots \ldots 0]$, where C_j is an $m_{j-1} \times m_j$ dimensional lower echelon form matrix (see Example 5.1).

Step 3: Compute

$$H_j' A_j H_j = \left[\begin{array}{ccc} A_{jj} & : & \underline{C}_{j+1} \\ \ldots & \ldots & \ldots \\ X & : & A_{j+1} \\ & : & \end{array}\right] \begin{array}{l} \} m_j \\ \\ \\ \\ \end{array} \quad (5.6)$$

$$\quad m_j$$

Step 4: Update matrix

$$H = H \left[\begin{array}{ccc} I_{j-1} & : & 0 \\ \ldots & \ldots & \ldots \\ 0 & : & H_j \end{array}\right]$$

where I_{j-1} is an (n_{j-1})-dimensional identity matrix.

Step 5: $n_j = n_{j-1} + m_j$. If $n_j = n$ or if $\underline{C}_{j+1}$ of (5.6) equals 0, then let $\nu = j$ and go to Step 7.

Step 6: Let $j = j + 1$ so that $(A_{j+1}, \underline{C}_{j+1})$ of (5.6) becomes $(A_j, \underline{C}_j)$. Then return back to Step 2.

Step 7: The final result is

$$
\begin{bmatrix} CH \\ \cdots\cdots\cdots \\ H'AH \end{bmatrix} = \left[\begin{array}{cccccc:ccc}
C_1 & 0 & \cdots & \cdots & \cdots & 0 & 0 & \cdots & 0 \\
\cdots & \cdots & \cdots & \cdots & \cdots & \cdots & & & \\
A_{11} & C_2 & 0 & \cdots & \cdots & 0 & 0 & \cdots & 0 \\
\vdots & A_{22} & C_3 & 0 & \cdots & 0 & & & \\
\vdots & \vdots & & \ddots & & \vdots & & & \\
\vdots & & & & \ddots & 0 & & & \\
\vdots & \vdots & & & & C_\nu & & & \\
A_{\nu 1} & \cdots & \cdots & \cdots & \cdots & A_{\nu\nu} & 0 & \cdots & 0 \\
\cdots & \cdots & \cdots & \cdots & \cdots & \cdots & \cdots & \cdots & \cdots \\
 & & & X & & & & \bar{A}_o &
\end{array}\right] \}m
$$

$$
\equiv \left[\begin{array}{c:c}
C_o & 0 \\
\cdots & \vdots \\
A_o & 0 \\
\cdots & \cdots \\
X & \bar{A}_o
\end{array}\right] \tag{5.7}
$$

The matrix pair (A_o, C_o) of (5.7) is in block-observable Hessenberg form (5.5), and is the observable part of the system with dimension n_ν $(=m_1+\ldots+m_\nu)$. This result will be used in the later analysis and design as (A, C) and with dimension n.

The main computation of this algorithm is at Step 3. According to Section A.2, the order of computation of this algorithm (based on Step 3) is about $4n^3/3$.

Definition 5.1: Observability Indices

Each system output is associated with one of the m rows of system matrix C. From the description of Algorithm 5.2, each of these m outputs will be linked to one more system state, if the corresponding row of this output remains linearly independent in the next block C_{j+1}. Thus, we can define the number of observable system states that are linked to the i-th output as the i-th observability index ν_i, $i=1, \ldots, m$.

In other words, $\nu_i=j$ if the row corresponding to the i-th output becomes linearly dependent in matrix block C_{j+1}. Thus, all observability indices can be determined using Algorithm 5.2, and $\nu \equiv \max\{\nu_i, i=1, \ldots, m\}$ can be found in Step 5 of Algorithm 5.2.

It is also clear that $\nu_1+\ldots+\nu_m=n$.

Another set of dimensional parameters m_j, $j=1, \ldots, \nu$ can also be found along with Algorithm 5.2. From the description of block-observable Hessenberg form (5.5) and Definition 5.1, parameter m_j equals the number of observability indices that are $\geq j$.

Example 5.3

Let the block-observable Hessenberg form of a 9-th order and four-output system be

$$\begin{bmatrix} C \\ \cdots \\ A \end{bmatrix} = \begin{bmatrix} C_1 & 0 & 0 & 0 \\ \cdots & \cdots & \cdots & \cdots \\ A_{11} & C_2 & 0 & 0 \\ A_{21} & A_{22} & C_3 & 0 \\ A_{31} & A_{32} & A_{33} & C_4 \\ A_{41} & A_{42} & A_{43} & A_{44} \end{bmatrix}$$

$$= \left[\begin{array}{ccccccccccc} * & 0 & 0 & 0 & : & 0 & 0 & 0 & 0 & 0 \\ x & + & 0 & 0 & : & 0 & 0 & 0 & 0 & 0 \\ x & x & @ & 0 & : & 0 & 0 & 0 & 0 & 0 \\ x & x & x & \# & : & 0 & 0 & 0 & 0 & 0 \\ \cdots & \cdots & \cdots & \cdots & \cdots & \cdots & \cdots & \cdots & \cdots & \cdots \\ x & x & x & x & : & * & 0 & 0 & 0 & 0 \\ x & x & x & x & : & x & + & 0 & 0 & 0 \\ x & x & x & x & : & x & x & 0 & 0 & 0 \\ x & x & x & x & : & x & x & \# & 0 & 0 \\ \cdots & \cdots & \cdots & \cdots & \cdots & \cdots & \cdots & \cdots & \cdots & \cdots \\ x & x & x & x & : & x & x & x & 0 & 0 \\ x & x & x & x & : & x & x & x & + & 0 \\ x & x & x & x & : & x & x & x & x & 0 \\ \cdots & \cdots & \cdots & \cdots & \cdots & \cdots & \cdots & \cdots & \cdots & \cdots \\ x & x & x & x & : & x & x & x & x & + \\ \cdots & \cdots & \cdots & \cdots & \cdots & \cdots & \cdots & \cdots & \cdots & \cdots \\ x & x & x & x & : & x & x & x & x & x \end{array}\right] \begin{array}{l} \}m_0 = m = 4 \\ \}m_1 = 4 \\ \}m_2 = 3 \\ \}m_3 = 1 \\ \}m_4 = 1 \end{array} \quad (5.8a)$$

From Definition 5.1, corresponding to the four system outputs, which are represented by the nonzero elements with symbols *, +, @, and #, respectively, the four observability indices are $\nu_1=2$, $\nu_2=4$, $\nu_3=1$, and $\nu_4=2$. Each index equals the number of appearances of the corresponding output/symbol in the state matrix.

We can also verify that $\nu_1+\nu_2+\nu_3+\nu_4=9=n$ and $\nu=\nu_2=4$. We can also verify that $m_1=4$ means four observability indices≥ 1; $m_2=3$ means three observability indices≥ 2, $m_3=1$ means one observability index≥ 3, and $m_4=1$ means one observability index≥ 4.

In Chen (1984), the block-observable Hessenberg form (5.8a) can be further transformed to the block-observable canonical form (1.16) by elementary similarity transformation:

$$\begin{bmatrix} CE \\ \cdots\cdots\cdots \\ E^{-1}AE \end{bmatrix} = \begin{bmatrix} C_1 I_1 & 0 & 0 & 0 \\ \cdots & \cdots & \cdots & \cdots \\ A_{11} & I_2 & 0 & 0 \\ A_{21} & 0 & I_3 & 0 \\ A_{31} & 0 & 0 & I_4 \\ A_{41} & 0 & 0 & 0 \end{bmatrix}$$

$$= \left[\begin{array}{cccccccccc} 1 & 0 & 0 & 0 & : & 0 & 0 & 0 & 0 & 0 \\ x & 1 & 0 & 0 & : & 0 & 0 & 0 & 0 & 0 \\ x & x & 1 & 0 & : & 0 & 0 & 0 & 0 & 0 \\ x & x & x & 1 & : & 0 & 0 & 0 & 0 & 0 \\ \cdots & \cdots & \cdots & \cdots & \cdots & \cdots & \cdots & \cdots & \cdots & \cdots \\ x & x & x & x & : & 1 & 0 & 0 & 0 & 0 \\ x & x & x & x & : & 0 & 1 & 0 & 0 & 0 \\ x & x & x & x & : & 0 & 0 & 0 & 0 & 0 \\ x & x & x & x & : & 0 & 0 & 1 & 0 & 0 \\ \cdots & \cdots & \cdots & \cdots & \cdots & \cdots & \cdots & \cdots & \cdots & \cdots \\ x & x & x & x & : & 0 & 0 & 0 & 0 & 0 \\ x & x & x & x & : & 0 & 0 & 0 & 1 & 0 \\ x & x & x & x & : & 0 & 0 & 0 & 0 & 0 \\ \cdots & \cdots & \cdots & \cdots & \cdots & \cdots & \cdots & \cdots & \cdots & \cdots \\ x & x & x & x & : & 0 & 0 & 0 & 0 & 1 \\ \cdots & \cdots & \cdots & \cdots & \cdots & \cdots & \cdots & \cdots & \cdots & \cdots \\ x & x & x & x & : & 0 & 0 & 0 & 0 & 0 \end{array}\right] \begin{array}{l} \\ \}m_1 = m = 4 \\ \\ \\ \\ \\ \}m_2 = 3 \\ \\ \\ \\ \\ \}m_3 = 1 \\ \\ \\ \}m_4 = 1 \\ \\ \}m_4 = 1 \end{array} \tag{5.8b}$$

where matrix E is an elementary matrix operator (Chen, 1984) and is usually not a unitary matrix.

The comparison of (5.8a) and (5.8b) shows that the block-observable canonical form is a special case of the block-observable Hessenberg form, in the sense that all nonzero elements or symbols in the C_j $(j \geq 1)$ blocks become 1 and that all other x elements of (5.8a) except those in the left m_1 columns become 0.

Although the parameters of a block-observable canonical form system matrix can be substituted directly into the polynomial matrix fraction description (MFD) $G(s) = C_1\ D(s)^{-1}N(s)$ (see Example 1.7), this advantage is offset by the unreliability of its similarity transformation computation, which is usually ill-conditioned (Wilkinson, 1965; Laub, 1985). For this reason, block-observable Hessenberg forms (5.5) and (5.8a) is recommended as the basis of future analysis and design computation, as described at the beginning of this section.

5.2 Computation of the Solution of Matrix Equation $TA-FT=LC$

In this matrix equation, system matrices (A, C) are given and are observable, and observer parameters (F, T, L) are the solution to be computed. From Chapter 4, it is critical for a really good design that the rows of this solution be completely decoupled so that the number of rows of this solution can be freely designed. We presume the number of rows of this solution, r, which is also the observer order, to be its maximal possible value $n-m$, although the actual value of r can be freely designed, reduced, and adjusted later.

To simplify the computation of solution of (4.1), block-observable Hessenberg form system matrices $(H'AH, CH)$ are computed using Algorithm 5.2. Thus, Eq. (4.1) becomes $T(H'AH)-F\ T=L(CH)$, although we still consider this new equation as (4.1) in this section. This means that after solution T of this new equation is computed, only TH' is the solution corresponding to the original Eq. (4.1), which corresponds to the original system matrices (A, C).

Mathematically, the eigenvalues $(\lambda_i, i=1, \ldots, n-m)$ of matrix F of (4.1) can be arbitrarily chosen and given. However, in actual design, we need to select these eigenvalues with the following four considerations. These four considerations are more comprehensive and wiser than other existing designs, which only require these eigenvalues to be stable.

1. These eigenvalues must have negative and sufficiently negative real parts, in order to guarantee not only observer stability and estimation convergence, but also sufficiently fast convergence (Theorem 3.2).
2. Each plant system's stable transmission zero must be matched by one of the eigenvalues of F. This is the necessary condition for the corresponding rows of matrix T to satisfy (4.3) ($TB=0$) if $m \le p$ (Conclusion 6.1).
3. These eigenvalues cannot differ too much from the eigenvalue of matrix A, because according to root locus, this would cause large observer gain L with very serious consequences (Example 3.2 and Subsection 4.3.6).
4. All eigenvalues of matrix F will be transmission zeros of the observer feedback system (Patel, 1978), and should be selected with the properties of these transmission zeros in mind (see Section 1.4).

There have been some other suggestions for the selection of eigenvalues of matrix F, but they have been proven wrong. For example, the suggestion that the eigenvalues of F other than those matching the plant stable transmission zeros have infinite magnitude with the Butterworth pattern, is criticized by Sogaard-Anderson (1987). Another suggestion that all $n-m$ eigenvalues of F be clustered around the plant stable transmission zeros (Tsui, 1988b), will cause a very high gain $\underline{K}$ while trying to satisfy (4.2) ($K=\underline{K}\ \underline{C}$).

Once the eigenvalues of F are selected, the matrix F is required in this book to be in its Jordan form, with all multiple eigenvalues forming a single Jordan block (1.10). This is because only in the Jordan form are all eigenvalues decoupled (or most decoupled), and only the decoupled eigenvalues of F can make the corresponding rows of solution (T, L) decoupled.

In other words, the solution (T, L) can be separately derived and computed corresponding to each Jordan block of matrix F. We will divide Jordan block size equals one (Case A) and greater than one (Case B) in the following.

5.2.1 Eigen-Structure Case A

For distinct and real eigenvalues λ_i, $i=1, 2, \ldots$ of matrix F, Eq. (4.1) can be partitioned as

$$\mathbf{t}_i A - \lambda_i \mathbf{t}_i = \mathbf{l}_i C, \quad i = 1, 2, \ldots \tag{5.9}$$

where $\mathbf{t}_i$ and $\mathbf{l}_i$ are the i-th row of matrices T and L respectively, corresponding to λ_i.

Based on the block-observable Hessenberg form (5.5) where $C=[\,C_1 : 0 \ldots 0]$ and C_1 has m columns, Eq. (4.1) can be partitioned as the left m columns

$$\mathbf{t}_i (A - \lambda_i I)[\, I_m : 0\,]' = \mathbf{l}_i C[\, I_m : 0\,]' = \mathbf{l}_i C_1 \tag{5.10a}$$

and the right $n-m$ columns:

$$\mathbf{t}_i (A - \lambda_i I)[\,0 : I_{n-m}\,]' = \mathbf{l}_i C[\,0 : I_{n-m}\,]' = 0 \tag{5.10b}$$

Because C_1 of (5.5) is full column rank and $\mathbf{l}_i$ is free to design, (5.10a) can always be satisfied by this $\mathbf{l}_i$ for whatever $\mathbf{t}_i$ at the other side of (5.10a). Therefore, the problem of (5.9) is simplified to satisfy (5.10b) only, which has only $n-m$ columns instead of n columns of (5.9).

Because of the form of (5.5) and observability criterion (Definition 1.3), the matrix at the left side of Eq. (5.10b) always has m linearly dependent rows, so the nonzero and linearly independent solution $\mathbf{t}_i$ of (5.10b) always exists. Furthermore, from (5.5),

$$(A - \lambda_i I)[0 \,:\, I_{n-m}]' = \begin{bmatrix} C_2 & 0 & \cdots & \cdots & 0 \\ A_{22} - \lambda_i I & C_3 & 0 & \cdots & 0 \\ \vdots & & \ddots & & \vdots \\ \vdots & & & \ddots & 0 \\ \vdots & & & & C_v \\ A_{v2} & \cdots & \cdots & \cdots & A_{vv} - \lambda_i I \end{bmatrix} \tag{5.11}$$

According to the description of this form in Examples 5.1–5.3, these m linearly dependent rows of the $n\times(n-m)$ matrix of (5.11) are clearly marked. Each of these m linearly dependent rows of (5.11) can be expressed as a linear combination of the linearly independent rows above it.

Conclusion 5.1

The solution $\mathbf{t}_i$ of Eq. (5.10b) has m linearly independent basis vectors $\mathbf{d}_{ij}$, $j=1, \ldots, m$. If system matrices (A, C) are in the block-observable Hessenberg form, then these m $\mathbf{d}_{ij}$ vectors are formed by the linear combination coefficients for each of the m linearly dependent rows in (5.11), and these $\mathbf{d}_{ij}$ vectors can be computed by simple back substitution operation from (5.11) (see Section A.2).

Example 5.4: For Single-Output Cases ($m=1$)

From (5.2),

$$(A-\lambda_i I)[0 \ : \ I_{n-1} \ ]' = \begin{bmatrix} * & 0 & \cdots & \cdots & 0 \\ x & * & 0 & \cdots & 0 \\ \vdots & & \ddots & & \vdots \\ \vdots & & & \ddots & 0 \\ \vdots & & & & * \\ x & & & & x \end{bmatrix}$$

This matrix has only one ($m=1$) linearly dependent row, which is the last row (without *). There is only one ($m=1$) vector $\mathbf{t}_i$ that can satisfy (5.10b) for this matrix.

Example 5.5: For a Multi-Output Case ($m=4>1$)

Consider the matrix of (5.11) of Example 5.3 ($m=4$). For each λ_i, the solution $\mathbf{t}_i$ of (5.10b) has m (= 4) basis vectors as:

$$\mathbf{d}_{i1} = \begin{bmatrix} x & x & 0 & x & : & 1 & 0 & 0 & : 0 : 0 \end{bmatrix}$$

$$\mathbf{d}_{i2} = \begin{bmatrix} x & x & 0 & x & : & 0 & x & 0 & : x : 1 \end{bmatrix}$$

$$\mathbf{d}_{i3} = \begin{bmatrix} x & x & 1 & 0 & : & 0 & 0 & 0 & : 0 : 0 \end{bmatrix}$$

$$\mathbf{d}_{i4} = \begin{bmatrix} x & x & 0 & x & : & 0 & x & 1 & : 0 : 0 \end{bmatrix}$$

These vectors are computed by back substitution. In each of the above $\mathbf{d}_{ij}$ vectors, the position of 1 is corresponding to the linearly dependent row of the j-th output, and the positions of x element are corresponding to the linearly independent rows (those with nonzero symbols like *) above that dependent row in the matrix.

Hence, at the position of each element 1, the elements of other $\mathbf{d}_{ij}$ vectors are all 0. Therefore, all m (= 4) $\mathbf{d}_{ij}$ vectors are linearly independent of each other.

The actual solution $\mathbf{t}_i$ of (5.10b) or (5.9) can be an arbitrary linear combination of its m basis vectors such as, in this example, $\mathbf{t}_i=[c_{i1}\ c_{i2}\ c_{i3}\ c_{i4}][\mathbf{d}_{i1}'\ \mathbf{d}_{i2}'\ \mathbf{d}_{i3}'\ \mathbf{d}_{i4}']' \equiv \mathbf{c}_i D_i$, $i=1, \ldots, n-m$. The m-dimensional vector $\mathbf{c}_i$ is free and is the remaining freedom of (4.1). This remaining freedom will be fully used to achieve other observer design objectives such as $TB=0$ (4.3) (see Section 6.1) or function observer order reduction (see Example 4.3 and Section 7.1).

The dual of Eq. (5.9), $A'\mathbf{v}_i-\lambda_i\mathbf{v}_i=C'\mathbf{k}_i$, is the state feedback control design problem for assigning eigenvalue λ_i (see (5.18)). The only general additional design objective is to make eigenvector $\mathbf{v}_i$ as linearly independent (or as decoupled) as possible from the other eigenvectors. It is well known that for multiple eigenvalues of λ_i, the corresponding generalized eigenvectors are not decoupled and should be avoided (1.10b). To achieve that, the above property that vector $\mathbf{t}_i$ has m linearly independent and decoupled basis vectors $\mathbf{d}_{ij}$ ($j=1, \ldots, m$), can be used. That is to assign up to m corresponding $\mathbf{t}_i$ eigenvectors equal its m basis vectors $\mathbf{d}_{ij}$ (with up to m different values of j). This idea forms the foundation of the analytical decoupling eigenvector assignment rules (Section 9.2).

By replacing the block-observable Hessenberg form (5.8a) with its special case – the block-observable canonical form (5.8b), the corresponding four basis vectors for $\mathbf{t}_i$ become:

$$\mathbf{d}_{i1} = \begin{bmatrix} \lambda_i & 0 & 0 & 0 & : & 1 & 0 & 0 & : 0 : & 0 \end{bmatrix}$$

$$\mathbf{d}_{i2} = \begin{bmatrix} 0 & \lambda_i^3 & 0 & 0 & : & 0 & \lambda_i^2 & 0 & : \lambda_i : & 1 \end{bmatrix}$$

$$\mathbf{d}_{i3} = \begin{bmatrix} 0 & 0 & 1 & 0 & : & 0 & 0 & 0 & : 0 : & 0 \end{bmatrix}$$

$$\mathbf{d}_{i4} = \begin{bmatrix} 0 & 0 & 0 & \lambda_i & : & 0 & 0 & 1 & : 0 : & 0 \end{bmatrix}$$

Another property is revealed by these four basis vectors as follows.

Conclusion 5.2

For a fixed parameter j (j can be 1 to m), and for up to ν_j different eigenvalues λ_i (ν_j different values of i), the corresponding ν_j basis vectors $\mathbf{d}_{ij}$ are linearly independent, because they form an extended Vandermonde matrix with ν_j rows. For the example above, if $j=2$, then ν_2 (=4) different $\mathbf{d}_{i2}$ vectors (four different values of i) are linearly independent of each other; if $j=1$, then ν_1 (=2) different $\mathbf{d}_{i1}$ vectors are linearly independent. The generalized version of this property will be proved in Theorem 9.1.

It is also clear from the above example that for a fixed value of j, there are ν_j-1 different $\mathbf{d}_{ij}$ vectors that are linearly independent of the rows of matrix C.

5.2.2 Eigen-Structure Case B

For complex conjugate or multiple eigenvalues of matrix F, larger size Jordan blocks of (1.10) will be formed, and the results of Subsection 5.2.1 will be generalized.

If λ_i and λ_{i+1} are a complex conjugate pair $a \pm jb$, the corresponding Jordan block (1.10) is

$$F_i = \begin{bmatrix} a & b \\ -b & a \end{bmatrix}$$

Then the corresponding (5.9), (5.10a) and (5.10b) will become

$$\begin{bmatrix} \mathbf{t}_i \\ \mathbf{t}_{i+1} \end{bmatrix} A - F_i \begin{bmatrix} \mathbf{t}_i \\ \mathbf{t}_{i+1} \end{bmatrix} = \begin{bmatrix} \mathbf{l}_i \\ \mathbf{l}_{i+1} \end{bmatrix} C \tag{5.12}$$

$$\begin{bmatrix} \mathbf{t}_i \\ \mathbf{t}_{i+1} \end{bmatrix} A \begin{bmatrix} I_m \\ 0 \end{bmatrix} - F_i \begin{bmatrix} \mathbf{t}_i \\ \mathbf{t}_{i+1} \end{bmatrix} \begin{bmatrix} I_m \\ 0 \end{bmatrix} = \begin{bmatrix} \mathbf{l}_i \\ \mathbf{l}_{i+1} \end{bmatrix} C_1 \tag{5.13a}$$

and

$$\begin{bmatrix} \mathbf{t}_i \\ \mathbf{t}_{i+1} \end{bmatrix} A \begin{bmatrix} 0 \\ I_{n-m} \end{bmatrix} - F_i \begin{bmatrix} \mathbf{t}_i \\ \mathbf{t}_{i+1} \end{bmatrix} \begin{bmatrix} 0 \\ I_{n-m} \end{bmatrix} = 0 \tag{5.13b}$$

respectively.

Because in (5.13a), C_1 is full column rank and $\mathbf{l}_i$ and $\mathbf{l}_{i+1}$ are completely free, (5.13a) can always be satisfied for whatever at the left side of (5.13a). Thus, we only need to satisfy (5.13b), which is equivalent of

$$[\mathbf{t}_i : \mathbf{t}_{i+1}] \left(\begin{bmatrix} A\begin{bmatrix} 0 \\ I_{n-m} \end{bmatrix} & : & 0 \\ \cdots & \cdots & \cdots \\ 0 & : & A\begin{bmatrix} 0 \\ I_{n-m} \end{bmatrix} \end{bmatrix} - \begin{bmatrix} a\begin{bmatrix} 0 \\ I_{n-m} \end{bmatrix} & : & -b\begin{bmatrix} 0 \\ I_{n-m} \end{bmatrix} \\ \cdots & \cdots & \cdots \\ b\begin{bmatrix} 0 \\ I_{n-m} \end{bmatrix} & : & a\begin{bmatrix} 0 \\ I_{n-m} \end{bmatrix} \end{bmatrix} \right) = 0 \tag{5.13c}$$

where the two matrices inside the bracket have dimension $2n \times 2(n-m)$ and can be expressed as $I_2 \otimes A \begin{bmatrix} 0 \\ I_{n-m} \end{bmatrix}$ and $F_i' \otimes \begin{bmatrix} 0 \\ I_{n-m} \end{bmatrix}$, respectively.

Here the operator ⊗ stands for a "Kronecker product".

It is not difficult to verify that, like matrix (5.10b) of case A, the matrix of (5.13c) has $2m$ linearly dependent rows. Therefore, the solution $[\mathbf{t}_i : \mathbf{t}_{i+1}]$ of (5.13c) has $2m$ basis vectors $[\mathbf{d}_{ij} : \mathbf{d}_{i+1,j}]$, $j=1, \ldots, 2m$.

Example 5.6: Single-Output Case ($m=1$)

Let matrix A of (5.2) be $A=\begin{bmatrix} x & * & 0 \\ x & x & * \\ x & x & x \end{bmatrix}$ $(n=3)$, where the elements * are nonzero.

The matrix of (5.13c) will be
$$\begin{bmatrix} * & 0 & : & 0 & 0 \\ x & * & : & b & 0 \\ x & x & : & 0 & b \\ \cdots & \cdots & \cdots & \cdots & \cdots \\ 0 & 0 & : & * & 0 \\ -b & 0 & : & x & * \\ 0 & -b & : & x & x \end{bmatrix}$$

Clearly, this matrix has $2m$ (= 2) linearly dependent rows (those without *). Therefore, the solution $[\mathbf{t}_i : \mathbf{t}_{i+1}]$ of (5.13c) and (5.12) will have $2m$ (= 2) basis vectors as

$$[\mathbf{d}_{i1} : \mathbf{d}_{i+1,1}] = \begin{bmatrix} x & x & 1 & : & x & x & 0 \end{bmatrix}$$

and $$[\mathbf{d}_{i2} : \mathbf{d}_{i+1,2}] = \begin{bmatrix} x & x & 0 & : & x & x & 1 \end{bmatrix}$$

where the positions of 1 or 0 are corresponding to the positions of the two linearly dependent rows of the matrix. These two vectors can be computed by either a modified back substitution or by Givens rotational method (Tsui, 1986a).

For a multiple of q eigenvalues λ_i and its q-dimensional Jordan block,

$$F_i' = \begin{bmatrix} \lambda_i & 1 & 0 & \cdots & \cdots & 0 \\ 0 & \lambda_i & 1 & \ddots & & : \\ 0 & 0 & \ddots & \ddots & & : \\ : & & & \ddots & \ddots & 0 \\ : & & & 0 & \lambda_i & 1 \\ 0 & & & & 0 & \lambda_i \end{bmatrix}$$

The corresponding equations of (5.12) and (5.13a–c) are

$$\begin{bmatrix} \mathbf{t}_{i1} \\ \vdots \\ \mathbf{t}_{iq} \end{bmatrix} A - F_i \begin{bmatrix} \mathbf{t}_{i1} \\ \vdots \\ \mathbf{t}_{iq} \end{bmatrix} = \begin{bmatrix} \mathbf{l}_{i1} \\ \vdots \\ \mathbf{l}_{iq} \end{bmatrix} C \tag{5.14}$$

$$\begin{bmatrix} \mathbf{t}_{i1} \\ \vdots \\ \mathbf{t}_{iq} \end{bmatrix} A \begin{bmatrix} I_m \\ 0 \end{bmatrix} - F_i \begin{bmatrix} \mathbf{t}_{i1} \\ \vdots \\ \mathbf{t}_{iq} \end{bmatrix} \begin{bmatrix} I_m \\ 0 \end{bmatrix} = \begin{bmatrix} \mathbf{l}_{i1} \\ \vdots \\ \mathbf{l}_{iq} \end{bmatrix} C_1 \tag{5.15a}$$

$$\begin{bmatrix} \mathbf{t}_{i1} \\ \vdots \\ \mathbf{t}_{iq} \end{bmatrix} A \begin{bmatrix} 0 \\ I_{n-m} \end{bmatrix} - F_i \begin{bmatrix} \mathbf{t}_{i1} \\ \vdots \\ \mathbf{t}_{iq} \end{bmatrix} \begin{bmatrix} 0 \\ I_{n-m} \end{bmatrix} = 0 \tag{5.15b}$$

$$\text{and} \begin{bmatrix} \mathbf{t}_{i1} : \ \ldots \ : \mathbf{t}_{iq} \end{bmatrix} \left(I_q \otimes A \begin{bmatrix} 0 \\ I_{n-m} \end{bmatrix} - F_i' \otimes \begin{bmatrix} 0 \\ I_{n-m} \end{bmatrix} \right) = 0 \tag{5.15c}$$

where $\mathbf{t}_{ij}$ ($j=1, \ldots, q$) are the q rows of solution T corresponding to the Jordan block F_i of dimension q.

Because C_1 is full column rank, (5.15a) can always be satisfied by the free vectors $\mathbf{l}_{ij}$ for whatever $\mathbf{t}_{ij}$ at the left side of (5.15a), $j=1, \ldots, q$.

It is not difficult to verify that in (5.15c), the whole matrix in the bracket of (5.15c) has qm linearly dependent rows. Thus, the solution [$\mathbf{t}_{i1}$: … : $\mathbf{t}_{iq}$] of (5.15c) has qm basis vectors [$\mathbf{d}_{i1}$: … : $\mathbf{d}_{i,q}$].

Because of the simplicity of the bi-diagonal form of Jordan block F_i, Eq. (5.15b) or (5.15c) can be spread as

$$\mathbf{t}_{ij}\left(A - \lambda_i I\right) \begin{bmatrix} 0 \\ I_{n-m} \end{bmatrix} = \mathbf{t}_{i,j-1} \begin{bmatrix} 0 \\ I_{n-m} \end{bmatrix}, \ j = 1, \ldots, q, \ \mathbf{t}_{i0} = 0 \tag{5.15d}$$

Equation (5.15d) shows that all $\mathbf{t}_{ij}$ vectors (except $\mathbf{t}_{i1}$) are calculated based on other vector $\mathbf{t}_{i,j-1}$. These vectors are called "generalized" or "defective" eigenvectors (Golub and Wilkinson, 1976b).

The above case A and case B can be summarized into the following general algorithm for the computation of solution of (4.1).

Algorithm 5.3: Computation of Solution of Matrix Equation *TA–FT = LC*

Step 1: Based on each of the three eigenvalue cases and different dimensions of the Jordan block, compute the *m*, 2*m*, and *qm* basis vectors for the solution vectors $\mathbf{t}_i$, [$\mathbf{t}_i$: $\mathbf{t}_{i+1}$], and [$\mathbf{t}_{i1}$:… : $\mathbf{t}_{iq}$], and for eigenvalue λ_i, $i=1, \ldots, n$, and according to (5.10b), (5.13c), and (5.15c), respectively.

Step 2: Design the linear combination coefficient vector $\mathbf{c}_i$ for each of the solution vectors $\mathbf{t}_i$, [$\mathbf{t}_i$: $\mathbf{t}_{i+1}$], and [$\mathbf{t}_{i1}$:... : $\mathbf{t}_{iq}$], $i=1, \ldots, n$. The dimensions of these $\mathbf{c}_i$ vectors of these three cases are m, $2m$, and qm, respectively. For $q=1$, each row of solution matrix T,

$$\mathbf{t}_i = \mathbf{c}_i D_i \tag{5.16a}$$

where D_i is an $m \times n$ matrix stacked with all basis vectors corresponding to λ_i.

Step 3: After matrix T is determined, then the solution matrix L is determined uniquely from the left m columns of equation $TA-FT=LC$ [or (5.10a), (5.13a), and (5.15a)]

$$(TA - FT)\begin{bmatrix} I_m \\ 0 \end{bmatrix} C_1^{-1} = L \tag{5.16b}$$

Conclusion 5.3

Algorithm 5.3 computes solution (F, T, L) that satisfies (4.1). Obviously, the first two steps of the algorithm compute solution matrix T so that the right $n-m$ columns of (4.1) are satisfied. The left m columns of (4.1) are satisfied by solution matrix L from (5.16b) in Step 3.

This solution is general without any additional restrictions or conditions.

The remaining freedom of this solution of (4.1) is the free row vectors $\mathbf{c}_i$ for each vector of (5.16a). This is the full, simplest, and most explicit expression of the remaining freedom of (4.1). □

Computational reliability and efficiency of Algorithm 5.3:

Because the computational error of Step 1 of an algorithm propagates through all steps of that algorithm, and because most of the computation of this algorithm is at Step 1, this analysis concentrates on Step 1 only.

This step can be carried out by back substitution operation (see Section A.2), which is itself numerically stable (Wilkinson, 1965). However, because this operation requires repeated divisions by those * nonzero elements of the Hessenberg form matrix (5.5) (see Examples 5.1, 5.5, and 5.6), this operation can be ill-conditioned if these nonzero elements do not have large enough magnitude (see Definition 2.5 and Example 2.3).

According to Section A.2 and Step 2 of Algorithm 5.2, the magnitude of these nonzero elements concerns with the determination of whether the corresponding state is observable or not (see Examples 1.5–1.7 and Algorithms 5.1 and 5.2). Therefore, requiring large enough nonzero elements implies admitting only the strongly observable enough states into the observable part of the system.

However, reducing the number of observable states or reducing the order of the system implies reducing the information and the accuracy of the model that describes the original system. This is because every state of the original state space model is generically observable no matter how weak, just like every real-world number is nonzero no matter how small. This tradeoff between model accuracy and computation reliability is studied in depth by Lawson and Hanson (1974) and Golub et al. (1976a).

To measure more accurately the number of linearly dependent rows of matrix $\underline{C}_j$, the SVD method (see Section A.3) can be used instead of the Householders method in Step 2 of Algorithm 5.2 (Van Dooren, 1981; Patel, 1981). However, the SVD method cannot determine at that step the linear dependency of each individual row among the m_{j-1} rows of matrix $\underline{C}_j$, and thus cannot determine the very important analytical system parameters such as observability indices. The SVD method cannot make matrix $\underline{C}_j$ into its lower echelon form either, so that the simple back substitution operation of Algorithm 5.3 cannot be implemented.

The distinct advantage of the computational efficiency of Algorithm 5.3 over other algorithms for the solution of (4.1), is that every basis vector between different Jordan blocks of matrix F can be computed in complete parallel. This advantage is uniquely enabled by the fact that the basis vectors between different Jordan blocks are completely decoupled. In addition, the back substitution operation is itself very efficient.

The basic reason for the overall very good computational reliability and efficiency of Algorithm 5.3, is that matrix F is set in its Jordan form in this algorithm. The Jordan form is the most decoupled and the simplest form. This is why the Jordan form is the most thought-after form in the first place. In this particular problem of solving (4.1), matrix F is freely designed with the given (not to-be-calculated) eigenvalues. Obviously, F should be set in the best possible form and the most thought-after form – the Jordan form.

Conclusion 5.4

The computation of Algorithm 5.3 is very reliable and very efficient, as compared with other algorithms for solving Eq. (4.1).

The much more important advantages of the solution of (4.1) of Algorithm 5.3 are at the analytical properties.

Analytical properties of the results of Algorithm 5.3.

Equation (4.1) is the most important equation of observer design (see Theorems 3.2, 4.1, Section A.4 and Tsui (1987a, 1993a, 2015)). In addition, the dual of (4.1) with matrix F set in Jordan form Λ is

$$AV - V\Lambda = B\underline{K} \tag{5.17}$$

which implies $A-BK = V\,\Lambda\,V^{-1}$, where $K = \underline{K}\,V^{-1}$ is the state feedback control gain, and V is the right eigenvector matrix of $A-BK$ corresponding to Λ.

Therefore, (4.1) is also the key equation for state feedback control design for eigenstructure assignment.

Because of these reasons, if Lyapunov and Sylvester equations

$$AV - VA' = B \quad \text{and} \quad AV - V\Lambda = B \tag{5.18}$$

are the most important equations for system analysis, and if the algebraic Riccati equation is the most important equation for linear-quadratic optimal control design, then Eq. (4.1) or (5.17) should be the most important equation for the state space control system design.

However, the completely general and completely decoupled solution of (4.1) with its remaining freedom in the simplest and most explicit form (5.16a) was not derived until 1985 (Tsui, 1987a, 1993a, 2015).

For example, before 1985, the solution of (5.18) was generally used as the substitute for the solution of (5.17) (Tsui, 1986c). Because (5.18) lacks the free parameter $\underline{K}$ of (5.17) (or the free parameter L in the dual observer design (4.1)), it cannot be simplified to the form of (5.10b), (5.13c) or (5.15c). Thus, the existence of solution to (5.18) is questionable if matrices A and Λ share common eigenvalues (Gantmacher, 1959; Chen, 1984; Friedland, 1986). Thus, the solution of (5.18) is not general and is simply inferior to the solution of (5.17) (Tsui, 1986c).

From Conclusion 5.3, the completely general and completely decoupled solution of (4.1) with its remaining freedom in the simplest and most explicit form (5.16a) was derived and computed by Algorithm 5.3. Such a solution to such a most important equation to the overall state space control design will surely have a great impact on the state space control theory.

As will be shown in the rest of this book, only this solution has enabled the first general dynamic output feedback compensator design (Tsui, 1992, 1993b) (Chapters 3, 4, and 6), the simplest formulation and the best possible theoretical result of minimal order function observer design (Tsui, 2015) (Example 4.3, Section 7.1), the systematic design of robust fault detectors necessary for fault detection and isolation (Tsui, 1989, 2015) (Section 7.2), the really general and systematic generalized state feedback control design for eigenvalue assignment (Tsui, 1999a) (Chapter 8), and the really general and systematic state feedback control design for numerical eigenvector assignment (Kautsky et al., 1985) (Chapter 9).

Figure 5.1 outlines the sequential relationship of these design results and their Chapters and Sections in this book.

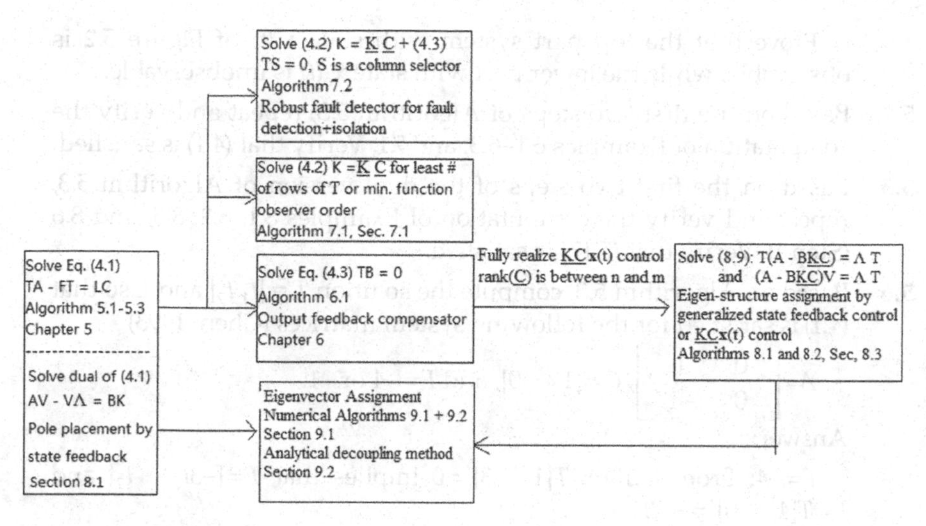

FIGURE 5.1
Sequence of design algorithms of this book.

Exercises

5.1 Repeat the computation of transformation to the block-observable Hessenberg form of Example 6.2, based on Algorithm 5.2.

5.2 Repeat the computation of similarity transformation to the block controllable Hessenberg form of Example 9.2.

5.3 Partition the states of the system of (5.7) as $[\mathbf{x}_o(t)' : \bar{\mathbf{x}}_o(t)']'$. Then the block diagram is

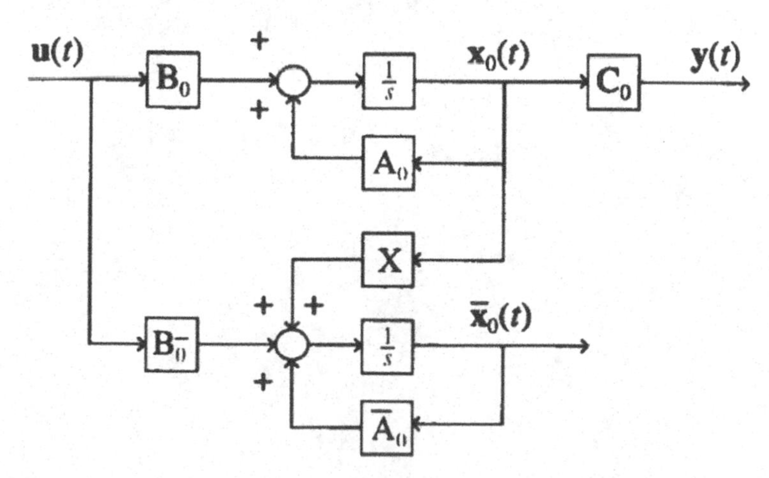

FIGURE 5.2
Block diagram of systems with partitioned observable and unobservable parts.

Prove that the top part system with state $\mathbf{x}_o(t)$ of Figure 5.2 is observable, while the lower part with state $\overline{\mathbf{x}}_o(t)$ is unobservable.

5.4 Based on the first two steps of Algorithm 5.3, repeat and verify the computation of Examples 6.1–6.3, and 7.1. Verify that (4.1) is satisfied.

5.5 Based on the first two steps of the dual version of Algorithm 5.3, repeat and verify the computation of Examples 8.1, 8.2, 8.5, and 8.6 (Step 1). Verify that (5.17) is satisfied.

5.6 Based on Algorithm 5.3, compute the solution $T \equiv [t_1\ t_2]$ and L so that (4.1) is satisfied for the following system matrices (Chen, 1993)

$$A=\begin{bmatrix} 0 & 1 \\ 0 & -1 \end{bmatrix},\ C=[1 \quad 0], \text{ and } F=-4 \text{ or } -1:$$

Answer:

$F=-4$: From (5.10b): $T[1 \quad 3]'=0$ implies that $T=[-3t_2 \quad t_2]$ and $L=T[4 \quad 0]'=-12t_2$.

$F=-1$: From (5.10b): $T[1 \quad 0]'=0$ implies that $T=[0 \quad t_2]$ and $L=T[1 \quad 0]'=0$. This solution does not exist in Chen (1993), which pre-excluded $L=0$ as a possible solution.

5.7 Repeat Exercise 5.6 for the following system:

$$A=\begin{bmatrix} -1 & 1 \\ 0 & 0 \end{bmatrix},\ C=[1 \quad 0], \text{ and } F=-4 \text{ or } -1$$

Answer: $F=-4$: $T=[-4t_2 \quad t_2]$ and $L=-12t_2$;

$F=-1$: $T=[t_1 \quad t_1]$, $L=0$.

6

Observer Design for Robust Realization

Step 2 of Algorithm 5.3 revealed the remaining freedom of Eq. (4.1). This freedom will be used fully for several design applications listed in Figure 5.1 at the end of Chapter 5.

This chapter describes the first of these applications – observer design for the guaranteed full realization of the robust properties of its state feedback control. Failure to realize this robust property has been the key reason that has limited successful practical applications of the modern control theory. This chapter will demonstrate that with the full use of the remaining freedom of (4.1), the robustness properties of generalized state feedback control can be fully realized for a great majority of open-loop plant systems! This result of Chapters 5 and 6 forms the core of the new design principle of this book.

This chapter consists of four sections.

Section 6.1 presents design Algorithm 6.1 that uses the remaining freedom of (4.1) to satisfy Eq. (4.3), $TB=0$, which is the key condition for the full realization of the robustness properties of the $\underline{KC}\mathbf{x}(t)$-control (Theorem 3.4).

Section 6.2 analyzes the generality of the solution of (4.1) and (4.3). It proves that the exact solution exists and can be computed by Algorithm 6.1, for a great majority of plant systems. It also illustrates Algorithm 6.1 using six examples. Under the separation principle, the solution to (4.1)–(4.3) does not exist for a great majority of plant systems.

Section 6.3 demonstrates a theoretical significance of the new generalized state feedback control $\underline{KC}\mathbf{x}(t)$ of this book (Definition 4.2) – it unifies completely and most explicitly, the existing state feedback control and static output feedback control as its two extremes.

Section 6.4 describes the selection and adjustment of observer order r and rank($\underline{C}$)$=r+m$. This rank determines the strength of the $\underline{KC}\mathbf{x}(t)$-control – the higher the observer order r, the less constraint and stronger the control to improve the performance and robustness and to meet the design requirement, but the more difficult to satisfy (4.3) and to fully realize this control.

DOI: 10.1201/9781003259572-6

6.1 Solution of Matrix Equation $TB = 0$

Let us summarize the result at Step (2) of Algorithm 5.1.

For distinct and real eigenvalue λ_i,

$$t_i = c_i D_i \tag{6.1a}$$

For complex conjugate λ_i and λ_{i+1}

$$\left[\mathbf{t}_i : \mathbf{t}_{i+1}\right] = \left[\mathbf{c}_i \mid \mathbf{c}_{i+1}\right]\left[D_i : D_{i+1}\right] \tag{6.1b}$$

For multiple of q λ_i's

$$\left[\mathbf{t}_{i1} : \ldots : \mathbf{t}_{iq}\right] = \left[\mathbf{c}_{i1} : \ldots : \mathbf{c}_{iq}\right]\left[D_{i1} : \ldots : D_{iq}\right] \tag{6.1c}$$

The dimensions of each row vectors $\mathbf{t}_i$ and $\mathbf{c}_i$ are n and m, respectively.

Algorithm 6.1: Compute the Solution of $TB = 0$

Step 1: Substitute (6.1a–c) into (4.3), we have

$$c_i\left[D_i B\right] = 0 \tag{6.2a}$$

$$\left[\mathbf{c}_i \mid \mathbf{c}_{i+1}\right]\left[D_i B : D_{i+1} B\right] = 0 \tag{6.2b}$$

$$\text{and } \left[\mathbf{c}_{i1} : \ldots : \mathbf{c}_{iq}\right]\left[D_{i1} B : \ldots : D_{iq} B\right] = 0 \tag{6.2c}$$

Step 2: Compute the solutions (the m-dimensional **c** row vectors) of (6.2a–c).

Equations of (6.2) are simple sets of linear equations. Nonetheless, there are two special cases, which need special treatments in the following. For simplicity of presentation, only (6.2a) is discussed.

Case A:

If the exact solution of (6.2a) does not exist (this usually happens when $m < p+1$), then compute the least-squares solution of (6.2a):

$$c_i = u_m \tag{6.3}$$

where $\mathbf{u}_m$ is the m-th column of matrix U and where $U\Sigma V' = D_i B$ is the singular value decomposition of $D_i B$ with m nonzero singular values (A.21). The corresponding $\mathbf{c}_i D_i B$ is $\sigma_m \mathbf{v}_m$, where σ_m and $\mathbf{v}_m$ are the m-th singular value and the m-th row of matrix V', respectively. The corresponding norm of $\mathbf{t}_i B$ is $\sigma_m > 0$ (see Example A.6).

The solution of Case A does not satisfy $TB=0$. Hence, the corresponding observer (3.16) is not an output feedback compensator (Definition 4.2), and the loop transfer function $L_{Kx}(s)$ and robustness properties of the $\underline{KC}x(t)$-control are not fully realized, even though the best possible approximate realization is attempted since the nonzero row of TB is minimized in a least-squares sense.

To summarize, Condition (4.3) or $TB=0$ is satisfied or best satisfied by Algorithm 6.1, using the remaining freedom of (4.1).

Case B:

If the exact solution of (6.2a) is not unique (this usually happens when $m>p+1$), the remaining freedom of (6.2a) and (4.1) exists. This freedom will be fully used to maximize the angles between the rows of matrix $\underline{C} \equiv [T' \vdots C']'$. The purpose is to maximize the span of range space of $\underline{C}$ or to best strengthen the corresponding $\underline{KC}x(t)$-control, since matrix $\underline{KC}$ is limited by this space. This operation is carried out by the following three sub-steps.

Sub-step 2.1: Compute all $(n-m)\times(m-p)$ linearly independent solutions $\mathbf{c}_{ij}$ ($j=1$ to $m-p$, $i=1, \ldots, n-m$) such that

$$c_{ij}[D_i B] = 0 \tag{6.4}$$

Sub-step 2.2: Compute all $(m-p)\times n$ dimensional matrices

$$\bar{D}_i = \begin{bmatrix} \mathbf{c}_{i1} & D_i \\ \vdots & \vdots \\ \mathbf{c}_{i,m-p} & D_i \end{bmatrix}, i = 1,\ldots,n-m \tag{6.5}$$

Sub-step 2.3: Compute the $(m-p)$ dimensional row $\mathbf{c}_i$ such that the angles between the rows

$$\mathbf{t}_i \begin{bmatrix} 0 \\ I_{n-m} \end{bmatrix} = \mathbf{c}_i \bar{D}_i \begin{bmatrix} 0 \\ I_{n-m} \end{bmatrix}, i = 1,\ldots,n-m \tag{6.6}$$

are maximized or as close to ±90° as possible. Two general and effective algorithms for this maximization are presented in Chapter 9. The linear independence between these rows (or the angles between these rows do not equal 0° or 180°) is easily and surely guaranteed, even if the angles between these vectors are not maximized. Because of the form of matrix C in (5.5), Sub-step 2.3 guarantees the maximal possible rank of matrix $\underline{C}$.

6.2 Analysis and Examples of This Design Solution

Design Algorithm 5.3 (for (4.1)) and design Algorithm 6.1 (for (4.3)) completely determine the dynamic part of observer (3.16a) and essentially define the new design principle of this book. The solution of this design is analyzed in this section and is illustrated by six examples.

Conclusion 6.1

A sufficient condition to satisfy (4.1) and (4.3) is $m>p$ (more plant output measurements than control inputs). This is clear from Conclusion 5.3 and Step 2 of Algorithm 6.1.

For plant systems with $m \le p$, there is another sufficient condition to satisfy (4.1) and (4.3). That is, the plant has at least one stable transmission zero and that zero is matched by an eigenvalue of matrix F. According to (1.18), a defining property of transmission zero z_i is that rank $\begin{bmatrix} A - z_i I & : & B \\ -C & : & 0 \end{bmatrix} < n + m$

This condition implies the existence of nonzero vectors $\mathbf{t}_i$ and $\boldsymbol{l}_i$ such that

$$[\ \mathbf{t}_i : \boldsymbol{l}_i]S = [\ \mathbf{t}_i : \boldsymbol{l}_i] \begin{bmatrix} A - \lambda_i I & : & B \\ -C & : & 0 \end{bmatrix} = 0 \tag{6.7}$$

The left n columns of (6.7) are equivalent to (4.1) if the eigenvalue λ_i matches the (stable) transmission zero z_i, and the right p columns of (6.7) are equivalent to (4.3).

It should be noticed that the number of rows of our solution (F, T, L) of (4.1) is completely flexible and can be as low as one. Thus, the existence of at least one plant stable transmission zero z_i implies the existence of our solution to (4.1) and (4.3).

Definition 1.5 also implies that the existence of at least one plant (stable) transmission is also a necessary condition for the existence of our solution to (4.1) and (4.3), if the plant system has $m \le p$ (number of output measurements is not more than the number of control inputs).

Conclusion 6.2

It is obvious that Algorithms 5.3 and 6.1 fully used all freedom of the observer's dynamic part (F, T, L) to satisfy (4.1) and (4.3), and to minimize the constraint and maximize the strength, of the resulting $\underline{K}C\mathbf{x}(t)$-control (see Sub-step 2.3 of Algorithm 6.1). Here $\underline{K}$ consists of all free parameters of the observer's output part (3.16b).

Conclusion 6.3 (Kudva et al., 1980)

If the plant system (A, B, C) satisfies all of the following three conditions: (a) minimum-phase; (b) $m \geq p$; and (c) rank $(CB)=p$, then exact solution of (4.1), (4.3), and rank$(\underline{C})=n$ exists. From Conclusion 6.2, Algorithms 5.3 and 6.1 can result in an observer that satisfies (4.1), (4.3), and rank$(\underline{C})=n$ if the plant system satisfies these three conditions. Observers satisfying (4.1), (4.3), and rank$(\underline{C})=n$ are also called "unknown input observers" (Wang et al., 1975) if matrix B is the system gain to the unknown inputs.

Proof

For systems with $m=p$: According to Davison and Wang (1974), conditions $m=p$ and rank $(CB)=p$ imply that the plant system has $n-m$ transmission zeros. Thus, the condition of minimum-phase (all transmission zeros are stable) implies that the system has $n-m$ stable transmission zeros. Then, Conclusions 6.1 and 6.2 imply that the result of Algorithms 5.3 and 6.1 not only satisfies (4.1) and (4.3) but also has n linearly independent rows in matrix $\underline{C}$.

For systems with $m>p$: There are a number of other proofs besides that of Kudva et al. (1980), such as those of Hou and Muller (1992), Syrmos (1993). We present here the proof of Hou and Muller (1992) because it is simpler and clearer, though longer.

This proof is as follows.

Let a nonsingular matrix

$$Q=\left[B:\underline{B}\right] \tag{6.8}$$

where $\underline{B}$ is an arbitrary matrix that makes matrix Q nonsingular. Then make a similarity transformation on the plant (A, B, C) (see the end of Section 1.2):

$$\underline{x}(t)=Q^{-1}x(t)\equiv\left[\underline{x}_1(t)':\underline{x}_2(t)'\right]' \tag{6.9}$$

and $$\left(Q^{-1}AQ, Q^{-1}B, CQ\right)\triangleq\left(\begin{bmatrix} A_{11} & : & A_{12} \\ A_{21} & : & A_{22}\end{bmatrix}, \begin{bmatrix} I_p \\ 0\end{bmatrix}, [CB:C\underline{B}]\right) \tag{6.10}$$

From (6.9) and (6.10),

$$d\underline{\mathbf{x}}_2(t)/dt = A_{21}\underline{\mathbf{x}}_1(t)+A_{22}\underline{\mathbf{x}}_2(t) \tag{6.11a}$$

$$y(t)=CB\underline{\mathbf{x}}_1(t)+C\underline{B}\underline{\mathbf{x}}_2(t) \tag{6.11b}$$

Because $m>p$ and rank $(CB)=p$, we can set another nonsingular matrix

$$P = [CB : \underline{CB}]$$

where $\underline{CB}$ is an arbitrary matrix that makes matrix P nonsingular. By multiplying P^{-1} on the left side of (6.11b), we have

$$P^{-1}\mathbf{y}(t) \triangleq \begin{bmatrix} P_1 \\ P_2 \end{bmatrix} \mathbf{y}(t) = \begin{bmatrix} I_p & : & P_1 C\underline{B} \\ 0 & : & P_2 C\underline{B} \end{bmatrix} \begin{bmatrix} \underline{\mathbf{x}}_1(t) \\ \underline{\mathbf{x}}_2(t) \end{bmatrix} \quad (6.12)$$

From the first p rows of (6.12),

$$\underline{\mathbf{x}}_1(t) = P_1[y(t) - C\underline{B}\underline{\mathbf{x}}_2(t)] \quad (6.13)$$

By substituting (6.13) and (6.12) into (6.11), we have the following system of order $n-p$:

$$d\underline{\mathbf{x}}_2(t)/dt = (A_{22} - A_{21}P_1C\underline{B})\underline{\mathbf{x}}_2(t) + A_{21}P_1\mathbf{y}(t)$$
$$\equiv \tilde{A}\underline{\mathbf{x}}_2(t) + \tilde{B}\mathbf{y}(t) \quad (6.14a)$$

$$\underline{\mathbf{y}}(t) \equiv P_2 y(t) = P_2 C\underline{B}\underline{\mathbf{x}}_2(t) \equiv \tilde{C}\underline{\mathbf{x}}_2(t) \quad (6.14b)$$

Because system (6.14) does not involve original system input $\mathbf{u}(t)$, its corresponding state observer is an unknown input observer. In addition, if $\underline{\mathbf{x}}_2(t)$ is estimated by this observer, then $\underline{\mathbf{x}}_1(t)$ can also be determined by (6.13).

Thus, the sufficient condition for the existence of an unknown input observer of the original system is the detectability (unobservable part of the system is stable so that all unobservable states are known to be zero) of system (6.14). Conditions $m>p$ and rank $(CB)=p$ are also part of this sufficient condition because they are necessary to the derivation of system (6.14).

For the system (6.10),

$$\text{Rank}\begin{bmatrix} sI_p - A_{11} & -A_{12} & : & I_p \\ -A_{21} & sI_{n-p} - A_{22} & : & 0 \\ CB & C\underline{B} & : & 0 \end{bmatrix}$$

$$= p + \text{Rank}\begin{bmatrix} -A_{21} & sI_{n-p} - A_{22} \\ CB & C\underline{B} \end{bmatrix} \quad (6.15a)$$

$$= p + \text{Rank}\left(\begin{bmatrix} I_{n-p} & 0 \\ 0 & P^{-1} \end{bmatrix} \begin{bmatrix} -A_{21} & sI_{n-p} - A_{22} \\ CB & C\underline{B} \end{bmatrix} \begin{bmatrix} -P_1 C\underline{B} & I_p \\ I_{n-p} & 0 \end{bmatrix} \right)$$

$$= p + \text{Rank} \begin{bmatrix} sI_{n-p} - \tilde{A} & -A_{21} \\ 0 & I_p \\ \tilde{C} & 0 \end{bmatrix} \begin{matrix} \}n-p \\ \}p \\ \}m-p \end{matrix}$$

$$= 2p + \text{Rank} \begin{bmatrix} sI_{n-p} - \tilde{A} \\ \tilde{C} \end{bmatrix} \tag{6.15b}$$

A comparison of (6.15a) and (6.15b) shows that the transmission zero for system (6.10), or the value of s that makes rank (6.15a) less than $p+n$, is the value of s that makes rank (6.15b) less than $2p+(n-p)$, and equals the pole of the unobservable part of system $(\tilde{A}, \tilde{C})$ or of system (6.14) (see Definition 1.3).

Because the sufficient condition for the existence of an unknown input observer for the original system (6.10) is that the unobservable part of system (6.14) is stable, this sufficient condition becomes that all transmission zeros of system (6.10) are stable. Thus, the proof.

Conclusion 6.4

For the three conditions of Conclusion 6.3 sufficient for Algorithms 5.3 and 6.1 to yield an observer that satisfies (4.1), (4.3), and rank($\underline{C}$)=n, each is also necessary for this observer.

Proof

Because all transmission zeros must be matched by observer poles, all transmission zeros must be stable (minimum-phase or Condition (a)).

Because rank (CB) cannot have full-column rank if $m<p$, Condition (b) ($m \geq p$) is necessary.

For matrix $\underline{C} \equiv [\,T' \vdots C'\,]'$ to have n linearly independent rows and for $TB=0$, CB must have full-column rank (Condition (c)). See Example A.7.

Conclusions 6.3 and 6.4 are summarized in the following Table 6.1.

TABLE 6.1

Necessary and Sufficient Conditions for Exact LTR or for an Observer to Satisfy (4.1), (4.3), and rank($\underline{C}$)=n

Conditions:	$m<p$	$m=p$	$m>p$
A) Have $n-m$ stable transmission zeros	Necessary and sufficient	Necessary and sufficient	Sufficient
B) Minimum-phase and CB full-column rank	Necessary	Necessary and sufficient	Necessary and sufficient

Three points can be noticed in Table 6.1. First, Condition (A) is stricter than Condition (B) because it is not necessary for systems with $m>p$. Thus, for systems with $m>p$, some additional systems may satisfy the sufficient Condition (B) but not (A). Second, for systems with $m=p$, Conditions (A) and (B) are equivalent because systems with $m=p$ and rank(C)$=p$ always have $n-m$ transmission zeros. Third, for systems with $m<p$, both Conditions (A) and (B) are necessary, yet both conditions cannot be satisfied.

Table 6.1 and Conclusion 6.4 all show that the conditions of minimum-phase and rank(CB)$=p$ are both necessary. Yet they both are very rarely satisfied by plant systems. First, it is very hard to require every zero to be stable among many existing transmission zeros. Because almost all systems with $m=p$ have $n-m$ transmission zeros (Davison and Wang, 1974), Exercise 4.2 shows that almost all of them are non-minimum phase. Second, rank (CB)$=p$ is not satisfied by many (say half) of the plant systems such as airborne systems. Thus, Example 4.6 shows that at reasonable assumptions about m and p, about 85% of all plant systems, or the great majority of plant systems, *cannot* satisfy the two necessary conditions of Conclusion 6.4. This means that the great majority of plant systems *cannot* realize the robust properties of state feedback control (or unable to have exact LTR) under separation principle.

Conclusion 6.5

From Conclusion 6.1, the new design principle of this book only requires either $m>p$ or at least one stable transmission zero. Because it is equally very hard to have every zero to be unstable among many existing transmission zeros (see Exercises 4.2 and 4.6), and because almost all systems with $m=p$ have $n-m$ transmission zeros, Exercises 4.3 and 4.7 show that almost all systems with $m=p$ have at least one stable transmission zero! Thus, almost all systems with $m\geq p$, or the great majority of plant systems, satisfy the condition of Conclusion 6.1 or *can* fully realize a generalized state feedback control under the new design principle of this book.

Furthermore, Example 4.6 shows that under two reasonable assumptions about m and p, the great majority (at least 80% or 85%) of plant systems are *unable* to have exact LTR under separation principle, yet the great majority (90% or 80%) of plant systems *can* fully realize a generalized state feedback control which is at least very good under the new design principle of this book!

This is the overwhelming and decisive advantage of the new design principle of this book over the existing separation principle.

Example 6.1: Four Systems with $m > p$

This is an example of four plant systems that share a common system matrix pair (A, C)

$$A = \begin{bmatrix} x & x & x & : & 1 & 0 & 0 & : & 0 \\ x & x & x & : & 0 & 1 & 0 & : & 0 \\ x & x & x & : & 0 & 0 & 1 & : & 0 \\ \cdots & \cdots & \cdots & \cdots & \cdots & \cdots & \cdots & \cdots & \cdots \\ x & x & x & : & 0 & 0 & 0 & : & 1 \\ x & x & x & : & 0 & 0 & 0 & : & 0 \\ x & x & x & : & 0 & 0 & 0 & : & 0 \\ \cdots & \cdots & \cdots & \cdots & \cdots & \cdots & \cdots & \cdots & \cdots \\ x & x & x & : & 0 & 0 & 0 & : & 0 \end{bmatrix}$$

and

$$C = \begin{bmatrix} 1 & 0 & 0 & : & 0 & 0 & 0 & : & 0 \\ x & 1 & 0 & : & 0 & 0 & 0 & : & 0 \\ x & x & 1 & : & 0 & 0 & 0 & : & 0 \end{bmatrix} \tag{6.16}$$

where x elements are arbitrary. This is a very common system example among all seventh order and three-output and two-input systems.

The system matrix pair of (6.16) is in the block observable canonical form (1.16) or (5.8b). The four different plant systems are distinguished by their respective system matrix B:

$$B_1 = \begin{bmatrix} 1 & 0 \\ 1 & 1 \\ 1 & 0 \\ \cdots & \cdots \\ -1 & 1 \\ 1 & -1 \\ 1 & 2 \\ \cdots & \cdots \\ -2 & -2 \end{bmatrix}, \quad B_2 = \begin{bmatrix} 0 & 0 \\ 0 & 1 \\ 1 & 0 \\ \cdots & \cdots \\ 1 & 0 \\ 2 & 2 \\ 3 & 1 \\ \cdots & \cdots \\ 1 & 1 \end{bmatrix}, \quad B_3 = \begin{bmatrix} 1 & 0 \\ 1 & 0 \\ 1 & 0 \\ \cdots & \cdots \\ -1 & 1 \\ 1 & 2 \\ 1 & 1 \\ \cdots & \cdots \\ -2 & -2 \end{bmatrix}, \text{ and}$$

$$B_4 = \begin{bmatrix} 1 & 0 \\ 1 & 1 \\ 1 & 0 \\ \cdots & \cdots \\ -1 & 1 \\ -2 & -1 \\ -2 & 2 \\ \cdots & \cdots \\ -2 & -1 \end{bmatrix}$$

Using the method of Example 1.7, we can derive directly the polynomial matrix fraction description of these four systems' transfer function model $G(s)=C_1D(s)^{-1}N(s)$. Because these four systems share the same (A, C) matrices, they also share the same $C_1D(s)^{-1}$. Only the four different $N(s)$ matrices distinguish these four systems.

$$N_1(s)=\begin{bmatrix} (s+1) & (s-2) & : & (s-2) \\ (s+1) & & : & (s-1) \\ (s+1) & & : & 2 \end{bmatrix},$$

$$N_2(s)=\begin{bmatrix} (s+1) & : & 1 \\ 2 & : & (s+2) \\ (s+3) & : & 1 \end{bmatrix},$$

$$N_3(s)=\begin{bmatrix} (s+1) & (s-2) & : & (s-2) \\ (s+1) & & : & 2 \\ (s+1) & & : & 1 \end{bmatrix},$$

and

$$N_4(s)=\begin{bmatrix} (s-2) & (s+1) & : & (s-1) \\ (s-2) & & : & (s-1) \\ (s-2) & & : & 2 \end{bmatrix}$$

The four $N(s)$ matrices reveal that only the first and third systems have one common stable transmission zero –1, and the fourth system has an unstable transmission zero (+2).

Thus, the common observer state matrix F for all four plant systems will have an eigenvalue that matches that stable transmission zero –1.

$$F=\text{diag}\{-1,-2,-3,-4\}$$

The dimension of F is presumed to be $n-m=4$, though it can be freely adjusted and reduced later.

Because Step 1 of Algorithm 5.3 operates only on matrices (A, C) and F which are common for all four systems, the result of this step is also common for the four systems. For each of the $n-m$ (= 4) eigenvalues of matrix F, the corresponding $n-m$ basis vector matrices of dimension $m\times n$, are computed like the end of Example 5.5 as in the following

$$D_1=\begin{bmatrix} 1 & 0 & 0 & : & -1 & 0 & 0 & : & 1 \\ 0 & -1 & 0 & : & 0 & 1 & 0 & : & 0 \\ 0 & 0 & -1 & : & 0 & 0 & 1 & : & 0 \end{bmatrix},$$

$$D_2 = \begin{bmatrix} 4 & 0 & 0 & : & -2 & 0 & 0 & : & 1 \\ 0 & -2 & 0 & : & 0 & 1 & 0 & : & 0 \\ 0 & 0 & -2 & : & 0 & 0 & 1 & : & 0 \end{bmatrix}$$

$$D_3 = \begin{bmatrix} 9 & 0 & 0 & : & -3 & 0 & 0 & : & 1 \\ 0 & -3 & 0 & : & 0 & 1 & 0 & : & 0 \\ 0 & 0 & -3 & : & 0 & 0 & 1 & : & 0 \end{bmatrix},$$

and

$$D_4 = \begin{bmatrix} 16 & 0 & 0 & : & -4 & 0 & 0 & : & 1 \\ 0 & -4 & 0 & : & 0 & 1 & 0 & : & 0 \\ 0 & 0 & -4 & : & 0 & 0 & 1 & : & 0 \end{bmatrix}$$

The result of T matrices for each of the four different systems is computed using Algorithm 6.1 as follows:

$$T_1 = \begin{bmatrix} [0 & 1 & 1]D_1 \\ [1 & 4/5 & 16/5]D_2 \\ [1 & 5/6 & 25/6]D_3 \\ [1 & 6/7 & 36/7]D_4 \end{bmatrix} = \begin{bmatrix} 0 & -1 & -1 & 0 & 1 & 1 & 0 \\ 4 & -8/5 & -32/5 & -2 & 4/5 & 16/5 & 1 \\ 9 & -15/6 & -75/6 & -3 & 5/6 & 25/6 & 1 \\ 16 & -24/7 & -144/7 & -4 & 6/7 & 36/7 & 1 \end{bmatrix},$$

$$T_2 = \begin{bmatrix} [0 & 1 & -1]D_1 \\ [1 & 1 & -1]D_2 \\ [1 & 1 & 0]D_3 \\ [0 & 1 & 2]D_4 \end{bmatrix} = \begin{bmatrix} 0 & -1 & 1 & 0 & 1 & -1 & 0 \\ 4 & -2 & 2 & -2 & 1 & -1 & 1 \\ 9 & -3 & 0 & -3 & 1 & 0 & 1 \\ 0 & -4 & -8 & 0 & 1 & 2 & 0 \end{bmatrix}$$

$$T_3 = \begin{bmatrix} [0 & 1 & -2]D_1 \\ [1 & 0 & 4]D_2 \\ [1 & 0 & 5]D_3 \\ [1 & 0 & 6]D_4 \end{bmatrix} = \begin{bmatrix} 0 & -1 & 2 & 0 & 1 & -2 & 0 \\ 4 & 0 & -8 & -2 & 0 & 4 & 1 \\ 9 & 0 & -15 & -3 & 0 & 5 & 1 \\ 16 & 0 & -24 & -4 & 0 & 6 & 1 \end{bmatrix}, \text{ and}$$

$$T_4 = \begin{bmatrix} [2 & -1 & 1]D_1 \\ [1 & -1/5 & 6/5]D_2 \\ [1 & 0 & 2]D_3 \\ [1 & 1/7 & 20/7]D_4 \end{bmatrix} = \begin{bmatrix} 2 & 1 & -1 & -2 & -1 & 1 & 2 \\ 4 & 2/5 & -12/5 & -2 & -1/5 & 6/5 & 1 \\ 9 & 0 & -6 & -3 & 0 & 2 & 1 \\ 16 & -4/7 & -80/7 & -4 & 1/7 & 20/7 & 1 \end{bmatrix}$$

It can be easily verified that the above four matrices satisfy (4.3) ($TB=0$) and the right $n-m$ (= 4) columns of (4.1). The left m (= 3) columns of (4.1) can always be satisfied by matrix L in (5.16) and are omitted here. Thus, the above solutions satisfying (4.1) and (4.3) for the four systems are correctly computed. Many fractional numbers in this solution indicate that this large amount of state of art results are computed simply by hand!

These four examples demonstrate part 1 of Conclusion 6.1 that $m>p$ is a sufficient condition for the existence of exact solutions to (4.1) and (4.3).

Except for the observer order, the observer dynamic part (3.16a) (F, T_i, L_i) for the four plant systems are designed, $i=1, \ldots, 4$.

Let us analyze now the observer output part $\underline{K}_i$, $i=1, \ldots, 4$, for these four systems. The generalized state feedback control $\underline{K}_i\underline{C}_i\mathbf{x}(t)$ will be designed for these four systems, $i=1, \ldots, 4$.

Because systems 1 and 2 satisfy the three conditions of Conclusion 6.3, rank($\underline{C}_i=[\ T_i':C']'$)$=n=7$ for these two systems, $i=1, 2$, as can be easily verified.

Because systems 3 and 4 do not satisfy the two necessary conditions of Conclusion 6.4 – system 3 has rank(CB)$=1<p$ and system 4 has an unstable transmission zero (2), rank($\underline{C}_i$)$=6<7=n$, for these two systems, $i=3, 4$. In other words, of the seven rows of matrix $\underline{C}_i$, or the four rows of T_i, one row is linearly dependent on the other and is redundant.

Let us randomly eliminate the last row of T_i, $i=3$ and 4, then the new observer order of these two systems is reduced from 4 to 3, and the new observer state matrix for these two systems will be $F=\text{diag}\{-1, -2, -3\}$, and the new generalized state feedback control $\underline{K}_i\underline{C}_i\,\mathbf{x}(t)$ will be:

$$K_3 = \underline{K}_3\left[\overline{T}_3' : C'\right]' = \underline{K}_3 \begin{bmatrix} 0 & -1 & 2 & 0 & 1 & -2 & 0 \\ 4 & 0 & -8 & -2 & 0 & 4 & 1 \\ 9 & 0 & -15 & -3 & 0 & 5 & 1 \\ & & & & & & \\ 1 & 0 & 0 & 0 & 0 & 0 & 0 \\ x & 1 & 0 & 0 & 0 & 0 & 0 \\ x & x & 1 & 0 & 0 & 0 & 0 \end{bmatrix}$$

and

$$K_4 = \underline{K}_4\left[\overline{T}_4' : C'\right]' = \underline{K}_4 \begin{bmatrix} 2 & 1 & -1 & -2 & -1 & 1 & 2 \\ 4 & 2/5 & -12/5 & -2 & -1/5 & 6/5 & 1 \\ 9 & 0 & -6 & -3 & 0 & 2 & 1 \\ & & & & & & \\ 1 & 0 & 0 & 0 & 0 & 0 & 0 \\ x & 1 & 0 & 0 & 0 & 0 & 0 \\ x & x & 1 & 0 & 0 & 0 & 0 \end{bmatrix}$$

These two generalized state feedback controls are not as strong as the direct state feedback control or the controls for systems 1 and 2. However, they are corresponding to rank $(\underline{C}_i)$ (=6) output measurements and are therefore much stronger than static output feedback control, which corresponds to only m (= 3) output measurements. For instance, these two controls can assign all arbitrarily given poles and $n-p$ (=5) eigenvectors because rank $(\underline{C}_i)+p=8>7=n$ (see Algorithm 8.2), while the static output feedback control cannot assign arbitrarily given poles and cannot guarantee feedback system stability because $m\times p=6\leq 7=n$ (Wang, 1996).

Even much more important, the controls for all four systems are all guaranteed of full realization especially their loop transfer functions and robust properties, because $T_iB_i=0, i=1, \ldots, 4$.

For the systems as challenging as the non-minimum phase (system 4) or as rank $(CB)<p$ (system 3), there exists no other control design methodology that can design such a strong control that is fully realizable!

Example 6.2: The Case When the Eigenvalues of F Are Complex Conjugates

$$\text{Let}\begin{bmatrix} A & : & B \\ C & : & 0 \end{bmatrix} = \begin{bmatrix} 1.0048 & -0.0068 & -0.1704 & -18.178 & : & 39.611 \\ -7.7779 & 0.8914 & 10.784 & 0 & : & 0 \\ 1 & 0 & 0 & 0 & : & 0 \\ 0 & 0 & 0 & 0 & : & 1 \\ & & & & & \\ 1 & 0 & 0 & 0 & : & 0 \\ 0 & 1 & 0 & 0 & : & 0 \end{bmatrix}$$

This is the state space model of a real combustion engine system (Liubakka, 1987). The four state variables are manifold pressure, engine rotation speed, manifold pressure of the previous rotation, and throttle position, respectively. Its control input is the throttle position for the next rotation, and its two output measurements are manifold pressure and engine rotation speed, respectively.

To transform this model into block observable Hessenberg form, we apply the operation at Steps 2 and 3 ($j=2$) of Algorithm 5.2, where the operator matrix

$$H = \begin{bmatrix} 1 & 0 & 0 & 0 \\ 0 & 1 & 0 & 0 \\ 0 & 0 & -0.0093735 & 0.999956 \\ 0 & 0 & -0.99996 & -0.0093735 \end{bmatrix}$$

is determined by the elements [−0.1704 −18.178] of the first row of matrix A. The resulting block observable Hessenberg form system matrices are

$$\begin{bmatrix} H'AH & : & H'B \\ CH & : & 0 \end{bmatrix}$$

$$= \begin{bmatrix} 1.0048 & -0.0068 & 18.1788 & 0 & : & 39.6111 \\ -7.7779 & 0.8914 & -0.1011 & 10.7835 & : & 0 \\ -0.00937 & 0 & 0 & 0 & : & -1 \\ 1 & 0 & 0 & 0 & : & -0.0093735 \\ \\ 1 & 0 & 0 & 0 & : & 0 \\ 0 & 1 & 0 & 0 & : & 0 \end{bmatrix}$$

The large magnitude difference of the above matrix elements and the long significant digits of the above operation all imply, the importance of numerical computation reliability and the advantage of unitary matrix H in this regard. Because this system does not have any stable transmission zeros to match, we set matrix

$$F = \begin{bmatrix} -1 & 1 \\ -1 & -1 \end{bmatrix} \text{ with eigenvalues } -1 \pm j.$$

By substituting matrices $H'AH$ and F into (5.13b) of Algorithm 5.3, we have

$$[D_1 : D_2] \times \left(I_2 \otimes H'AH \begin{bmatrix} 0 \\ I_2 \end{bmatrix} - F' \otimes \begin{bmatrix} 0 \\ I_2 \end{bmatrix} \right)$$

$$= \begin{bmatrix} -0.05501 & 0 & 1 & 0 & : & -0.05501 & 0 & 0 & 0 \\ -0.0005157 & -0.092734 & 0 & 1 & : & -0.0005157 & -0.092734 & 0 & 0 \\ \\ 0.05501 & 0 & 0 & 0 & : & -0.05501 & 0 & 1 & 0 \\ 0.0005157 & 0.092734 & 0 & 0 & : & -0.0005157 & -0.092734 & 0 & 1 \end{bmatrix}$$

$$\qquad\qquad D_1 \qquad\qquad\qquad\qquad\qquad D_2$$

$$\begin{bmatrix} 18.1788 & 0 & : & 0 & 0 \\ -0.1011 & 10.7835 & : & 0 & 0 \\ 1 & 0 & : & 1 & 0 \\ 0 & 1 & : & 0 & 1 \\ \\ 0 & 0 & : & 18.1788 & 0 \\ 0 & 0 & : & -0.1011 & 10.7835 \\ -1 & 0 & : & 1 & 0 \\ 0 & -1 & : & 0 & 1 \end{bmatrix} = 0$$

By substituting this $[D_1 \vdots D_2]$ into (6.2b) of Algorithm 6.1, we have

$$\underset{\mathbf{c}}{[-0.0093735 \ 1 \ 0 \ 0]} \underset{\quad D_1B \qquad\qquad\qquad D_2B}{\begin{bmatrix} -3.179 & : & -2.179 \\ -0.029801 & : & -0.02043 \\ & & \\ 2.179 & : & -3.179 \\ 0.02043 & : & -0.029801 \end{bmatrix}} = 0$$

Thus, the result of Step 2 of Algorithm 5.3 is:

$$T = \begin{bmatrix} \mathbf{c}D_1 \\ \mathbf{c}D_2 \end{bmatrix} = \begin{bmatrix} 0 & -0.092734 & -0.0093735 & 1 \\ 0 & -0.092734 & 0 & 0 \end{bmatrix}$$

The matrices F and T satisfy the right $n-m$ columns of (4.1) and (4.3) ($TB=0$).

Now we need to transform this matrix T back to correspond to the original system (A, B, C) (see the beginning of Section 5.2):

$$T = TH' = \begin{bmatrix} 0 & -0.092734 & 1 & 0 \\ 0 & -0.092734 & 0 & 0 \end{bmatrix}$$

By substituting this T into (5.16) of Step 3 of Algorithm 5.3, we have

$$L = (TA - FT)\begin{bmatrix} I_2 \\ 0 \end{bmatrix} = \begin{bmatrix} 1.721276 & -0.082663 \\ 0.721276 & -0.26813 \end{bmatrix}$$

Now the left m (=2) columns of (4.1) are also so satisfied.

However, the resulting rank ($\underline{C}=[T' \vdots C']'$)$=3<n$. This is caused by the fact that this engine system has an unstable zero 0.4589 that the eigenvalue of F cannot match. Nonetheless, the generalized state feedback control $\underline{KC}\mathbf{x}(t)$ of our design is still much stronger than the static output feedback control, because rank ($\underline{C}$) is higher than rank (C).

In addition to Examples 6.1 and 6.2, Example 7.1 of Chapter 7 provides a numerical example about satisfying (4.1) at the multiple eigenvalue case. Thus, all different kinds of eigenvalues are presented in numerical examples in this book.

Example 6.3: A Case of Approximate Solution to (4.1) and (4.3)

Let the system matrices be

$$(A,B,C) = \left(\begin{bmatrix} x & x & 1 & 0 \\ x & x & 0 & 1 \\ x & x & 0 & 0 \\ x & x & 0 & 0 \end{bmatrix}, \begin{bmatrix} 1 & 3 \\ 1 & 2 \\ 2 & 6 \\ -1 & -2 \end{bmatrix}, \begin{bmatrix} 1 & 0 & 0 & 0 \\ 0 & 1 & 0 & 0 \end{bmatrix} \right)$$

where x's are arbitrary elements. Because this system has the same number of outputs and inputs and satisfies rank $(CB)=p$, this system has $n-m=4-2=2$ transmission zeros. Using the method of Example 1.7 on this block observable canonical form system matrices, we have matrix $N(s)$ of the polynomial MFD $C_1D^{-1}(s)N(s)$ (1.17a) of this system as:

$$N(s)=\begin{bmatrix} s+2 & 3(s+2) \\ s-1 & 2(s-1) \end{bmatrix}$$

It is clear from $N(s)$ that this system has two transmission zeros, as predicted. They are –2 and 1. The zero 1 is unstable and cannot be matched by an eigenvalue of matrix F, and this system is non-minimum phase.

Set $F=\text{diag}\{-2,-1\}$, which matches the stable transmission zero –2 of the plant system. Then by solving (5.10b), we have

$$D_1=\begin{bmatrix} -2 & 0 & 1 & 0 \\ 0 & -2 & 0 & 1 \end{bmatrix} \quad \text{and} \quad D_2=\begin{bmatrix} -1 & 0 & 1 & 0 \\ 0 & -1 & 0 & 1 \end{bmatrix}$$

Substituting this result into (6.2a),

$$\mathbf{c}_1D_1B=\mathbf{c}_1\begin{bmatrix} 0 & 0 \\ -3 & -6 \end{bmatrix}=0 \quad \text{and} \quad \mathbf{c}_2D_2B=\mathbf{c}_2\begin{bmatrix} 1 & 3 \\ -2 & -4 \end{bmatrix}=0 \qquad (6.17)$$

The first equation of (6.17) has an exact solution $\mathbf{c}_1=[x \quad 0]$ $(x\neq 0)$, because the corresponding zero –2 is matched by the first eigenvalue of F. The second equation of (6.17) does not have an exact solution, because the corresponding zero (1) is unmatched by the second eigenvalue of F. This is a proof that the minimum-phase is a necessary condition for the system to have a solution satisfying (4.1), (4.3), and rank $(\underline{C})=n$. This is also a proof of the second part of Conclusion 6.1 that for systems with $m=p$, an eigenvalue of F must match a zero of the system, for the corresponding row of T to satisfy (4.1) and (4.3).

To minimize $\mathbf{c}_2D_2B$ of (6.17) in the least-squares sense, we use (6.3) of Algorithm 6.1 such that

$$\mathbf{c}_2=\mathbf{u}_2{}'=\begin{bmatrix} 0.8174 & 0.576 \end{bmatrix}$$

Here, $\mathbf{u}_2$ is the normalized right eigenvector of matrix $[D_2B][D_2B]'$ and is corresponding to that matrix's smallest eigenvalue σ_2^2 (= 0.13393), or is the second column of unitary matrix U of the singular value decomposition of matrix D_2B. It can be verified that the least-squares norm $\|\mathbf{c}_2D_2B\|=\sigma_2=0.366$.

All of the above results provide us with two possible observers. Observer 1 satisfies (4.1) and (4.3) completely, but does not include the pole –1 and is of order $r=1$ (so that rank $(\underline{C}_1)=r+m=3$). Observer 2 includes the pole –1 and is of order $r=2$ (so that rank $(\underline{C}_2)=n$). It satisfies (4.1), but satisfies (4.3) only approximately in the least-squares sense.

The dynamic part (3.16a) of these two observers are

$$(F_1, T_1) = (-2, \mathbf{c}_1 D_1 = [-2\mathbf{c}_1 : \mathbf{c}_1])$$

$$\text{and } (F_2, T_2) = \left(\begin{bmatrix} -2 & 0 \\ 0 & -1 \end{bmatrix}, \begin{bmatrix} & -2\mathbf{c}_1 & : & \mathbf{c}_1 & \\ -0.8174 & -0.576 & : & 0.8174 & 0.576 \end{bmatrix} \right)$$

Therefore, Observer 1 can fully realize (including loop transfer function and robustness) a generalized state feedback control $\underline{K}_1\underline{C}_1x(t)$. Although this control is not as ideal as the direct state feedback control because rank$(\underline{C}_1)=3<4=n$ (because the plant is non-minimum phase), this control is still excellent because rank$(\underline{C}_1)+p=5>n$ Tsui (2005). This control is also much stronger than static output feedback control because rank$(\underline{C}_1)>$rank$(C)=2$.

Therefore, Observer 2 can generate an ideal state feedback control signal because rank $(\underline{C}_2=[T_2':C']')=n$. Although the loop transfer function and robustness properties of this control can only be approximately realized (because the plant is a non-minimum phase), this approximate realization is far more effective than the existing approximate LTR (see Example 4.5) because this approximation is achieved by minimizing TB.

For such a non-minimum phase plant system, there exists no other design methodology that can design systematically the control results as excellent, as rich of choices/options, and as simple as presented here.

6.3 Complete Unification of Two Existing Basic Modern Control System Structures

As mentioned in Section 4.4, besides the guaranteed full realization of robust properties of generalized state feedback control for a great majority of plant systems, another major theoretical significance of the new design principle of this book is that, this generalized state feedback control unifies completely the existing state feedback control and static feedback control, under a unified frame that the loop transfer function and robust properties of these controls are fully realized.

These two existing controls are very basic especially the state feedback control, which existed even before the start of the modern control theory in 1960. However, no attempt has been made to unify the huge difference between these two basic controls. See 4.3.7 and 4.3.8 of Section 4.3.

The generalized state feedback control (Definition 4.2) of the new design principle of this book, can completely unify these two existing controls as its two extremes. This unification can be very clearly seen in Figure 6.1 and Table 6.2.

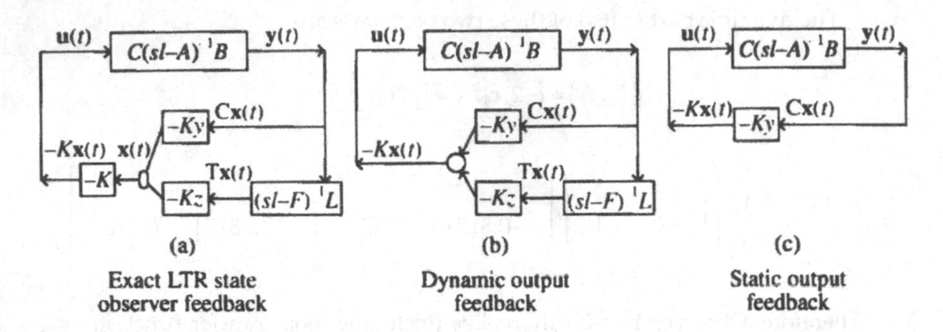

FIGURE 6.1
Three modern control structures capable of realizing generalized state feedback control and their robustness properties.

TABLE 6.2
Three Output Feedback Control Systems of the Modern Control Theory

Control Structure	**(a)**	**(b)**	**(c)**
Control signal	$-K\mathbf{x}(t)$	$-\underline{K}\,\underline{C}\,\mathbf{x}(t)$	$-K_yC\mathbf{x}(t)$
Gain on $\underline{C}\mathbf{x}(t)$	$K[\,K_z : K_y\,]=K\underline{C}^{-1}$	$[\,K_z : K_y\,]$	K_y
Matrix $\underline{C}$	$[\,T' : C'\,]'$	$[\,T' : C'\,]'$	C
Rank $(\underline{C})=r+m$	n	$n \geq \text{rank}\,(\underline{C}) \geq m$	m
Controller order r	$n-m$	$n-m \geq r \geq 0$	0
Gain K on state $\mathbf{x}(t)$	No constraint	No constraint to most constrained	Most constrained
State matrix	$A-BK$	$A-B\underline{K}[\,T' : C'\,]'$	$A-BK_yC$
Loop transfer function	$-K(sI-A)^{-1}B$	$-\underline{K}\underline{C}(sI-A)^{-1}B$	$-K_yC(sI-A)^{-1}B$
Existence condition on plant system	All zeros are stable $m \geq p$ and rank $(CB)=p$	At least one stable zero or $m>p$	None

The properties of these three feedback structures (a), (b), and (c) are summarized in Table 6.2 above.

Conclusion 6.6

Table 6.2 clearly shows that the new feedback control structure (b) of this book completely unified the existing two feedback control structures (a) and (c), as its two extremes in all of nine aspects. Structure (a) realizes the strongest control but is available to the least plant systems, while structure (c) implements the weakest control but is general to all plant systems. The two common features that formed the foundation of this unification are that all control signals are constant gains on system state $\mathbf{x}(t)$, and that the loop transfer functions and robust properties of these controls are fully realized.

A direct consequence of this unification is, that the design of the generalized state feedback control $\underline{K}\underline{C}\mathbf{x}(t)$ of this book can readily use the existing design results: the direct state feedback control design results if rank $(\underline{C})=n$, and the static output feedback control design results if rank ($\underline{C}$)$<n$. In both Chapters 8 (eigenstructure assignment) and 10 (LQ optimal control), the design results are presented in the sequence of rank ($\underline{C}$)$=n$ and then followed by rank($\underline{C}$)$<n$.

6.4 Observer Order Adjustment to Tradeoff between Performance and Robustness

One of the main and unique design features of the new design principle of this book is that the controller/observer order r is completely flexible, while the order of the existing controllers of modern control design theory is fixed. For example, the existing state observer order r is fixed at either n or $n-m$ (see Examples 4.1 and 4.2), and the existing static output feedback control order r is fixed at 0 (see Table 6.2).

Also because of this unique feature, the new output feedback compensator of this book can unify completely these two existing controls as shown in the middle of Table 6.2.

The unique technical reason that enabled this unique feature is that the rows of the observer dynamic part (3.16a) (F, T, L), are completely decoupled (Tsui, 2000a, 2003a, 2004a, 2015). This unique feature is also enabled by the basic design concept of Section 4.1 that the whole system state vector $\mathbf{x}(t)$ needs not to be estimated explicitly.

Example 6.4

The third and fourth observers of Example 6.1 have order 3, which is less than $n-m$ (=4). The first observer of Example 6.3 has order 1, while $n-m=2$.

This section will discuss how to finally decide the observer/controller order r. This decision is based on the following two mutually contradictory properties:

Property 1 The control gain $K=\underline{K}\underline{C}=\underline{K}\ [T':C']'$ is constrained when rank($\underline{C}$)$=r+m<n$. The higher the rank($\underline{C}$) and the observer order r, the less the constraint, and the stronger the control, and the more effective the improvement of feedback system performance and robustness. Thus, the observer/controller order r should be high enough to satisfy the feedback system performance and robustness requirement.

Property 2 Condition (4.3) ($TB=0$) is the necessary and sufficient condition to realize the robust properties of the $\underline{K}\underline{C}\mathbf{x}(t)$-control (Theorem 3.4). Because matrix B is given, the lower the observer order r or the lower the number (r) of rows of matrix T, the easier it is to satisfy (4.3). In addition, the lower-order observer/controller is much easier to simulate. Thus, the observer/controller order r should be low enough to satisfy or best satisfy (4.3).

Therefore, it is certainly plausible to set the value of order r at the maximum possible value, while (4.1) and (4.3) are satisfied.

Definition 6.1

Let parameter $\bar{r}$ be the maximal rank($\underline{C}$=[T': C']')$-m$, where matrix T satisfies (4.1) and (4.3).

Based on Conclusion 6.1, for plant systems with $m \leq p$, this $\bar{r}$ simply equals the number of stable transmission zeros of that plant system.

But for plant systems with $m>p$, this order $\bar{r}$ is very difficult to predict. This is because for this kind of systems, the value of $\bar{r}$ depends on too many system parameters such as n, m, p, the number of system stable and unstable zeros, rank (CB), etc.

Although Examples 6.1 to 6.3 show that if a plant system either has rank (CB)=p–1 or has one unstable transmission zero, then $\bar{r}$=maximal rank ($\underline{C}$)–m –1=(n–m) –1, this value of $\bar{r}$ is less predictable if the plant system has both of these two conditions.

For example, the third system of Example 6.1 also has a stable zero (–1) in addition to the condition rank (CB)=p–1. It appears that even if this stable zero is not matched by an observer pole (say the three observer poles are –2, –3, and –4), then maximal rank ($\underline{C}_3$) is still at n–1=6. The same situation repeats even if this stable zero –1 becomes unstable (see Exercises 6.5 and 6.6). If the number of zeros of this system is increased to two, then maximal rank ($\underline{C}_3$)=n–2 if only one of these two zeros is unmatched by an observer pole (see Exercise 6.7), while maximal rank ($\underline{C}_3$)=n–3 if both zeros are unmatched by observer poles (see Exercise 6.8).

Even though the value of $\bar{r}$ is hard to predict when $m>p$, the design Algorithms 5.3 and 6.1 can guarantee to yield an observer/controller that satisfies (4.1) and (4.3) and that has a maximal possible rank ($\underline{C}$) or a maximal possible value of $\bar{r}$ (Conclusion 6.2).

There is another numerical method to compute directly the value of $\bar{r}$ (Saberi et al., 1993). This method computes the similarity transformation of the plant's state space model to a "special coordinate basis" in which the system is completely decoupled into five blocks. The dimensions of these blocks are the number of system's stable and unstable zeros, etc. Because some of these decoupled subsystems can have unknown input observers, the value of $\bar{r}$ is claimed to be predictable from the dimensions of these blocks.

The main problem of this method is that the computation of this similarity transformation is very ill-conditioned – it should be much more ill-conditioned than computing the rank of a matrix. Although the numerical stability of this computational algorithm has been improved somehow later (Chu, 2000), it is well known that the original ill-conditioning of a problem itself cannot be improved by the improved computation algorithm that solves that problem.

Although $\bar{r}$ of Definition 6.1 is a theoretically plausible choice of observer order r, the actual value of r may still need to be adjusted from actual design requirements on performance and robustness.

One design requirement that all observer designs must comply is the guarantee of feedback system stability, or the guarantee that matrix $A-B\underline{KC}$ is stable.

Because the stability of a matrix requires that all eigenvalues of that matrix have negative real parts, this requirement is much easier to satisfy than the requirement of arbitrary and exact eigenvalue assignment. According to Wang (1996), the condition to generically guarantee arbitrary and exact eigenvalue assignment is rank($\underline{C}$)$\times p > n$ or $(r+m)\times p > n$. Therefore, we require that

$$r > n/p - m \tag{6.18}$$

This lower bound of the actual observer order r should be more than enough to guarantee feedback system stability.

Because the observer designed in this book is an output feedback compensator, such an observer with order satisfying (6.18) fits the definition of "strong stabilization" (Youla et al., 1974; Vidyasagar, 1985).

A practical upper (sufficient) bound for the actual observer order may be derived as follows. Because system poles can fully and far more effectively guarantee system performance, while eigenvector assignment can improve the robustness of the corresponding eigenvalues, the pole assignment and partial eigenvector assignment for some key eigenvalues should have excellent and satisfactory feedback system performance and robustness. Because such an assignment can be guaranteed by rank($\underline{C}$)$+p > n$ or $(r+m)+p > n$ (Algorithms 8.1 and 8.2 and Tsui (2005)), we require

$$r > n - p - m \tag{6.19}$$

Conclusion 6.7

Based on Exercises 4.5 and 4.8, almost all plant systems with $m=p$ satisfy $\bar{r} > n/p - m$ (6.18). Based on Exercises 4.4 and 4.7, at least half of the plant systems with $m=p$ satisfy $\bar{r} > n-p-m$ (6.19). Because plant systems with $m>p$ are better than plant systems with $m=p$ (more measurement information and less input generation requirement), more percentage of plant systems with $m>p$ can satisfy (6.18) and (6.19) than plant systems with $m=p$. To summarize, the new generalized state feedback control of this book can achieve a very good level of feedback system performance and robustness for the great majority of plant systems. See Examples 4.6 and 4.7.

The actual determination and adjustment of observer/controller order r from the suggested value $\bar{r}$ can be based on the actual design requirements and their associated bounds such as (6.18) and (6.19). See Exercises 8.13–8.15 and Tsui (1999c).

If $\bar{r}$ is higher than (6.19), which usually satisfies sufficiently the design requirement, or if $\bar{r}$ is higher than any value of r sufficient for a given design requirement, then the actual value of r may be reduced from $\bar{r}$. A reduced controller order r can make the controller easier to simulate. See Exercises 6.11–6.13.

If $\bar{r}$ is lower than (6.18), which is almost necessary for guaranteed stability, or if $\bar{r}$ is lower than any value of r required by a given design requirement, then the actual value of r should be higher than $\bar{r}$, or that some rows of matrix T cannot satisfy (4.3) ($TB=0$). Of the $(n-m)-\bar{r}$ rows of T not satisfying (4.3), we will select those with the least values of σ_m (see (6.3)) to be added to the $\bar{r}$ rows of matrix T. The purpose of this selection is to minimize the actual $\|TB\|$ in order to best approximate $L(s)$ to $L_{Kx}(s)$. See Exercises 6.14–6.16.

Exercises

6.1 Verify the computation of Algorithm 6.1 to satisfy (4.3), of the four plant systems of Example 6.1.

6.2 Verify the computation of Algorithm 6.1 to satisfy (4.3), of Example 6.2.

6.3 Verify the computation of Algorithm 6.1 to satisfy (4.3), of Example 6.3.

6.4 For plant systems with $m=3$ and $p=2$, the matrix D_i of Eq. (6.2a) will have m rows ($\mathbf{d}_{ij}$, $j=1, 2, 3$). Verify that the solution $\mathbf{c}_i$ of Eq. (6.2a) can be computed using the following three different possible ways:

$$\mathbf{c}_1=[1 \quad c_2 \quad c_3] \text{ implies that } [c_2 \quad c_3]=-[\mathbf{d}_1 B]\left(\begin{bmatrix} d_2 \\ d_3 \end{bmatrix} B\right)^{-1}$$

$$\mathbf{c}_2=[c_1 \quad 1 \quad c_3] \text{ implies that } [c_1 \quad c_3]=-[\mathbf{d}_2 B]\left(\begin{bmatrix} d_1 \\ d_3 \end{bmatrix} B\right)^{-1}$$

$$\mathbf{c}_3=[c_1 \quad c_2 \quad 1] \text{ implies that } [c_1 \quad c_2]=-[\mathbf{d}_3 B]\left(\begin{bmatrix} d_1 \\ d_2 \end{bmatrix} B\right)^{-1}$$

The above computation is very simple because the 2×2 matrix inverse (if exists) is very simple.

6.5 Change matrix B_3 and $N(s)$ of system 3 of Example 6.1 to

$$B_3{}'=\begin{bmatrix} 1 & 1 & 1 & : & -3 & -1 & -1 & : & 2 \\ 0 & 0 & 0 & : & 1 & 2 & 1 & : & -2 \end{bmatrix}',$$

$$N(s)=\begin{bmatrix} (s-1)(s-2) & (s-1) & (s-1) \\ (s-2) & 2 & 1 \end{bmatrix}'$$

Repeat the design of Example 6.1 using the same four eigenvalues of F. Verify that even if the original stable zero −1 is changed to an unstable 1, while rank(CB) remains at p−1, then the new max. rank($\underline{C}$) remains unchanged at n−1 (=6).

Partial answer:

For the four eigenvalues of F: $\mathbf{c}_1=[1\ 0\ 3]$, $\mathbf{c}_2=[1\ 0\ 4]$, $\mathbf{c}_3=[1\ 0\ 5]$, and $\mathbf{c}_4=[1\ 0\ 6]$.

6.6 Change matrix B_3 and $N(s)$ of system 3 of Example 6.1 to

$$B_3' = \begin{bmatrix} 1 & 1 & 1 & : & 3 & -1 & -1 & : & 2 \\ 0 & 0 & 0 & : & 1 & 0 & 1 & : & -2 \end{bmatrix}',$$

$$N(s) = \begin{bmatrix} (s-1)(s-2) & (s-1) & (s-1) \\ (s-2) & 0 & 1 \end{bmatrix}'$$

Repeat the design of Example 6.1 using the same four eigenvalues of F. Verify that even if the original stable zero –1 is changed to an unstable 1, then the new rank($\underline{C}$) remains unchanged at 6.

6.7 Change matrix B_3 and $N(s)$ of system 3 of Example 6.1 to

$$B_3' = \begin{bmatrix} 1 & 1 & 1 & : & -1 & 1 & 1 & : & -2 \\ 0 & 0 & 0 & : & 1 & 0 & 0 & : & -2 \end{bmatrix}',$$

$$N(s) = \begin{bmatrix} (s+1)(s-2) & (s+1) & (s+1) \\ (s-2) & 0 & 0 \end{bmatrix}'$$

Repeat the design of Example 6.1 using the same four eigenvalues of F. Verify that if the increase in the number of transmission zeros is from 1 to 2, while the number of unmatched zeros remains at one and rank (CB) remains at $p-1$, then the new rank($\underline{C}$) is reduced from $n-1$ (=6) to $n-2$ (=5).

Partial answer:

For the four eigenvalues of F: $\mathbf{c}_1=[0\ 1\ 1]$, $\mathbf{c}_2=\mathbf{c}_3=\mathbf{c}_4=[0\ \ 1\ \ -1]$.

6.8 Change matrix B_3 and $N(s)$ of system 3 of Example 6.1 to

$$B_3' = \begin{bmatrix} 1 & 1 & 1 & : & -3 & -1 & -1 & : & 2 \\ 0 & 0 & 0 & : & 1 & 0 & 0 & : & -2 \end{bmatrix}',$$

$$N(s) = \begin{bmatrix} (s-1)(s-2) & (s-1) & (s-1) \\ (s-2) & 0 & 0 \end{bmatrix}'$$

Repeat the design of Example 6.1 using the same four eigenvalues of F. Verify that if the increase in the number of unstable transmission zeros is from 0 to 2 so that the number of unmatched zeros increased to two, while rank (CB) remains at $p-1$, then the new max. rank ($\underline{C}$) is reduced from $n-1$ (=6) to $n-3$ (=4).

Partial answer:

For the four eigenvalues of F: $\mathbf{c}_1=\mathbf{c}_2=\mathbf{c}_3=\mathbf{c}_4=[0\ \ 1\ \ -1]$.

6.9 For the system of Example 6.3, change the stable transmission zero from –2 to –3 by changing the B matrix. Then, repeat the design of Example 6.3 using matrix $F=\text{diag}\{-3\ \ -1\}$.

Partial answer: $B=\begin{bmatrix} 1 & 1 & 3 & -1 \\ 2 & 3 & 6 & -3 \end{bmatrix}$

6.10 Condition rank $(CB)=p$ is the sufficient condition for the plant system with $m=p$ to have $n-m$ transmission zeros. Analyze the effect of this condition on the maximal rank($\underline{C}$) for systems with $m>p$. Say rank$(CB)=p-1$, or $p-2$, ….

6.11 Let a plant system have parameters $n=10$, $m=p=3$, and 6 stable transmission zeros. Derive the following: (a) The total number of transmission zeros. (b) The value of $\overline{r}$ of Definition 6.1. (c) lower bound of observer order r of (6.18) for guaranteed stability. (d) Lower bound of observer order r of (6.19) for an excellent level of performance and robustness. (e) Compare the results of (b)–(d). Answer:

a)	b)	c)	d)	e)
$n-m=7$	$\overline{r}$=# of stable zeros=6	$n/p-m=0$	$n-p-m=4$	Excellent control is fully achievable since 6>4.

6.12 Repeat Exercise 6.11 for systems with $n=12$, $m=p=3$, and eight stable transmission zeros.

Answer:

a)	b)	c)	d)	e)
9	8	1	6	Excellent control is fully achievable since 8>6

6.13 Repeat Exercise 6.11 for systems with $n=9$, $m=p=3$, and 5 stable transmission zeros Answer:

a)	b)	c)	d)	e)
6	5	0	3	Excellent control is fully achievable since $\overline{r}>3$

6.14 Repeat Exercise 6.11 for systems with $n=9$, $m=p=3$, and 3 stable transmission zeros.

Answer:

a)	b)	c)	d)	e)
6	3	0	3	Excellent control not fully achievable since $\overline{r}$ (= 3)≤3. Very good control is fully achievable since $\overline{r}>0$.

6.15 Repeat Exercise 6.11 for systems with $n=12$, $m=p=2$, and 4 stable transmission zeros.

Answer:

a)	b)	c)	d)	e)
10	4	4	8	Even very good control is not fully achievable because $\overline{r} \leq 4$ Order r is recommended to be 5 (> 4) to guarantee stability, though this control is not fully realizable because this $r=5>\overline{r}$.

6.16 Repeat Exercise 6.11 for systems with $n=10$, $m=p=3$, and 4 stable transmission zeros.

Answer:

a)	b)	c)	d)	e)
7	4	0	4	Like 6.14: Excellent control is not fully realizable but very good control is, since $\overline{r}$ (= 4) is ≤ 4, but is > 0.

Exercises 6.5–6.16 demonstrated the following. There are infinitely many different situations of plant system conditions (such as different n, m, p, and $\overline{r}$) and design requirements. Only the new design principle of this book can provide general, wise, and specific rules and guidance of design (including the controller/observer order), and only the new design principle of this book can fully realize a new generalized state feedback control, which is much more effective than other existing fully realizable controls and which is sufficiently effective for a great majority of plant systems.

7

Observer Design for Other Special Purposes

The purpose of the observer design of Chapter 6 is not only to generate a good control signal $\underline{K}\underline{C}\mathbf{x}(t)$ but also to guarantee that the loop transfer function and robustness property of this control is fully realized. Robustness is the foremost purpose and design requirement of feedback control.

This chapter discusses observer design for other two different purposes.

Section 7.1 is about minimal order function observer design (Definition 4.2). The purpose of this design is to generate a $K\mathbf{x}(t)$ signal for a previously and separately designed and arbitrarily given constant K, while minimizing the function observer order. Mathematically, from Theorem 3.2, this design is to satisfy (4.1) and then fully use the remaining freedom of solution T of (4.1), to minimize the number of rows of T needed to satisfy (4.2): $K = \underline{K}[T' : C']'$.

This design problem has received continuous attention since its proposition by Luenberger (1971). The simplest design formulation and the best possible theoretical result of this design problem are derived, and this design problem is therefore essentially solved, in 1986. However, the last two decisive conclusions were formally published only around 2012 (Tsui, 2015), 26 years after 1986.

This happening of control theory development has another odd implication. That is, the situation that the new and overwhelmingly superior design principle of this book has not yet been accepted by the mainstream control research community, is not surprising.

The practical significance of the problem of minimal order function observer design is at observer order reduction. For example, Example 7.4 shows that under very simple and common plant system conditions, Algorithm 7.1 guarantees reducing the existing state observer order from 100 to only 15 of the function observer order! The simulation of a 100th order controller is much harder than that of a 15th order controller, even for very powerful computers.

Section 7.2 deals with the design of fault detection, isolation, and accommodation control observers. Here, each control input fault is rare and severe, represented by an additional unknown input signal, and is totally different from the normal and minor control disturbances discussed in the other chapters of this book. Thus, input faults should be treated specially, like detecting and treating the severely infected patients is different from building general immunity against virus infection. This is a very important problem since the 1980s and has been treated exclusively in some monographs. This system can complement the normal observer feedback system so that all kinds of disturbance and faults are handled comprehensively.

DOI: 10.1201/9781003259572-7

Of the three tasks of fault detection, isolation, and accommodation control, the task of fault isolation is particularly important and pivotal – fault isolation/diagnosis is much more difficult than merely detecting the fault occurrence, and accurate fault isolation information is also key to any effective fault control. This is just like an accurate diagnosis of the disease is much more difficult than just feeling sick, and is also the key to any effective cure treatment.

The input fault detection and isolation are accomplished by special observers called robust fault detectors, and Chapters 5 and 6 present the far more general and simpler design method for such special observers (Tsui, 1989, 2015). Based on this far better result of fault detection and isolation, far more effective adaptive fault control with general treatment of model uncertainty and output measurement noise, are also developed and presented. Thus, the fault detection, isolation, and control of this book is far more general and complete, even with a run throughout example which is also very rare in other comparable books.

7.1 Minimal-Order Linear Functional Observer Design

7.1.1 Simplest Possible Design Formulation – Most Significant Theoretical Development

A minimal order function observer is an observer (3.16) that can generate signal $K\mathbf{x}(t)$ (K is arbitrarily given) and has minimized observer order (Definition 4.2). For example, a state observer estimates signal $\mathbf{x}(t) = I\mathbf{x}(t)$ and is therefore a special case of function observer when K has a special value I.

Physically speaking, a function observer can have an order lower than the state observer order, because it generates only p input control signals $K\mathbf{x}(t)$ while a state observer must estimate all n signals of $\mathbf{x}(t)$, and p is usually much less than n.

From Theorem 3.2, this design is to satisfy Eq. (4.1) first, and then use the remaining freedom of the solution T of (4.1) to minimize the number of rows of T needed to satisfy Eq. (4.2) (or $K = \underline{K}[\,T' \colon C']'$, "′" stands for the transpose operation).

Mathematically speaking, function observer order or the number of rows of T can be lower than a state observer order, because on the left side of its design requirement (4.2), rank $(K) = p$ while rank $(I) = n$, and p is usually much less than n.

If in the output part of the observer system (3.16b), $K_y = 0$, then the corresponding observer is called "strictly proper", and the observer is called "proper" if $K_y \neq 0$. Both types of observers will be discussed in this chapter. However, the proper-type observers use more information of m elements of $\mathbf{y}(t)$ to help generate signal $K\mathbf{x}(t)$ (or used m more rows of matrix C to help satisfy (4.2)). The proper-type observer always has a lower order than its strictly proper counterparts (i.e., strictly proper state observer order $= n$, while proper state observer order $= n - m$, see Examples 4.1 and 4.2). We will concentrate on the proper-type function observer design and then easily extend the result to strictly proper-type function observers.

Because in a real practical observer design, not only the estimation convergence but also the convergence rate and smoothness must be satisfied, it is required that observer poles be arbitrarily given in almost all existing observer design formulations. This book is no exception.

Because Algorithm 5.3 computed the solution T of (4.1) and all its remaining freedom (see Step 2), we will use this remaining freedom fully to satisfy (4.2) ($K = \underline{K}[T'\!: C']'$) while minimizing the number of rows of matrix T (= observer order). According to the overview of Example 4.3, only based on the decoupled property of the rows of matrix T by Tsui (1985), can the number of rows of T be systematically computed and can the minimal order function observer design computation algorithm be general and systematic (Tsui, 1993a, 1998a, 2000b, 2003a, 2015).

Like Algorithm 5.3 for Eq. (4.1), based on the form of matrix C in a block observable Hessenberg form (5.5), Eq. (4.2) can be separated into two parts. The left m columns of (4.2) are

$$K\begin{bmatrix} I_m \\ 0 \end{bmatrix} = \begin{bmatrix} K_Z & : & K_y \end{bmatrix}\begin{bmatrix} T \\ C \end{bmatrix}\begin{bmatrix} I_m \\ 0 \end{bmatrix} = K_Z T\begin{bmatrix} I_m \\ 0 \end{bmatrix} + K_y C_1 \tag{7.1a}$$

and the right $n-m$ columns of (4.2):

$$K\begin{bmatrix} 0 \\ I_{n-m} \end{bmatrix} \triangleq \bar{K} = K_Z T\begin{bmatrix} 0 \\ I_{n-m} \end{bmatrix} \triangleq K_Z \bar{T} \tag{7.1b}$$

Because matrix C_1 is full column rank and because K_y is free to design in (7.1a), Eq. (7.1a) can always be satisfied by parameter K_y. Hence, the problem is simplified to satisfy (7.1b) only.

For simplicity of presentation, we will let observer poles be distinct and real only. These poles are supposed to be arbitrarily given after all, and there is no commanding reason for other kinds of observer poles. The corresponding solution of (4.1) is (5.16a): $\mathbf{t}_i = \mathbf{c}_i D_i$, $i = 1, \ldots, n-m$.

Substitute this solution into (7.1b):

$$\bar{K} = K_Z \bar{T} = K_Z \underbrace{\begin{bmatrix} \mathbf{c}_1 & & \\ & \ddots & \\ & & \mathbf{c}_r \end{bmatrix}}_{m \;\cdots\; m} \underbrace{\begin{bmatrix} \bar{D}_1 \\ \vdots \\ \bar{D}_r \end{bmatrix}}_{n-m} \tag{7.1c}$$

where $\bar{K}$, $\bar{T}$, and $\bar{D}_i$ are the right $n-m$ columns of matrices K, T, and D_i, respectively, and r ($\leq n-m$) is the minimal number of rows of T needed to satisfy (7.1c).

Equation (7.1c) is essentially a set of linear equations, because all unknown variables of (7.1c) are at the same side of the equation. Although these unknown variables have a separate data structure of two parts (K_Z and diag$\{\mathbf{c}_1, \ldots, \mathbf{c}_r\}$), K_Z is the full remaining freedom of (3.16b) after K_y of (7.1a), while $\{\mathbf{c}_1, \ldots, \mathbf{c}_r\}$ is the full remaining freedom of (3.16a) after (4.1). Therefore, (7.1c) is obviously the simplest possible formulation with all available observer design freedom, of minimal order function observer design problem (Tsui, 1998a, 2003a).

Therefore, Eq. (7.1c) is also the most significant theoretical development of the minimal order function observer design problem (Tsui, 2015).

7.1.2 Really Systematic Design Algorithm and Guaranteed Observer Order Upper Bound

Because we only need to satisfy Eq. (7.1c) for a minimal value of r, and because (7.1c) is only a set of linear equations, we can use the most standardized way to compute the solution of this problem – lower triangularizing or echelonizing at the right side of the given matrix of (7.1c) (Section A.2; Tsui, 1983b; Lay, 2006).

Algorithm 7.1: Minimal-Order Function Observer Design (Tsui, 1985)

Step 1: Operate (see Section A.2) on the right side of the following matrix S until it becomes the echelon form of (7.2):

$$SH \triangleq \begin{bmatrix} \bar{D}_1 \\ \vdots \\ \bar{D}_{n-m} \\ \bar{K} \end{bmatrix} H = \begin{bmatrix} * & & 0 & \vdots & & & \\ & \ddots & & \vdots & & 0 & \\ X & & * & \vdots & & & \\ \cdots & \cdots & \cdots & \cdots & \cdots & \cdots & \cdots \\ & & & X & & & \\ \cdots & \cdots & \cdots & \cdots & \cdots & \cdots & \cdots \\ & & & \bar{D}_{r_1+1}H & & & \\ & & & \vdots & & & \\ & & & \bar{D}_{n-m}H & & & \\ & & & X & & & \\ x & \cdots & x & \vdots & 0 & \cdots & 0 \\ & & & X & & & \end{bmatrix} \begin{matrix} \left.\right\} (r_1-1)m+1 \text{ to } r_1 m \text{ rows} \quad \left.\right\} r_1 m \text{ rows} \\ \\ \left.\right\} m \\ \\ \left.\right\} m \\ \leftarrow \text{the } q_1\text{th row } \boldsymbol{k}_1 \end{matrix}$$

$$\triangleq \bar{S} \tag{7.2}$$

Step 2: The form of $\bar{S}$ of (7.2) indicates that the q_1-th row of $\bar{K}$ is a linear combination of the rows of $\bar{D}_i$ $(i = 1, \ldots, r_1)$ or $\mathbf{k}_1 = \mathbf{c}_1 \bar{D}_1 H + \ldots + \mathbf{c}_{r1} \bar{D}_{r1} H$. Compute these linear combination coefficients $\mathbf{c}_i$ $(i = 1, \ldots, r_1)$ by back substitution. Let $\underline{t}_i = \mathbf{c}_i \bar{D}_i$ be the i-th row of $\bar{T}$, $i = 1, \ldots, r_1$. Let the q_1-th row of K_z be [1 … 1: 0 … … 0] with r_1 "1" elements.

Step 3: Operate (see Section A.2) on the right side of the following matrix S_1 until it becomes the echelon form of (7.3):

$$S_1 H_1 \triangleq \begin{bmatrix} \mathbf{c}_1 \bar{D}_1 H \\ \vdots \\ \mathbf{c}_{r1} \bar{D}_{r1} H \\ \bar{D}_{r1+1} H \\ \vdots \\ \bar{D}_{n-m} H \\ \bar{K} H \end{bmatrix} H_1$$

$$= \begin{bmatrix} * & & 0 & \vdots & & & \\ & \ddots & & \vdots & & 0 & \\ X & & * & \vdots & & & \\ \cdots & \cdots & \cdots & \cdots & \cdots & \cdots & \cdots \\ & & & X & & & \\ \cdots & \cdots & \cdots & \cdots & \cdots & \cdots & \cdots \\ & & & \bar{D}_{r1+r2+1} H H_1 & & & \\ & & & \vdots & & & \\ & & & \bar{D}_{n-m} H H_1 & & & \\ & & & X & & & \\ x & \cdots & x & \vdots & 0 & \cdots & 0 \\ & & & X & & & \end{bmatrix} \begin{matrix} \}\, r_1 + (r_2 - 1)m + 1 \text{ to } r_1 + r_2 m \text{ rows} \\ \\ \\ \\ \}\, m \\ \vdots \\ \}\, m \\ \leftarrow \text{the } q_2\text{th row } \boldsymbol{k}_2 \\ (q_2 \neq q_1) \end{matrix}$$

$$\triangleq \bar{S}_1 \qquad (7.3)$$

Step 4: The form of $\bar{S}_1$ of (7.3) indicates that the q_2-th row of $\bar{K}$ is a linear combination of the rows of $\mathbf{c}_i \bar{D}_i$ $(i = 1, \ldots, r_1)$ and the rows of $\bar{D}_i$ $(i = r_1 + 1, \ldots, r_1 + r_2)$. Or

$$\mathbf{k}_2 = k_{21}\mathbf{c}_1\bar{D}_1HH_1 + \cdots + k_{2,r1}\mathbf{c}_{r1}\bar{D}_{r1}HH_1 + \mathbf{c}_{r1+1}\bar{D}_{r1+1}HH_1 + \cdots + \mathbf{c}_{r1+r2}\bar{D}_{r1+r2}HH_1$$

Compute $\mathbf{k}_{Z2} = [k_{21} \ldots k_{2,r1}]$ and $\mathbf{c}_i$ $(i = r_1+1, \ldots, r_1+r_2)$ by back substitution. Then, let the q_2-th row of

$$K_Z = [\mathbf{k}_{Z2} : 1 \ldots\ldots 1 : 0 \ldots\ldots\ldots 0] \tag{7.4a}$$

with r_2 "1" elements.

Steps 3 and 4 are repeated until all p rows of $\bar{K}$ are expressed as the linear combinations of the rows of $\bar{D}_i$ $(i = 1, \ldots\ldots, r_1 + \ldots + r_p \equiv r)$, where this r is the observer order.

At any stage of the algorithm, if the calculated parameter $\mathbf{c}_i = 0$, then the corresponding matrix $\bar{D}_iH\ldots$ will be removed and redeployed at the lower part of matrix S, to express another row of $\bar{K}$ (Tsui, 1986b).

Without loss of generality, we assume $q_i = i$ $(i = 1, \ldots, p)$. Then from (7.4a) matrix K_Z will be

$$K_Z = \begin{bmatrix} 1\ldots1 & : & 0 & : & \cdots & \cdots & : & 0 \\ \mathbf{k}_{Z2} & : & 1\ldots1 & : & 0 & : \cdots & : & 0 \\ - \quad \mathbf{k}_{Z3} & & - & : & 1\ldots1 & : \ddots & : & \vdots \\ & & \ddots & & & \ddots & : & 0 \\ & & - & \mathbf{k}_{Zp} & - & & : & 1\ldots1 \end{bmatrix} \tag{7.4b}$$
$$\quad r_1 \qquad\qquad r_2 \qquad\qquad \cdots \qquad \cdots \qquad\qquad r_p$$

Finally, parameters T and L are determined by Steps 2 and 3 of Algorithm 5.3, and parameter K_y is determined by Eq. (7.1a).

Based on Conclusion 5.2 and the general assumption

$$v_1 \geq v_2 \geq \cdots \geq v_m, \quad \text{and} \quad r_1 \geq r_2 \geq \cdots \geq r_p \tag{7.5}$$

It is proven by Tsui (1986b) that in Algorithm 7.1,

$$r_i \leq v_i - 1, \quad i = 1, \ldots, p \tag{7.6a}$$

Hence, the observer order is

$$r = (r_1 + \cdots + r_p) \leq (v_1 - 1) + \cdots + (v_p - 1) \tag{7.6b}$$

It is proven in Tsui (1986b) that observer order r also satisfies

$$r \leq n - m \tag{7.7}$$

The actual observer order upper bound will take the minimum of (7.6b) and (7.7).

For strictly proper function observers ($K_y = 0$), Eq. (7.1c) becomes

$$K = K_Z T = K_Z \begin{bmatrix} \mathbf{c}_1 & & \\ & \ddots & \\ & & \mathbf{c}_r \end{bmatrix} \begin{bmatrix} D_1 \\ \vdots \\ D_r \end{bmatrix} \tag{7.8}$$

$$\quad m \quad \cdots \quad m \qquad n$$

which has n columns instead of $n-m$ columns as in (7.1c). Therefore, Algorithm 7.1 can be used directly and the results of (7.6a, 7.6b, and 7.7) become

$$r_i \le \nu_i, i = 1,\ldots,p \tag{7.9a}$$

$$r = r_1 + \cdots + r_p \le \nu_1 + \cdots + \nu_p \tag{7.9b}$$

$$\text{and } r \le n \tag{7.10}$$

respectively. The actual observer order upper bound will take the minimum of (7.9b) and (7.10).

Table 7.1 summarizes the lower and upper bounds of the function observer order guaranteed by Algorithm 7.1, for both state observer and function observer, and for both strictly proper and proper observers. These observers have arbitrarily given real and distinct poles.

TABLE 7.1

Lower and Upper Bounds of All Observer Orders Guaranteed by Algorithm 7.1

	State Observers	**Function Observers**
Observer Type	$K = I, p = n$	**Arbitrary** $K, p \le n$
Strictly proper	$r = n$	$1 \le r \le \min\{n, \nu_1 + \ldots + \nu_p\}$
Proper	$r = n-m$	$0 \le r \le \min\{n-m, (\nu_1-1) + \ldots + (\nu_p-1)\}$

Table 7.1 shows that the function observer order of Algorithm 7.1 varies between its upper and lower bounds. The actual value of this order depends on the actual values of K and D_i matrices in (7.1c) and (7.8). For example, if the actual value of $K = I$ (state observers), then the matrix $[T' \vdots C']'$ of (7.1a) and the matrix T of (7.8) must be nonsingular. Thus, the number of rows of matrix T in these two cases must be $n-m$ and n, respectively (the lowest possible state observer order upper bounds). In another example, if the actual value of K equals 0 in (7.1c) or K equals a linear combination of the rows of only one D_i matrix in (7.8), then the number of rows of T needed to satisfy (7.1c) and (7.8) can be 0 and 1, respectively (the lowest possible functional observer order lower bounds).

Example 7.1: Tsui (1985)

Let the system's block observable Hessenberg form matrices be

$$A = \begin{bmatrix} -1 & 0 & 0 & : & 1 & 0 & 0 & : & 0 \\ 2 & 0 & 1 & : & -1 & 1 & 0 & : & 0 \\ 0 & 3 & 0 & : & 0 & 1 & 1 & : & 0 \\ \cdots & \cdots & \cdots & \cdots & \cdots & \cdots & \cdots & \cdots & \cdots \\ 0 & 0 & 0 & : & -3 & 0 & 1 & : & 1 \\ 0 & 0 & 0 & : & 0 & 1 & 0 & : & -1 \\ 1 & 0 & 0 & : & 0 & 0 & -1 & : & 0 \\ \cdots & \cdots & \cdots & \cdots & \cdots & \cdots & \cdots & \cdots & \cdots \\ 0 & 1 & 0 & : & 0 & 1 & 0 & : & -2 \end{bmatrix}$$

and

$$C = \begin{bmatrix} 1 & 0 & 0 & : & 0 & 0 & 0 & : & 0 \\ 1 & 1 & 0 & : & 0 & 0 & 0 & : & 0 \\ -1 & 0 & 1 & : & 0 & 0 & 0 & : & 0 \end{bmatrix}$$

Based on Definition 5.1, the three observability indices of this system are $\nu_1 = 3$, $\nu_2 = 2$, and $\nu_3 = 2$.

Before using Algorithm 7.1 to design function observers, we need to calculate basis vector matrices D_i each has m (= 3) rows, for each row of solution T of (4.1). There are $n-m$ (= 4) D_i matrices, each is for one of the four observer poles, which are selected as –1, –2, –3, and –1. From (5.10b) and (5.15d) of Algorithm 5.3, these four D_i matrices are

$$\begin{bmatrix} D_1 \\ \vdots \\ D_4 \end{bmatrix} = \begin{bmatrix} 2 & 0 & -1 & : & 1 & 0 & 0 & : & 1 \\ 1 & -1 & -1 & : & 1 & 1 & 0 & : & 0 \\ 0 & 0 & 0 & : & 0 & 0 & 1 & : & 0 \\ \cdots & \cdots & \cdots & \cdots & \cdots & \cdots & \cdots & \cdots & \cdots \\ -1 & -1 & 0 & : & 0 & 0 & 0 & : & 1 \\ -1 & -2 & -1 & : & 1 & 1 & 0 & : & 0 \\ 1 & 1 & -1 & : & 0 & 0 & 1 & : & 0 \\ \cdots & \cdots & \cdots & \cdots & \cdots & \cdots & \cdots & \cdots & \cdots \\ -2 & -2 & 1 & : & -1 & 0 & 0 & & 1 \\ -3 & -3 & -1 & : & 1 & 1 & 0 & & 0 \\ 2 & 2 & -2 & : & 0 & 0 & 1 & : & 0 \\ \cdots & \cdots & \cdots & \cdots & \cdots & \cdots & \cdots & \cdots & \cdots \\ 6 & 1 & -2 & : & 2 & 0 & 0 & : & 1 \\ 3 & 0 & -1 & : & 1 & 1 & 0 & : & 0 \\ -1 & -1 & 1 & : & 0 & 0 & 1 & : & 0 \end{bmatrix}$$

We will now design three different function observers, each is for one of the following three different and given K matrices, using Algorithm 7.1.

<u>For K_1</u> $= \begin{bmatrix} 3 & -2 & -2 & : & 1 & 2 & 1 & : & 0 \\ 2 & 0 & -1 & : & 1 & 1 & 0 & : & 0 \end{bmatrix}$

Step 1:

$$SH = \begin{bmatrix} \bar{D}_1 \\ \vdots \\ \bar{D}_4 \\ ---- \\ \bar{K}_1 \end{bmatrix} \begin{bmatrix} 1 & 0 & 0 & -1 \\ 0 & 1 & 0 & 1 \\ 0 & 0 & 1 & 0 \\ 0 & 0 & 0 & 1 \end{bmatrix} = \begin{bmatrix} 1 & 0 & : & 0 & 0 \\ 1 & 1 & : & 0 & 0 \\ \cdots & \cdots & \cdots & \cdots & \cdots \\ 0 & 0 & & 1 & 0 \\ \cdots & \cdots & \cdots & \cdots & \cdots \\ 0 & 0 & & 0 & 1 \\ 1 & 1 & & 0 & 0 \\ 0 & 0 & & 1 & 0 \\ -1 & 0 & & 0 & 2 \\ 1 & 1 & & 0 & 0 \\ 0 & 0 & & 1 & 0 \\ 2 & 0 & & 0 & -1 \\ 1 & 1 & & 0 & 0 \\ 0 & 0 & & 1 & 0 \\ & & & & \\ 1 & 2 & & 1 & 1 \\ 1 & 1 & : & 0 & 0 \end{bmatrix} \quad (7.11)$$

The last row of the above matrix is k_1, which is the second row of $\bar{K}_1 H$ $(q_1 = 2)$.

Step 2: $r_1 = 1$ and $\mathbf{c}_1 = \begin{bmatrix} 0 & 1 & 0 \end{bmatrix}$ such that $\mathbf{c}_1 \bar{D}_1 H = k_1$.

Step 3:

$$S_1 H_1 = \begin{bmatrix} 1 & 1 & & 0 & 0 \\ 0 & 0 & & 0 & 1 \\ 1 & 1 & & 0 & 0 \\ 0 & 0 & & 1 & 0 \\ -1 & 0 & & 0 & 2 \\ 1 & 1 & & 0 & 0 \\ 0 & 0 & & 1 & 0 \\ & & : & & \\ 1 & 2 & & 1 & 1 \\ 1 & 1 & & 0 & 0 \end{bmatrix} \begin{bmatrix} 1 & -1 & 0 & 0 \\ 0 & 1 & 0 & 0 \\ 0 & 0 & 1 & 0 \\ 0 & 0 & 0 & 1 \end{bmatrix}$$

$$= \begin{bmatrix} 1 & 0 & & 0 & 0 \\ 0 & 0 & & 0 & 1 \\ 1 & 0 & & 0 & 0 \\ 0 & 0 & & 1 & 0 \\ -1 & 1 & & 0 & 2 \\ \cdots & \cdots & \cdots & \cdots & \cdots \\ & & : & & \\ & & : & & \\ 1 & 1 & & 1 & 1 \\ x & x & & x & x \end{bmatrix} \begin{matrix} \\ \\ \\ \\ \\ \\ \\ \\ \leftarrow k_2 \\ \\ \end{matrix}$$

Step 4: $r_2 = 2$, $\mathbf{c}_2 = [-1\ 0\ 1]$, $\mathbf{c}_3 = [1\ 0\ 0]$, and $\mathbf{k}_{Z2} = 2$ such that

$$k_2 = \mathbf{k}_{Z2}\left(\mathbf{c}_1\bar{D}_1HH_1\right)+\mathbf{c}_2\left(\bar{D}_2HH_1\right)+\mathbf{c}_3\left(\bar{D}_3HH_1\right)$$

Finally, for K_1: the observer order $r = r_1 + r_2 = 1 + 2 = 3$, $F = \text{diag}\{-1, -2, -3\}$, and

$$T = \begin{bmatrix} \mathbf{c}_1D_1 \\ \mathbf{c}_2D_2 \\ \mathbf{c}_3D_3 \end{bmatrix} = \begin{bmatrix} 1 & -1 & -1 & 1 & 1 & 0 & 0 \\ 2 & 2 & -1 & 0 & 0 & 1 & -1 \\ -2 & -2 & 1 & -1 & 0 & 0 & 1 \end{bmatrix}$$

$$\text{From (7.4a), } K_Z = \begin{bmatrix} \mathbf{k}_{Z2} & : & 1 & 1 \\ 1 & : & 0 & 0 \end{bmatrix} = \begin{bmatrix} 2 & : & 1 & 1 \\ 1 & : & 0 & 0 \end{bmatrix}$$

$$\text{From (5.10a), } L = (TA - FT)\begin{bmatrix} I_3 \\ 0 \end{bmatrix}C_1^{-1} = \begin{bmatrix} 0 & -4 & -2 \\ 7 & 0 & 0 \\ -5 & -2 & 1 \end{bmatrix}$$

$$\text{and from (7.1a), } K_y = (K - K_ZT)\begin{bmatrix} I_3 \\ 0 \end{bmatrix}C_1^{-1} = \begin{bmatrix} 1 & 0 & 0 \\ 0 & 1 & 0 \end{bmatrix}.$$

$$\underline{\textit{For } K_2} = \begin{bmatrix} 2 & 0 & 2 & : & 1 & 0 & 1 & : & 1 \\ -3 & -3 & -2 & : & 1 & 2 & 0 & : & 0 \end{bmatrix}:$$

Step 1: The result of this step is similar to that of K_1 in (7.11), except for the lower part of (7.11)

$$\bar{K}_2H = \begin{bmatrix} 1 & 0 & 1 & 0 \\ 1 & 2 & 0 & 1 \end{bmatrix} \begin{matrix} \leftarrow q_1 = 1, \boldsymbol{k}_1 \\ \ \end{matrix}$$

Step 2: $r_1 = 1$ and $\mathbf{c}_1 = [1\ 0\ 1]$ such that $\mathbf{c}_1(\bar{D}_1H) = \boldsymbol{k}_1$.
Step 3:

$$S_1H_1 = \begin{bmatrix} 1 & 0 & 1 & 0 \\ 0 & 0 & 0 & 1 \\ 1 & 1 & 0 & 0 \\ 0 & 0 & 1 & 0 \\ & \vdots & & \\ & \vdots & & \\ 1 & 0 & 1 & 0 \\ 1 & 2 & 0 & 1 \end{bmatrix} \begin{bmatrix} 1 & 0 & -1 & 0 \\ 0 & 1 & 0 & 0 \\ 0 & 0 & 1 & 0 \\ 0 & 0 & 0 & 1 \end{bmatrix} = \begin{bmatrix} 1 & 0 & 0 & 0 \\ 0 & 0 & 0 & 1 \\ 1 & 1 & -1 & 0 \\ 0 & 0 & 1 & 0 \\ \cdots & \cdots & \cdots & \cdots \\ & & \vdots & \\ x & x & x & x \\ 1 & 2 & -1 & 1 \end{bmatrix} \begin{matrix} \\ \\ \\ \\ \\ \\ \\ \leftarrow \boldsymbol{k}_2 \end{matrix}$$

Step 4: $r_2 = 1$, $\mathbf{c}_2 = [1\ 2\ 1]$, and $\mathbf{k}_{Z2} = -1$ such that

$$k_2 = \mathbf{k}_{Z2}\left(\mathbf{c}_1\bar{\mathrm{D}}_1HH_1\right)+\mathbf{c}_2\left(\bar{\mathrm{D}}_2HH_1\right)$$

Finally, the minimal order function observer for K_2 is: $r = r_1 + r_2 = 2$, and

$$F = \begin{bmatrix} -1 & 0 \\ 0 & -2 \end{bmatrix}$$

$$T = \begin{bmatrix} \mathbf{c}_1 D_1 \\ \mathbf{c}_2 D_2 \end{bmatrix} = \begin{bmatrix} 2 & 0 & -1 & 1 & 0 & 1 & 1 \\ -2 & -4 & -3 & 2 & 2 & 1 & 1 \end{bmatrix}$$

$$K_Z = \begin{bmatrix} 1 & : & 0 \\ \mathbf{k}_{Z2} & : & 1 \end{bmatrix} = \begin{bmatrix} 1 & : & 0 \\ -1 & : & 1 \end{bmatrix}$$
$$\qquad\qquad\qquad\quad r_1 \qquad r_2$$

$$L = (TA - FT)\begin{bmatrix} I_3 \\ 0 \end{bmatrix} C_1^{-1} = \begin{bmatrix} 2 & -2 & -1 \\ -3 & -16 & -10 \end{bmatrix}$$

$$\text{and} \quad K_y = (K - K_Z T)\begin{bmatrix} I_3 \\ 0 \end{bmatrix} C_1^{-1} = \begin{bmatrix} 1 & 0 & 1 \\ 0 & 1 & 0 \end{bmatrix}$$

$$\underline{\text{For } K_3} = \begin{bmatrix} 0 & 2 & 3 & : & 0 & 0 & 0 & : & 0 \\ 1 & -1 & -1 & : & 1 & 1 & 0 & : & 0 \end{bmatrix}$$

Because the first row of K_3 is already a linear combination of rows of C (because the first row of $\bar{K}_3 = 0$), we let the linear combination coefficients [1 2 3] be $\mathbf{k}_{y1}$ so that the first row of K_3 equals $\mathbf{k}_{y1}C$.

Because the remaining second row of $\bar{K}_3$ equals the q_1-th row of K_1, we will take the result of Steps 1 and 2 (for the q_1-th row) of K_1, that is

$$r = r_1 = 1, F = -1, T = \begin{bmatrix} 1 & -1 & -1 & 1 & 1 & 0 & 0 \end{bmatrix}$$

The remaining parameters of the function observer for K_3 are then:

$$K_Z = \begin{bmatrix} 0 \\ 1 \end{bmatrix}, \quad K_y = \begin{bmatrix} 1 & 2 & 3 \\ 0 & 0 & 0 \end{bmatrix} = \begin{bmatrix} & \mathbf{k}_{y1} & \\ 0 & 0 & 0 \end{bmatrix}, \text{and}$$

$$L = \begin{bmatrix} 0 & -4 & -2 \end{bmatrix}$$

To summarize, the order of the three function observers is 3, 2, and 1, respectively. All of these three orders comply completely with the upper bound $((\nu_1-1)+(\nu_2-1)=3)$ of Table 7.1. The corresponding observer order for each of the two rows of K_3 also complies with the two lower bounds (0 or 1) respectively, of Table 7.1.

The three different function observer orders of this example also demonstrate quite convincingly, that the actual value of function observer order depends on the actual value of each of the many given data of the problem (such as each of the $p \times n$ entries of matrix K). Therefore, only the bounds of the function observer order of Table 7.1, not the actual value of function observer order, is the final form of the general theoretical result of the function observer order.

This means that if the bounds of Table 7.1 are the lowest possible, as has already been demonstrated by the published design results of the past 35 years, and as theoretically proved rigorously by Tsui (2012a, 2015), then the best possible theoretical result of the whole design problem is derived (Tsui, 2012a, 2015). This decisive and final conclusion will be claimed in the next subsection.

The entire design of these three function observers is very general, systematic, comprehensive, and simple! This kind of example actually demonstrates that the problem of minimal order function observer design is essentially solved (Tsui, 2015).

Example 7.2: Single-Output Case ($m = 1$)

In this case, the number of rows of matrix D_i for (4.1), m, becomes 1, and D_i becomes a row vector $= \mathbf{t}_i$, and there is only one observability index $\nu_1 = n$. Hence, (7.8) and (7.1b) become

$$K = K_Z \underbrace{\begin{bmatrix} \mathbf{t}_1 \\ \vdots \\ \mathbf{t}_r \end{bmatrix}}_{n} \quad \text{and} \quad \bar{K} = K_Z \underbrace{\begin{bmatrix} \mathbf{t}_1 \\ \vdots \\ \mathbf{t}_r \end{bmatrix} \begin{bmatrix} 0 \\ I_{n-m} \end{bmatrix}}_{n-m}, \text{ respectively.}$$

Because the number of columns of these two equations is n and $n-m$, respectively, the upper bound of observer order r is n $(=\nu)$ and $n-1$ $(=\nu-1)$, respectively. The lower bound of observer order r is 1 and 0 (when $\bar{K}=0$), respectively. All of these bounds are the lowest possible bounds, and all comply with Table 7.1.

Example 7.3: Single-Input Case ($p = 1$)

Because $p = 1$, the upper bound of function observer order of Table 7.1 becomes ν_1 and ν_1-1, for the strictly proper and proper-type observers, respectively. Because the highest possible value of ν_1 equals n (see Example 7.2), the upper observer order bound becomes n and $n-1$ of Table 7.1 respectively, as expected. The lowest possible number of ν_1 is 1; thus, the lower observer order bound becomes 1 and 0 of Table 7.1 respectively, also as expected.

As parameter p increments from 1, or as rank (K) and the number of signals of $K\mathbf{x}(t)$ increments, the corresponding observer order r must increase. However, in any case when $m = 1$, the observer order cannot exceed ν_1 and ν_1-1 for the two types of observers if $p = 1$ and cannot exceed $\nu_1 + \ldots + \nu_m = n$ and $(\nu_1 + 1) + \ldots + (\nu_m-1) = n-m$ for the two types of observers when $p = m$.

To summarize, (1) Eqs. (7.1c) and (7.8) are the simplest possible general design computation formulations of function observer design, and for proper and strictly proper-type observers, respectively. (2) Algorithm 7.1 is a modified version (more freedom $\{\mathbf{c}_i, i = 1, 2, \ldots\}$ is involved) of the well-standardized algorithm for computational problem (7.1c) and (7.8), and (3) the guaranteed results of Algorithm 7.1 (Table 7.1) unify all cases such as $K_y = 0$ or $\neq 0$, $m = 1$ or >1, $p = 1$ or >1, and $K = I$ or $\neq I$.

Let us notice again that the result of this section is only for distinct and real observer poles. Nonetheless, the result of this section is general because the observer poles have been presumed freely designable (Subsection 7.1.1).

7.1.3 The Lowest Possible Observer Order Upper Bound – The Best Possible Theoretical Result – The Whole Design Problem Is Essentially Solved

Theorem 7.1: The Function Observer Order Upper and Lower Bounds of Table 7.1, Are the Lowest Possible Bounds

Proof

Examples 7.1 to 7.3 already proved that the lower bounds of Table 7.1 are the lowest possible.

For upper bounds, the proof is achieved using the formal induction (Tsui, 2012a):

First prove that when $p = 1$, any upper bound lower than the one in Table 7.1 (ν_1-1) does not exist (see Examples 7.3).

Then prove, that if $(\nu_1-1) + \ldots + (\nu_p-1)$ is the lowest possible observer order upper bound for a value p, then for $p + 1$ $(\leq m)$, any upper bound lower than the one in Table 7.1, $(\nu_1-1) + \ldots + (\nu_{p+1}-1)$, does not exist (see the actual proof in the work of Tsui (2012a)).

Conclusion 7.1: Tsui (2012a, 2014, 2015)

Because generally an analytical formula for the actual value of minimal function observer order does not exist – the final form of the general theoretical formula for these orders are only their bounds (see argument at the end of Examples 7.1–7.3), Theorem 7.1 implies that the best possible theoretical result has already been derived in Algorithm 7.1 (Tsui, 1985, 1986b) and that the theoretical part of minimal order function observer design problem, proposed by Luenberger (1971), is finally solved.

The most important reason for deriving this best possible theoretical result is the simplest possible and far simpler design computation formulation (7.1c) or (7.8). This formulation is enabled only by the general and decoupled solution of (4.1) (Tsui, 1998a, 2003a, 2015).

Equation (7.1c) is only a set of linear equations added with more freedom $\{\mathbf{c}_i, i=1,2,\ldots\}$, see the end of Subsection 7.1.1. It is well known that in a normal set of $n-m$ linear equations $A\mathbf{x}=\mathbf{b}$, the minimal dimension of $\mathbf{x}$ that satisfies this equation depends on every given data in A and $\mathbf{b}$, and this minimal dimension itself obviously does not have a general formula except a bound $(n-m)$. See also the comments after Example 7.1. Hence, we also have the following conclusion.

Conclusion 7.2: Tsui (2015)

Like in the above $A\mathbf{x}=\mathbf{b}$ equation with arbitrarily given A and $\mathbf{b}$, the dimension of solution $\mathbf{x}$ will be generically near and close to its upper bound or the dimension of $\mathbf{b}$ $(=n-m)$, the minimal number of rows of T satisfying (7.1c) or the minimal function observer order (r) will be generically near and close to its upper bound. Therefore, since Algorithm 7.1 guarantees an upper bound of r that is the lowest possible (Theorem 7.1), the effort to further complicate the computation of Algorithm 7.1 in order to seek possible further reduction of the value of r is generically not worthwhile and not practical.

Therefore, the computational part of the minimal order function observer design problem is also solved. Together with Conclusion 7.1, the whole design problem is essentially solved.

Conclusion 7.2 is important because the effort to further reduce the functional observer order has continued well after 1986 (Darouach, 2000; Fernando et al., 2010).

In these newly proposed design algorithms, the computation is equivalent to repeating Algorithm 7.1 about $\sum_{j=1}^{r}[n-m:j]$ times! Here, $[n-m:j]$ is the combination of j $\bar{D}_i$ matrices out of $n-m$ $\bar{D}_i$ matrices (or trying these different combinations of $\bar{D}_i$ matrices into the S matrix of Algorithm 7.1). Thus, the amount of computation is about $(n-m)^3$ times that of Algorithm 7.1! Moreover, each time the computation formulation/problem is much more complicated than (7.1c)! On the other hand, Algorithm 7.1 runs only once and runs on only one matrix S, and Algorithm 7.1 itself is the most standardized and computational reliable way of solving the kind of computation problem of (7.1c) or (7.8), which are already the simplest computation formulation of the whole design problem. Therefore, this extreme further complication of Algorithm 7.1 is obviously worthless, when the actual minimal value of j or r is generically reaching the upper bound $(n-m)$, while the lowest possible upper bound is already guaranteed by Algorithm 7.1!

A relatively much simpler modification of Algorithm 7.1, which has already been mentioned in the first two editions of this book, is to try different sequence permutations of the $\bar{D}_i$ matrices in matrix S of (7.2). For example, if a $\bar{D}_1$–$\bar{D}_2$–$\bar{D}_3$ sequence in matrix S reveals that $\bar{K}$ is a linear combination of $\bar{D}_1$ and $\bar{D}_2$ (order $r = 2$), then a different $\bar{D}_3$–$\bar{D}_2$–$\bar{D}_1$ sequence may reveal that $\bar{K}$ is a linear combination of the rows of $\bar{D}_3$ only (order $r = 1$). But this kind of additional exhaustive sequence permutation is still not worthwhile, when the value of r is generically close to the guaranteed minimal possible upper bound, of Algorithm 7.1.

Conclusion 7.3

Besides the theoretical significance of Algorithm 7.1, there is another practical significance of Algorithm 7.1 – the observer order reduction. Order reduction has long been an important problem of control theory (see Example 4.3).

Based on Table 7.1, the function observer order is always lower than the state observer order whenever $m > p$, and is significantly lower than state observer order whenever $m >> p$ and the observability indices are not much different from each other. These situations are not rare at all in practice. Therefore, Algorithm 7.1 can guarantee significant observer order reduction from state observer order under many practical system conditions.

Example 7.4: Tsui (2015)

For all systems with 120th order, 20 outputs and 3 inputs, and observability indices $\nu_1 = \ldots = \nu_{20} = 6$, the state observer order is $n - m = 100$. But according to Table 7.1, using Algorithm 7.1, the function observer order is guaranteed not exceeding 15 for any parameter K!

The simulation of a 100th order controller is much harder than that of a 15th order controller, even for very powerful computers.

7.2 Fault Detection, Isolation, and Control Design

7.2.1 Fault Models and Design Formulation of Fault Detection and Isolation

System faults can belong to two categories: input control faults and output sensor faults (Frank, 1990; Gertler, 1991). Because the input control faults directly affect the system dynamics, they are far more difficult to detect and control than output measurement faults, which can usually be handled by redundant measurement devices. For example, to solve the recent and serious angle-of-attack sensor fault problem of Boeing 737MAX, the announced solution is simply adding one more such sensor to have a total of two such sensors.

This book will study the detection and control of input control faults only. These input faults are represented as additional unknown input signals $\mathbf{d}(t)$ in the state space model (1.1a):

$$d\mathbf{x}(t)/dt = A\mathbf{x}(t) + B\mathbf{u}(t) + B_1\mathbf{d}_1(t) + B_2\mathbf{d}_2(t) + \cdots + B_Q\mathbf{d}_Q(t)$$
$$\equiv A\mathbf{x}(t) + B\mathbf{u}(t) + B_d\mathbf{d}(t) \qquad (7.12)$$

where the fault signals $\mathbf{d}_i(t)$, $i = 1, 2, \ldots, Q$ all are zero when the fault is free, but some will become nonzero when the fault occurs, and matrices B_i, $i = 1, \ldots, Q$, are the given system gains to their corresponding fault signals. We generally let a vector $\mathbf{d}(t)$ stacked with all $\mathbf{d}_i(t)$ signals to represent all faults, and assign matrix $B_d = [B_1 : \ldots : B_Q]$.

An example throughout the entirety of this section has a special fault model of $Q = n$ and $B_d = I$. Thus, the corresponding state space model (7.12) will become $d\mathbf{x}(t)/dt = A\mathbf{x}(t) + B\mathbf{u}(t) + \mathbf{d}(t)$. In addition, because each fault signal $d_i(t)$ appears only at the i-th differential equation of state variable $x_i(t)$ in (7.12), the nonzero $d_i(t)$ can reasonably mean the failure of the system's state component $x_i(t)$ and will be called the i-th state fault in that throughout example of this section.

The problem of fault detection, or the detection of fault occurrence, can be achieved using a single-output fault detector. This single-output $e(t)$ equals zero at the normal and fault-free ($\mathbf{d}(t) = 0$) condition, but will become nonzero when fault occurs ($\mathbf{d}(t) \neq 0$). This $e(t)$ signal is called the "residual signal".

Therefore, a fault detector structure is a special case of general observer (3.16) as:

$$d\mathbf{z}(t)/dt = F\mathbf{z}(t) + L\mathbf{y}(t) + TB\mathbf{u}(t) \qquad (7.13a)$$

$$e(t) = \mathbf{n}\mathbf{z}(t) + \mathbf{m}\mathbf{y}(t) \qquad (7.13b)$$

where the fault detector parameters $\mathbf{n}$ and $\mathbf{m}$ are nonzero row vectors.

Based on Theorem 3.2, to meet the requirement that $e(t) = 0$ when fault signal $\mathbf{d}(t) = 0$, the fault detector (7.13) must satisfy

$$TA - FT = LC\,(F \text{ must be stable}) \qquad (7.14a)$$

$$\text{and } \mathbf{n}T + \mathbf{m}C = 0 \qquad (7.14b)$$

The problem of fault isolation and diagnosis is much more difficult than the problem of fault detection. It requires the detection of not only $\mathbf{d}(t) \neq 0$, but also which specific group of elements of $\mathbf{d}(t)$ is nonzero.

Obviously, the best way to achieve fault isolation is to have a band of fault detectors working together. Each of these fault detectors not only needs to have its output $e(t)$ become nonzero when $\mathbf{d}(t) \neq 0$, but also has its output $e(t)$ become nonzero only when a pre-specified group of elements of $\mathbf{d}(t)$ becomes nonzero. In other words, the fault detector output $e(t)$ must remain zero even if the (other) specific groups of (say q) elements of $\mathbf{d}(t)$ become nonzero, or the fault detector is insensitive or robust toward the occurrence of those q faults. These fault detectors are therefore called the "robust fault detector" (Ge and Fang, 1988).

This specific function of robust fault detector is very similar to that of the output feedback compensators (Definition 3.3). Therefore, in addition to (7.14a) and (7.14b), a robust fault detector must also satisfy

$$T \times \left[\text{a specific group of } q \text{ columns of } B_d \right] = 0 \tag{7.14c}$$

$$\text{and } T \times \left[\text{all other columns of } B_d \right] \neq 0 \tag{7.14d}$$

Because all columns are generically nonzero, the main design requirement and formulation of a robust fault detector is (7.14a–c).

This design requirement of (7.14a–c) is very similar to the main observer design requirement of this book (4.1–4.3), except that the number of faults q is replacing the number of control inputs p. From Conclusion 6.1, one sufficient condition to satisfy this requirement is $m > q$, and this condition is much simpler than the other sufficient condition of Conclusion 6.1. Therefore, we will concentrate only on the condition $m > q$.

It can be easily proven that after the value of q is chosen, a combination of [Q: q] robust fault detectors work together, can detect and isolate logically and instantly any group of up to q simultaneous faults, as will be shown in the following Example 7.5.

Example 7.5

A fourth-order plant system with three (m = 3) output measurements has a fault model (7.12) with $Q = n$ (= 4 fault signals), and the gain B_d equals I_n.

For q equal to either 1 or 2 (such that requirement $q < m$ is met), we will design two groups of [Q: q] (= $Q! \,/\, [(Q-q)!q!]$ = 4 or 6) robust fault detectors, respectively.

In the following Tables 7.2 and 7.3, we assign "x" as nonzero and as a logic "true", and "0" as zero and as a logic "false", for each residual signal $e(t)$. These two tables show how all groups of up to q simultaneous faults can be detected and isolated. "$\cap$" stands for "AND" logic operation.

TABLE 7.2

Isolation of One Fault of a Four-Possible-Fault System

Fault Situations	Residual Signals				Logic Determination of
$d(t) = [d_1\ d_2\ d_3\ d_4]'$	e_1	e_2	e_3	e_4	Fault Situation Isolation
$d_1(t)\neq 0$	0	x	x	x	$(d_1(t)\neq 0) = e_2 \cap e_3 \cap e_4$
$d_2(t)\neq 0$	x	0	x	x	$(d_2(t)\neq 0) = e_1 \cap e_3 \cap e_4$
$d_3(t)\neq 0$	x	x	0	x	$(d_3(t)\neq 0) = e_1 \cap e_2 \cap e_4$
$d_4(t)\neq 0$	x	x	x	0	$(d_4(t)\neq 0) = e_1 \cap e_2 \cap e_3$

TABLE 7.3

Isolation of up to Two Simultaneous Faults of a Four-Fault System

Fault Situations	Residual Signals						Logic Determination of
$d(t) = [d_1\ d_2\ d_3\ d_4]'$	e_1	e_2	e_3	e_4	e_5	e_6	Fault Situation Isolation
$d_1(t)\neq 0$	0	0	0	x	x	x	$(d_1(t)\neq 0) = e_4 \cap e_5 \cap e_6$
$d_2(t)\neq 0$	0	x	x	0	0	x	$(d_2(t)\neq 0) = e_2 \cap e_3 \cap e_6$
$d_3(t)\neq 0$	x	0	x	0	x	0	$(d_3(t)\neq 0) = e_1 \cap e_3 \cap e_5$
$d_4(t)\neq 0$	x	x	0	x	0	0	$(d_4(t)\neq 0) = e_1 \cap e_2 \cap e_4$

Notice that the logic determination of Table 7.3 can also isolate any combination of two simultaneous fault occurrence situations. For example, the fault situation that both $d_1 \cap d_2$ are nonzero, can be uniquely isolated and identified by the logic operation $e_2 \cap e_3 \cap e_4 \cap e_5 \cap e_6$.

7.2.2 Design Algorithm and Examples of Fault Detection and Isolation

From Subsection 7.2.1, the design formulation of fault detection and isolation system (7.14a–c) is very similar to the main observer design formulation (4.1–4.3) of this book. Thus, its design Algorithm 7.2 can be simply modified from Algorithms 5.3 and 6.1, which are for the solution of (4.1–4.3).

Algorithm 7.2: Tsui (1989)

Step 1: Based on Step 1 of Algorithm 5.3, compute the basis vector matrices D_i for each row of solution matrix T, $\mathbf{t}_i = \mathbf{c}_i D_i$, $i = 1, \ldots, r$ ($r = n-m+1$, see Step 3). Each of the matrices D_i will satisfy the $n-m$ columns of (7.14a) (like (5.10b)) which are corresponding to the $n-m$ zero columns of matrix C.

Step 2: Use the remaining freedom of matrix T, the m-dimensional rows $\mathbf{c}_i$, $i = 1, \ldots, r$, to satisfy (7.14c), or $\mathbf{c}_i D_i$(B_d's q selected columns) = 0. The simple sufficient condition to satisfy (7.14c) is $m>q$. After matrix T is designed, then the remaining m columns of (7.14a) corresponding to the m nonzero columns of matrix C can always be satisfied by parameter L.

Step 3: Because (7.14b) has n columns, the fault detector order r must exceed $n-m$ (or $r>n-m$) in general, in order for a nonzero solution $\mathbf{n}$ and $\mathbf{m}$ of (7.14b) ($\mathbf{n}\, T + \mathbf{m}\, C = 0$ or $[\ \mathbf{n} : \mathbf{m}\ ][T' : C']' = 0$) to exist.

In the following example, we will use Algorithm 7.2 to design a fault detection and isolation system, for a real system of the same kind as Example 7.5 ($Q = n = 4$ and $m = 3$). We will set $r = 2$ so that $r > n - m$. We will also choose $q = 2$, so that a combination of six (= [4: 2]) robust fault detectors will be designed, as specified in Table 7.3 of Example 7.5.

Example 7.6: Tsui (1993c)

Let the plant system be

$$(A, B, C) = \left(\begin{bmatrix} -20.95 & 17.35 & 0 & 0 \\ 66.53 & -65.89 & -3.843 & 0 \\ 0 & 1473 & 0 & -67420 \\ 0 & 0 & -0.00578 & -0.05484 \end{bmatrix}, \begin{bmatrix} 1 \\ 0 \\ 0 \\ 0 \end{bmatrix}, \right.$$

$$\left. \begin{bmatrix} 1 & 0 & 0 & 0 \\ 0 & 1 & 0 & 0 \\ 0 & 0 & 0 & 1 \end{bmatrix} \right)$$

This is the state space model of an actual automotive powertrain system (Cho and Paolella, 1990). The four state variables are engine speed, torque-induced turbine speed, driving axle torque (sum of both sides), and wheel rotation speed. The control input is an engine-induced torque. This example will be used throughout the rest of this section.

We will use Algorithm 7.2 to design all six robust fault detectors, according to Table 7.3. We first set a common matrix $F = \text{diag}\{-10 \;\; -20.7322\}$ for all six fault detectors.

Step 1: For each eigenvalue λ_i of matrix F, $i = 1, 2$, the two matrices D_i common to all six robust fault detectors are computed for the third column (corresponding to the zero column of matrix C) of (7.14a)

$$D_{i1} = \begin{bmatrix} 0.3587 & 0.8713 & 0.0002 & 0.3348 \\ -0.0005 & 0.0002 & 1 & -0.0005 \\ -0.9334 & 0.3348 & -0.0005 & 0.1287 \end{bmatrix}, \quad i = 1, \ldots, 6$$

and

$$D_{i2} = \begin{bmatrix} 0.1741 & 0.9697 & 0 & 0.1715 \\ -0.0003 & 0 & 1 & -0.0003 \\ -0.9847 & 0.1715 & -0.0003 & 0.0303 \end{bmatrix}, \quad i = 1, \ldots, 6$$

Step 2: The two $\mathbf{c}_{ji}$ rows ($j = 1, 2$) for each of the six robust fault detectors ($i = 1, \ldots, 6$) are computed to satisfy (7.14c) according to Table 7.3 as

$$\mathbf{c}_{11} = \begin{bmatrix} 0 & -1 & 0.0006 \end{bmatrix}, \qquad \mathbf{c}_{12} = \begin{bmatrix} 0 & -1 & 0.0003 \end{bmatrix}$$

$$\mathbf{c}_{21} = \begin{bmatrix} 0.0015 & 1 & 0 \end{bmatrix}, \qquad \mathbf{c}_{22} = \begin{bmatrix} 0.0015 & 1 & 0 \end{bmatrix}$$

$$\mathbf{c}_{31} = \begin{bmatrix} 0.9334 & 0 & 0.3587 \end{bmatrix}, \qquad \mathbf{c}_{32} = \begin{bmatrix} 0.9847 & 0 & 0.1741 \end{bmatrix}$$

$$\mathbf{c}_{41} = \begin{bmatrix} -0.3587 & 0.0005 & 0.9334 \end{bmatrix}, \ \mathbf{c}_{42} = \begin{bmatrix} -0.1741 & 0.0003 & 0.9847 \end{bmatrix}$$

$$\mathbf{c}_{51} = \begin{bmatrix} -0.3587 & 0.0005 & 0.9334 \end{bmatrix}, \ \mathbf{c}_{52} = \begin{bmatrix} -0.1741 & 0.0003 & 0.9847 \end{bmatrix}$$

$$\text{and } \mathbf{c}_{61} = \begin{bmatrix} -0.3587 & 0.0005 & 0.9334 \end{bmatrix}, \mathbf{c}_{62} = \begin{bmatrix} -0.1741 & 0.0003 & 0.9847 \end{bmatrix}$$

The last three robust fault detectors share the same parameters and are redundant. Thus, these three robust fault detectors will be reduced to only one. In addition, the two $\mathbf{c}_{ji}\ D_{ji}$ (i = 1 and 2) vectors are the same for the second and fourth robust fault detectors. Thus, we will reduce these two vectors to only one ($\mathbf{c}_{21}D_{21}$ and $\mathbf{c}_{41}D_{41}$), and we will set matrices $F_j = -10$, $T_j = \mathbf{c}_{j1}\ D_{j1}$, $j = 2, 4$.

The T matrices are computed, and then the corresponding matrices L are computed from (7.14a) and for the m nonzero columns of system matrix C, for the remaining four robust fault detectors:

$$T_1 = \begin{bmatrix} 0 & 0 & 0.0006 & 1 \\ 0 & 0 & 0.0003 & 1 \end{bmatrix} \quad L_1 = \begin{bmatrix} 0 & 0.8529 & -29.091 \\ 0 & 0.3924 & 3.715 \end{bmatrix},$$

$$T_2 = \begin{bmatrix} 0 & 0.0015 & 0 & -1 \end{bmatrix} \quad L_2 = \begin{bmatrix} 0.1002 & -0.0842 & -9.9451 \end{bmatrix},$$

$$T_3 = \begin{bmatrix} 0 & 0.9334 & 0.3587 & 0 \\ 0 & 0.9847 & 0.1741 & 0 \end{bmatrix} \quad L_3 = \begin{bmatrix} 62 & 476 & -24{,}185 \\ 66 & 213 & -11{,}740 \end{bmatrix},$$

$$\text{and } \ T_4 = \begin{bmatrix} -1 & -0 & 0 & -0 \end{bmatrix} \quad L_4 = \begin{bmatrix} 10.95 & -17.35 & 0 \end{bmatrix}.$$

It can be verified that both (7.14a) and (7.14c) (based on Table 7.3) are satisfied.

Step 3: We solve (7.14b) for the remaining four robust fault detectors,

$$[\mathbf{n}_1 : \mathbf{m}_1] = \begin{bmatrix} 0.3753 & -0.8156 & : & 0 & 0 & 0.4403 \end{bmatrix}$$

$$[\mathbf{n}_2 : \mathbf{m}_2] = [\ 0.7071:\ 0 \quad -0.0011 \quad 0.7071\ ]$$

$$[\mathbf{n}_3 : \mathbf{m}_3] = [\ 0.394 \quad -0.8116\ :\ 0 \quad 0.4314 \quad 0\ ]$$

$$[\mathbf{n}_4 : \mathbf{m}_4] = [1:\ 1 \quad 0 \quad 0\ ]$$

It can be verified that (7.14b) is satisfied for these four-fault detectors.

However, condition (7.14d) is satisfied only by the first three T_i matrices, $i = 1, 2, 3$, while T_4 does not satisfy (7.14d) because it has three columns equal to zero (should have only $n-q$ (= 2) zero columns). That means the e_4 signal will remain 0 when three fault signals d_2, d_3, and d_4 (corresponding to those zero columns of T_4) become nonzero.

All design results of these four robust fault detectors are summarized in Table 7.4.

TABLE 7.4

Fault Isolation Capability of the Actual Design Result of Example 7.6

Fault Situations $\mathbf{d}(t) = [d_1\ d_2\ d_3\ d_4]'$	Residual Signals e	e_2	e_3	e_4	Logic Determination of Fault Situation Isolation
$d_1(t) \neq 0$	0	0	0	x	$(d_1(t) \neq 0) = e_4$
$d_2(t) \neq 0$	0	x	x	0	$(d_2(t) \neq 0) = e_2 \cap e_3$
$d_3(t) \neq 0$	x	0	x	0	$(d_3(t) \neq 0) = e_1 \cap e_3$
$d_4(t) \neq 0$	x	x	0	0	$(d_4(t) \neq 0) = e_1 \cap e_2$

In addition to the isolation of four single fault situations of Table 7.3, this result can also isolate the first three two-fault situations of Table 7.3: d_1 and d_2, d_1 and d_3, and d_1 and d_4. These three fault situations are uniquely identified by the residual patterns $e_2 \cap e_3 \cap e_4$, $e_1 \cap e_3 \cap e_4$, and $e_1 \cap e_2 \cap e_4$, respectively.

However, the other three two-fault situations (all corresponding to $d_1 = 0$) cannot be isolated and identified, because the corresponding four residual signals are all the same: e_1, e_2, and e_3 are all nonzero, and only e_4 is zero. This pattern of residual signals only tells that d_1 is zero, but cannot tell which one of the other three fault signals is also zero.

Nonetheless, it is already remarkable to be able to isolate up to seven different fault situations, for this four-fault and three-measurement system.

In general, the value of q should be chosen to be near $Q/2$ so that the number of combinations of $[Q: q]$ is maximized, or q is maximized (as long as $q<m$), even though not all $[Q: q]$ robust fault detectors can be designed, like Example 7.6.

7.2.3 Adaptive Fault Control and Accommodation (Tsui, 1997)

As the purpose of diagnosing a disease is for treatment and cure of the disease, the purpose of fault detection and isolation is to control the identified fault situation.

The fault control signal of this book is not only based on the information of fault isolation and identification, but also based on the $k = [n-m: q]$ fault detector states $\mathbf{z}_i(t)$, $i = 1, \ldots, k$ and plant system output measurement $\underline{\mathbf{y}}(t)$. The control signal is a constant gain $[Kz: Ky]$ on these signals, just like the feedback control signals of (3.16b) of other chapters. This is because $\mathbf{y}(t) = C\mathbf{x}(t)$ and $\mathbf{z}_i(t) = T_i\mathbf{x}(t)$ just before fault occurrence, which are linear combinations of system state $\mathbf{x}(t)$ already.

Hence, the fault control of this book is based on exceptionally rich information and is therefore very effective. In addition, the constant control gains are usually much easier to design and realize.

Furthermore, this kind of control gain $[Kz: K_y]$ can be pre-designed based on the understanding of all different possible fault situations, and a specific gain of $[Kz: K_y]$ is prepared for each specific fault situation. The specific gain will be automatically and instantly switched on, as soon as its corresponding fault situation is identified.

Hence, the fault control of this book is adaptive to each of the detected fault situations and is therefore very focused and very timely, and thus very effective.

For example, if we know a-priori that under a certain fault situation what state variable is failed and unreliable, and since we know a-priori what fault detector state $\mathbf{z}_i(t)$ and what element of output $\mathbf{y}(t)$ are relatively more influenced by those failed system states, then we will design that fault situation's corresponding control gain $[Kz: K_y]$ not based on those unreliable state signals.

Example 7.7

Based on Table 7.3 of Example 7.5, there are ten identifiable fault situations labeled from S_1 to S_{10}. For each of these ten situations, the unfailed states and the control signals that include only those unfailed states are listed in the second and third columns of the following Table 7.5, respectively.

TABLE 7.5

Adaptive Control Signal for the Ten Identifiable Fault Situations of Example 7.5

Fault Situations	Unfailed System States	Adaptive Control Signal
S_1: $d_1(t) \neq 0$	$x_2(t)$, $x_3(t)$, $x_4(t)$	$K_1\,[\mathbf{z}_1(t)':\mathbf{z}_2(t)':\mathbf{z}_3(t)':\underline{\mathbf{y}}_1(t)']'$
S_2: $d_2(t) \neq 0$	$x_1(t)$, $x_3(t)$, $x_4(t)$	$K_2\,[\mathbf{z}_1(t)':\mathbf{z}_4(t)':\mathbf{z}_5(t)':\underline{\mathbf{y}}_2(t)']'$
S_3: $d_3(t) \neq 0$	$x_1(t)$, $x_2(t)$, $x_4(t)$	$K_3\,[\mathbf{z}_2(t)':\mathbf{z}_4(t)':\mathbf{z}_6(t)':\underline{\mathbf{y}}_3(t)']'$
S_4: $d_4(t) \neq 0$	$x_1(t)$, $x_2(t)$, $x_3(t)$	$K_4\,[\mathbf{z}_3(t)':\mathbf{z}_5(t)':\mathbf{z}_6(t)':\underline{\mathbf{y}}_4(t)']'$
S_5: $d_1(t) \neq 0$ and $d_2(t) \neq 0$	$x_3(t)$, $x_4(t)$	$K_5\,[\mathbf{z}_1(t)':\underline{\mathbf{y}}_5(t)']'$
S_6: $d_1(t) \neq 0$ and $d_3(t) \neq 0$	$x_2(t)$, $x_4(t)$	$K_6\,[\mathbf{z}_2(t)':\underline{\mathbf{y}}_6(t)']'$
S_7: $d_1(t) \neq 0$ and $d_4(t) \neq 0$	$x_2(t)$, $x_3(t)$	$K_7\,[\mathbf{z}_3(t)':\underline{\mathbf{y}}_7(t)']'$
S_8: $d_2(t) \neq 0$ and $d_3(t) \neq 0$	$x_1(t)$, $x_4(t)$	$K_8\,[\mathbf{z}_4(t)':\underline{\mathbf{y}}_8(t)']'$
S_9: $d_2(t) \neq 0$ and $d_4(t) \neq 0$	$x_1(t)$, $x_3(t)$	$K_9\,[\mathbf{z}_5(t)':\underline{\mathbf{y}}_9(t)']'$
S_{10}: $d_3(t) \neq 0$ and $d_4(t) \neq 0$	$x_1(t)$, $x_2(t)$	$K_{10}[\mathbf{z}_6(t)':\underline{\mathbf{y}}_{10}(t)']'$

In Table 7.5, each of the feedback control gains K_i ($\equiv [Kz_i : Ky_i]$) is designed based on the understanding of the corresponding fault situation S_i, $i = 1, \ldots, 10$. The signal $\underline{\mathbf{y}}_i(t)$ consists of signals of $\mathbf{y}(t)$ that are not related to the failed states (determined by matrix C) of fault situation S_i, $i = 1, \ldots, 10$. The actual design of K_i will be introduced in Chapters 8–10.

In actual practice, all six robust fault detectors will run simultaneously, and their states equal $T_i\mathbf{x}(t)$, $i = 1, \ldots, 6$. Once a fault situation S_i is detected and identified, the corresponding control signal with gain K_i, $i = 1, \ldots, 10$, according to Table 7.5, will be switched on automatically.

In case not all $[Q: q]$ robust fault detectors can be designed (see Example 7.6), then the above Table 7.5 will be modified as follows.

Example 7.8: Four General Rules of Modification of Adaptive Fault Control of This Book

We will use the actual design results of Example 7.6 to illustrate the four general rules for modifying the adaptive fault control listed in Table 7.5 of Example 7.7.

First, the last two robust fault detectors will not be available, and their states $\mathbf{z}_5(t)$ and $\mathbf{z}_6(t)$ will also be removed from Table 7.5. Second, the last three fault situations of Table 7.5 cannot be isolated and identified and hence cannot be adaptively controlled either.

Now based on these two modifications, the $\underline{\mathbf{y}}_i(t)$ signals for the remaining seven identifiable fault situations will be:

$\underline{\mathbf{y}}_1(t)$	$\underline{\mathbf{y}}_2(t)$	$\underline{\mathbf{y}}_3(t)$	$\underline{\mathbf{y}}_4(t)$	$\underline{\mathbf{y}}_5(t)$	$\underline{\mathbf{y}}_6(t)$	$\underline{\mathbf{y}}_7(t)$
$[y_2(t)\ y_3(t)]'$	$[y_1(t)\ y_3(t)]'$	$\mathbf{y}(t)$	$[y_1(t)\ y_2(t)]'$	$y_3(t)$	$[y_2(t)\ y_3(t)]'$	$y_2(t)$

Third, among the signals used for feedback control for each fault situation S_i, we also want to use and list only the signals that are not linearly dependent (or redundant) on each other. If there are dependent and redundant signals among the output measurement signal $\underline{\mathbf{y}}_i(t)$ and the fault detector states $\mathbf{z}_j(t)$, then the $\underline{\mathbf{y}}_i$ signals will be used instead of the signals of $\mathbf{z}_j(t)$, because $\underline{\mathbf{y}}_i(t)$ is more reliable than the $\mathbf{z}_j(t)$ signals.

For example, $\mathbf{z}_4(t) = T_4\,\mathbf{x}(t) = -x_1(t)$ is redundant with $y_1(t) = x_1(t)$, and thus will not be used for feedback if $y_1(t)$ is used.

Fourth and finally, if there are enough unfailed plant system states available for feedback control, then an additional modification can be made to Table 7.5 as follows. This modification is more elaborate than the first three modifications but can be very effective and very important.

Among the unfailed plant system states some may be more strongly influenced than the others, by the failed states. These unfailed plant states are therefore less reliable than the others for fault control and thus may be treated like the failed states and be removed from Table 7.5.

The strength of coupling influence between the plant system states is very clearly indicated by the magnitude of the corresponding element of plant system state matrix A. *For example,* $|a_{ij}|$ indicates how strongly state x_j influences state x_i.

Let us apply all above four modifications to the real design of Example 7.6, in the following.

Example 7.9

The state matrix of the plant system of Example 7.6 is

$$A = \begin{bmatrix} -20.95 & 17.35 & 0 & 0 \\ 66.53 & -65.89 & -3.843 & 0 \\ 0 & 1437 & 0 & -67,420 \\ 0 & 0 & -0.00578 & -0.0548 \end{bmatrix}$$

Matrix A indicates that state x_3 is very strongly influenced by state x_4 because $|a_{34}| = 67{,}420$ is large, yet x_4 is very weakly influenced by state x_3 because $|a_{43}| = 0.00578$ is small.

Let us arbitrarily set 10 as the threshold for $|a_{ij}|$ to be large or small. This threshold means that $|a_{12}|$ = 17.35 is large, while $|a_{23}|$ = 3.84 is small.

The completely modified Table 7.5 is shown in the following Table 7.6.

TABLE 7.6

Fault Accommodation Control of Example 7.6

Fault Situations	Unfailed System States	States Weakly Influenced by the Failed States	Adaptive Control Signal
S_1: $d_1(t) \neq 0$	x_2, x_3, x_4	x_3, x_4	$K_1\,[\mathbf{z}_1(t)' \vdots y_3(t)]'$
S_2: $d_2(t) \neq 0$	x_1, x_3, x_4	x_4	$K_2\,[y_3(t)]$
S_3: $d_3(t) \neq 0$	x_1, x_2, x_4	x_1, x_2, x_4	$K_3\,[\mathbf{y}(t)]$
S_4: $d_4(t) \neq 0$	x_1, x_2, x_3	x_1, x_2	$K_4\,[y_1(t)\; y_2(t)]'$
S_5: $d_1 \neq 0$ and $d_2 \neq 0$	x_3, x_4	x_4	$K_5\,[y_3(t)]$
S_6: $d_1 \neq 0$ and $d_3 \neq 0$	x_2, x_4	x_4	$K_6\,[y_3(t)]$
S_7: $d_1 {\neq} 0$ and $d_4 \neq 0$	x_2, x_3	x_2	$K_7\,[y_2(t)]$

For fault situation S_7, state x_2 is still considered weakly influenced by the failed states, even though the influence from the failed state x_1 is considered strong ($|a_{21}|$ = 66.53>10). This is because the only other unfailed state x_3 is even much more strongly influenced by a failed state x_4 ($|a_{34}| = 67{,}420$).

7.2.4 The Treatment of Model Uncertainty and Measurement Noise (Tsui, 1994b)

In the previous three subsections, a complete fault detection, isolation, and accommodation (DIA) control system is established. This subsection discusses the effect and the corresponding treatment of plant system model uncertainty and measurement noise. A threshold treatment will be proposed based on this analysis of these effects.

The effect of plant system model uncertainty and measurement noise (see Section 3.1) is studied for the overall combined feedback system of the observer system (3.16) and fault DIA system (7.13). The first system is for normal (fault-free) control and the second system is for accidental fault control. It is striking that these two systems (3.16) and (7.13) are very similar structure wise, and that these two systems implement the same form of $[K_z{:}K_y]$ $[T'{:}C']'\mathbf{x}(t)$-control.

Let the normal observer/output feedback commentator (3.16) parameters be denoted as $(F_0, T_0, L_0, K_z, K_y)$, and let the actual k robust fault detector (7.13) parameters be $(F_i, T_i, L_i, K_i, i = 1, \ldots, k)$. The combined feedback control system is depicted in Figure 7.1.

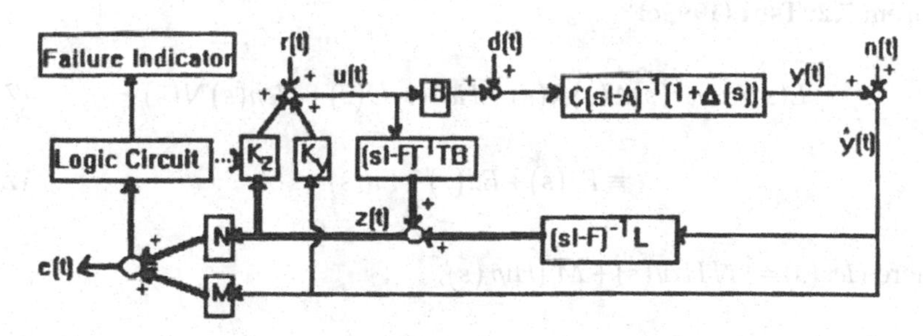

FIGURE 7.1
Block diagram of a combined observer and fault DIA feedback system.

In Figure 7.1, $\mathbf{r}(t) \equiv$ external reference signal with its Laplace transform $R(s)$,
$\mathbf{d}(t) \equiv$ fault signal with its Laplace transform $D(s)$,
$\mathbf{n}(t) \equiv$ measurement noise signal with its Laplace transform $N(s)$ and with $\bar{N}$ as the upper bound of its rms value,
$C(sI-A)^{-1}\Delta(s) \equiv$ plant system model uncertainty with $\bar{\Delta}$ as the upper bound of the scalar function $\Delta(s)$.
$\mathbf{e}(t) \equiv [e_1(t) \ \ldots \ e_k(t)]'$ is the residual signal with Laplace transform $E(s)$,
$\mathbf{z}(t) \equiv [\mathbf{z}_0(t)' : \mathbf{z}_1(t)' : \ldots : \mathbf{z}_k(t)']'$ with Laplace transform $Z(s)$,
$F \equiv \text{diag}\{F_0, F_1, \ldots, F_k\}$,
$N \equiv \text{diag}\{\mathbf{n}_1, \ldots, \mathbf{n}_k\}$, $(\mathbf{n}_0 = 0)$, and

$$T \triangleq \begin{bmatrix} T_0 \\ T_1 \\ \vdots \\ T_k \end{bmatrix}, L \triangleq \begin{bmatrix} L_0 \\ L_1 \\ \vdots \\ L_k \end{bmatrix}, M = \begin{bmatrix} \mathbf{m}_1 \\ \vdots \\ \mathbf{m}_k \end{bmatrix} (\mathbf{m}_0 \triangleq 0)$$

At normal operation before faults occurred and detected, the control signals are $[K_z : K_y][\mathbf{z}_0(t)' : \mathbf{y}(t)']' = [K_z : K_y]\,[T_0' : C']'\mathbf{x}(t)$. At fault situation S_i, the control signal will be switched to K_i[selected signals of $\mathbf{z}(t)$ and $\mathbf{y}(t)$ for S_i]', $i = 1, 2, \ldots$, (see Subsection 7.2.3).

Fault detection and isolation are achieved based on the zero and nonzero pattern of the residual signals $E(s)$, see Tables 7.2–7.4. Therefore, the effect of model uncertainty $C(sI-A)^{-1}\Delta(s)$ and measurement noise $N(s)$ is on the zero and nonzero determination of the $E(s)$ signals.

To analyze this effect, we need to establish the transfer function relationship among signals $R(s)$, $D(s)$, and $N(s)$ to $E(s)$, before a fault situation is detected.

Theorem 7.2: Tsui (1993c)

$$E(s) = Her(s)\Delta(s)R(s) + Hed(s)D(s) + Hen(s)N(s) \tag{7.15}$$

$$\equiv Er(s) + Ed(s) + En(s) \tag{7.16}$$

where $Her(s) = \left[N\,Hzy(s) + M\right]Hyr(s)$,

$$Hed(s) = \left[N\,Hzy(s) + M\right]Hyd(s),$$

$$Hen(s) = \left[N\,Hzy(s) + M\right]Hyn(s)$$

where $Hyr(s) = \left[I - Go(s)B\left(K_z Hzy(s) + K_y\right)\right]^{-1} \times Go(s)B\left(K_z Hzr(s) + 1\right)$

$$Hyd(s) = \left[I - Go(s)B\left(K_z Hzy(s) + K_y\right)\right]^{-1} \times Go(s)$$

$$Hyn(s) = \left[I - Go(s)B\left(K_z Hzy(s) + K_y\right)\right]^{-1}$$

And where (and because $T_0 B = 0$),

$$Hzyi(s) = \begin{cases} Gy_0(s) = (sI - F_0)^{-1} L_0, & \text{if } i = 0 \\ Gui(s) K_Z Gy_0(s) + Gui(s) K_y + Gyi(s), & \text{if } i \neq 0 \end{cases}$$

$$Hzr(s) = \left[I - Gu(s) K_z \right]^{-1} Gu(s),$$

and where $Gu(s) = (sI - F)^{-1} TB$, $Gy(s) = (sI - F)^{-1} L$, and $Go(s) = C(sI - A)^{-1}$

Proof

The proof or the derivation is drawn directly from Figure 7.1. It can also be found from the work of Tsui (1993c) and at the first two editions of this book (1996, 2004).

Theorem 7.2 shows that the effects of faults $D(s)$, of system model uncertainty $\Delta(s)$ (through the effect of system's reference input $R(s)$), and of measurement noise $N(s)$, on residual signal $E(s)$ are in separate terms $Ed(s)$, $Er(s)$, and $En(s)$. The harmful effect of $\Delta(s)$ and $N(s)$ is that, when fault is free ($D(s) = Ed(s) = 0$), the residual signal $Er(s) + En(s)$ can still be nonzero and thus can be misinterpreted as the fault occurrence and cause a false alarm.

The simple and practical way to prevent this false alarm is to set a threshold J_{th} on the actual $E(s)$. This threshold should be the estimated maximum value of $Er(s) + En(s)$. Only when $E(s)$ exceeds this threshold, then this nonzero $E(s)$ can be treated as caused by fault $D(s)$ instead of by model uncertainty $\Delta(s)$ and measurement noise $N(s)$.

This is like people usually use body temperature of 100°F as a threshold, to determine if a person is really sick or not. The body temperature above normal but still below 100° can be treated as not caused by sickness (like faults) but by other minor causes instead.

Let us establish this threshold J_{th}. From (7.16) of Theorem 7.2, when fault $D(s) = 0$,

$$E(s) = Er(s) + En(s) = Her(s)\Delta(s)R(s) + Hen(s)N(s) \tag{7.17}$$

$$\text{Hence, } J_{th} = \max\|Er(s) + En(s)\| \leq \max\|Er(j\omega)\| + \max\|En(j\omega)\|$$

$$\leq \max\{\sigma_1[Her(j\omega)]\|R(j\omega\|\}\,\bar{\Delta} + \max\{\sigma_1[Hen(j\omega)]\}\bar{N} \tag{7.18}$$

Although the threshold J_{th} of (7.18) is on $E(j\omega)$, which is in the frequency domain, its time domain equivalence can be derived using Parseval's theorem on the following value of $\mathbf{e}(t)$ over a window time length of τ (Emami-Naeini et al., 1988):

$$\|\mathbf{e}(t)\|\tau = \left[\left(\frac{1}{\tau}\right)\int_{t_0}^{t_0+\tau} \mathbf{e}(t)'\mathbf{e}(t)\,dt\right]^{1/2} \tag{7.19}$$

Only if this $\|\mathbf{e}(t)\|\tau$ is greater than J_{th}, can the nonzero residual $\mathbf{e}(t)$ be considered caused by fault $D(s)$ (or fault has occurred). In practice, the actual value of τ should be adjusted.

An extremely important improvement is to make this threshold for each of the k residual signals $e_i(t)$ $i = 1, \ldots, k$, instead of for the whole $\mathbf{e}(t)$. This treatment can greatly improve the resolution of the fault detection, by greatly reducing the threshold level from J_{th} to J_{thi}, $i = 1, \ldots, k$.

$$J_{thi} = \max\left\{\sigma_1\left[Her_i(j\omega)\right]\left\|R(j\omega)\right\|\right\}\bar{\Delta}$$
$$+\max\left\{\sigma_1\left[Hen_i(j\omega)\right]\right\}\bar{N},\ i = 1,\ldots,k \tag{7.20}$$

and $$e_{i\tau} = \left[\left(\frac{1}{\tau}\right)\int_{t_0}^{t_0+\tau} e_i^2(t)\,dt\right]^{1/2},\quad i = 1,\ldots,k \tag{7.21}$$

This improvement on the resolution by considering individual $e_i(t)$ instead of the whole $\mathbf{e}(t)$, is very much like improving the resolution of eigenvalue sensitivity by considering individual λ_i instead of the whole Λ. In addition, considering the sensitivity of the eigenvalues is much more focused than considering the whole system's sensitivity function. See Section 2.2. All of these are the key reasons for the superiority of the control analysis and design results, of this book.

After the threshold J_{th} (7.18) for the residual signals is established, it is useful to establish a theoretical lower bound for the fault signal $D(s)$, for guaranteed detection. Obviously, this lower bound of fault $D(s)$ must cause the residual signal level $\|E(s)\|$ to exceed J_{th}, or

$$\min\left\|E(s) = Er(s) + Ed(s) + En(s)\right\| \ge J_{th} \equiv \max\left\|Er(s) + En(s)\right\| \tag{7.22}$$

Theorem 7.3

For a sufficiently strong fault, such as that described by Emami-Naeini et al. (1988),

$$\min\left\|Ed(s)\right\| > \max\left\|Er(s) + En(s)\right\|, \tag{7.23}$$

the detectable fault $D(s)$ must satisfy

$$\|Ed(s) = Hed(s)D(s)\| > 2J_{th}/(1-\bar{\Delta}) \tag{7.24}$$

in order to be guaranteed detected.

Proof

From (7.23), $\min\|E(s) = Ed(s) + Er(s) + En(s)\| > \min\|Ed(s)\| - \max\|Er(s) + En(s)\|$

From (7.18), $\min\|E(s) = Ed(s) + Er(s) + En(s)\| > \min\|Ed(s)\| - J_{th}$

Therefore, (7.22) or $\min\|E(s) = Ed(s) + Er(s) + En(s)\| > J_{th}$ is guaranteed by

$$\min\|Ed(s)\| > 2J_{th} \tag{7.25}$$

In other words, $\min\|Ed(s)\| > 2J_{th}$ can guarantee $Ed(s)$ meet the detectable requirement (7.22).

Now from (7.16) and assuming $\bar{\Delta} < 1$,

$$\|Ed(s)\| = \|Hed(s)D(s) + \Delta(s)Hed(s)D(s)\| \geq (1-\bar{\Delta})\|Hed(s)D(s)\| \tag{7.26}$$

Combining (7.26) and (7.25), condition (7.24) is proved.

It should be emphasized again that the theoretical bounds of this subsection, derived from the norms of all transfer functions and signals and over all frequency ω, are understandably too conservative, not accurate enough and not practical enough, even though the result is relatively simple and general. For example, at the frequency ω_o when $\sigma_1(\text{Hen}(j\omega))$ of (7.18) is at the maximum, $N(j\omega_o)$ in reality is usually much less than its maximal level $\bar{N}$ of (7.18).

Exercises

7.1 Repeat Example 7.1 for a modified system of Example 6.1.

Let system matrix $C = \begin{bmatrix} 1 & 0 & 0 & : & 0 & 0 & 0 & 0 \\ 2 & 1 & 0 & : & 0 & 0 & 0 & 0 \\ 3 & 4 & 1 & : & 0 & 0 & 0 & 0 \end{bmatrix}$

Let matrices D_i ($i = 1, \ldots, 4$) of the four observer poles be the same as that of Example 6.1. The system has parameters $n = 7$, $m = 3$, $p = 2$, $\nu_1 = 3$, and $\nu_2 = \nu_3 = 2$.

a. $K = \begin{bmatrix} 1 & -1 & 0 & : & 1 & 2 & 3 & : & 1 \\ 0 & 0 & 1 & : & -4 & 3 & -2 & : & 0 \end{bmatrix}$

Answer: $r = 3$, $T = \begin{bmatrix} [\ 3 \quad 2 \quad 3\ ]D_1 \\ [\ -2 \quad 0 \quad 0\ ]D_2 \\ [\ 4 \quad 3 \quad -2\ ]D_3 \end{bmatrix}$,

$K_z = \begin{bmatrix} 1 & 1 & : & 0 \\ 0 & 2 & : & 1 \end{bmatrix}$, $K_y = \begin{bmatrix} 19 & -11 & 3 \\ -63 & 29 & -5 \end{bmatrix}$

b. $K = \begin{bmatrix} 1 & 0 & 0 & : & 1 & 2 & 3 & : & 1 \\ 0 & -1 & 1 & : & 2 & -4 & -6 & : & -4 \end{bmatrix}$

Answer: $r = 2$, $T = \begin{bmatrix} [\ 3 \quad 2 \quad 3\ ]D_1 \\ [\ -2 \quad 0 \quad 0\ ]D_2 \end{bmatrix}$, $K_z = \begin{bmatrix} 1 & 1 \\ -2 & -1 \end{bmatrix}$,

$K_y = \begin{bmatrix} 17 & -10 & 3 \\ -17 & 15 & -5 \end{bmatrix}$

c. $K = \begin{bmatrix} 11 & 7 & 1 & : & 0 & 0 & 0 & : & 0 \\ -6 & 2 & -4 & : & 6 & -2 & 4 & : & -6 \end{bmatrix}$

Answer: $r = 1$, $F = -1$, $T = [3 \ 1 \ -2]\ D_1$, $K_z = \begin{bmatrix} 0 \\ -2 \end{bmatrix}$,

$K_y = \begin{bmatrix} 2 & 3 & 1 \\ 0 & 0 & 0 \end{bmatrix}$

d. $K = \begin{bmatrix} 3 & 2 & -3 & : & -1 & -2/3 & 1 & : & 1/3 \\ 10 & 6 & 1 & : & 0 & 0 & 0 & : & 0 \end{bmatrix}$ and set matrix S of Step 1 of Algorithm 7.1 as $[\bar{D}_3' : \bar{D}_2' : \bar{D}_1' : \bar{K}']'$:

Answer: $r = 1$, $F = -3$, $T = [1 \ -2 \ 3]D_3$, $Kz = \begin{bmatrix} 1/3 \\ 0 \end{bmatrix}$,

$K_y = \begin{bmatrix} 0 & 0 & 0 \\ 3 & 2 & 1 \end{bmatrix}$

To the professors teaching this book: From your experience, can so many variations of Example 7.1 and Exercise 7.1 be so simply constructed, without really solving this design problem?

7.2 Let us practice more on part (d) of Exercise 7.1. Because the matrix $\bar{K}$ of (7.1c) has only one nonzero row, the equivalent parameter $p = 1$, and Steps 1 and 2 of Algorithm 7.1 can finish the design of this problem. The observer order upper bound for $p = 1$ is $\nu_1 - 1 = 2$ (see Table 7.1), so that only two $\bar{D}_i$ matrices are needed. Let matrix S of Step 1 be $[\bar{D}_1' : \bar{D}_2' : \bar{K}']'$ instead, and then compute Steps 1 and 2.

Answer:

Step 1:

$$S\begin{bmatrix} 1 & 0 & 0 & 1 \\ 0 & 1 & 0 & 0 \\ 0 & 0 & 1 & 0 \\ 0 & 0 & 0 & 1 \end{bmatrix} = \left[\begin{array}{cccc} -1 & 0 & 0 & 0 \\ 0 & 1 & 0 & 0 \\ 0 & 0 & 1 & 0 \\ \hline -2 & 0 & 0 & -1 \\ 0 & 1 & 0 & 0 \\ 0 & 0 & 1 & 0 \\ \hline -1 & -2/3 & 1 & -2/3 \end{array}\right] = \bar{S}$$

Step 2:

$$r = 2,\ F = \text{diag}\{-1\ -2\},\ T = \begin{bmatrix} [-1\ -2\ \ 3]D_1 \\ [\ 2\ \ \ 0\ \ \ 0\ ]D_2 \end{bmatrix},\ \text{and } K_z = \begin{bmatrix} 1/3 & 1/3 \\ 0 & 0 \end{bmatrix}.$$

Notice: if we set matrix $S = [\bar{D}_3' \vdots \bar{D}_1' \vdots \bar{K}']'$ in Step 1 (a different sequence of the $\bar{D}_i$ matrices in matrix S), then $r = 1$ instead of 2.

7.3 Repeat Exercise 7.2 for the same problem except setting matrix $S = [\bar{D}_1' \vdots \bar{D}_3' \vdots \bar{K}']'$:

Answer:

Step 1:

$$S\begin{bmatrix} 1 & 0 & 0 & 1 \\ 0 & 1 & 0 & 0 \\ 0 & 0 & 1 & 0 \\ 0 & 0 & 0 & 1 \end{bmatrix} = \left[\begin{array}{cccc} -1 & 0 & 0 & 0 \\ 0 & 1 & 0 & 0 \\ 0 & 0 & 1 & 0 \\ \hline -3 & 0 & 0 & -2 \\ 0 & 1 & 0 & 0 \\ 0 & 0 & 1 & 0 \\ \hline -3/3 & -2/3 & 1 & -2/3 \end{array}\right] = \bar{S}$$

Step 2: A mechanic execution of this step would yield a similar result to Exercise 7.2 ($r = 2$). However, an extra careful examination of the fourth to sixth rows (or the three rows corresponding to D_3) of matrix $\bar{S}$ will find that they alone can express $\bar{K}$. That means the following different and better answer: $r = 1$, $F = -3$, $T = [1 \quad -2 \quad 3]$ $D_3 = [9 \quad 6 \quad -9 : -3 \quad -2 \quad 3 \quad 1]$, and $Kz = [1/3 \quad 0]'$.

Furthermore, if K or $\bar{K}$ has additional rows, then all three rows of D_1 can be relocated below the three rows of D_3, for those additional rows of K or $\underline{K}$. This is the modification mentioned between (7.4a) and (7.4b).

7.4 Let the system matrices and state feedback control law be

$$(A,B,C,K) = \left(\begin{bmatrix} 0 & 0 & 5 \\ 1 & 0 & -5 \\ 0 & 1 & -2 \end{bmatrix}, \begin{bmatrix} 1 \\ 0 \\ 0 \end{bmatrix}, \begin{bmatrix} 0 & 0 & 1 \end{bmatrix}, \frac{\begin{bmatrix} 0 & -5 & 3 \end{bmatrix}}{8} \right)$$

Use Algorithm 7.1 and the following specially assigned observer poles:

a. Let $K_y = 0$ and the observer poles be -5.25, -2.44, and -4:

Answer: $r = 2$ ($< n$), but will be 3 if the first two poles are not as specified.

b. Let $K_y \neq 0$ and the observer poles be $-5/3$ and $-10/3$:

Answer: $r = 1$ ($< n-m$), but will be 2 if the first pole is not as specified as $-5/3$.

7.5 Derive the state observer order and function observer order upper bounds of Table 7.1, for plant systems with $n = 500$, $m = 100$, $p = 2$, $\nu_1 = \ldots = \nu_{100} = 5$, and with $K_y = 0$:

Answer: State observer order $r = 500$; function observer order $r \leq 10$!

7.6 Derive the state observer order and function observer order upper bounds of Table 7.1, for plant systems with $n = 25$, $m = 6$, $p = 2$, 3, and 4 (so that p is still less than m and $p \times m < n$), $\nu_1 = 5$, $\nu_2 = \ldots = \nu_6 = 4$, and with $K_y \neq 0$. Compare another existing function observer order bound $p(\nu_1 - 1)$ (Kaileth, 1980; Chen, 1984):

p	$n-m$	$p(\nu_1-1)$	$(\nu_1-1) + \ldots + (\nu_p-1)$ is the lowest
2	19	8	7
3	19	12	10
4	19	16	13

7.7 Repeat Exercise 7.6 for plant systems with $n = 25$, $m = 5$, $p = 2$–5 (so that $p \times m$ is still not exceeding n), $\nu_1 = 9$, $\nu_2 = \ldots = \nu_5 = 4$, and with $K_y \neq 0$.

p	$n-m$	$p(\nu_1-1)$	$(\nu_1-1)+\ldots+(\nu_p-1)$ is the lowest
2	20	16	11
3	20	24	14
4	20	32	17
5	20	40	20

7.8 Let $n = 5$, $m = 4$, and $Q = 5$.

a. Design fault detection and isolation systems for $q = 1$ and 2, like Tables 7.2 and 7.3.

b. Repeat part (a) for $q = 3$. Compare results with that for $q = 2$. Notice that although the same number of robust fault detectors ([5:3] = [5:2] = 10) are designed, the logic function for fault situation identification will be different.

c. Design the fault control function of Table 7.5, for $q = 1$, 2, and 3.

d. If we need to identify a fault situation of four simultaneous faults, is this possible? Why?

7.9 Knowing that matrix F is in Jordan form, what is the sufficient condition to satisfy (7.14d)? Hint: Refer to Examples 1.4 and 1.5.

7.10 Repeat the designs of Examples 7.6 and 7.9, for a new set of observer poles -1 (and -2 if needed).

Partial answer: $T_1 = \begin{bmatrix} 0 & 0 & 0.0058 & 1 \\ 0 & 0 & 0.0029 & 1 \end{bmatrix}$,

$$\begin{bmatrix} T_2 = & 0 & 0.0015 & 0 & -1 \end{bmatrix},$$

$$T_3 = \begin{bmatrix} 0 & 0.2518 & 0.9678 & 0 \\ 0 & 0.4617 & 0.8871 & 0 \end{bmatrix},$$

$$\text{and } T_4 = \begin{bmatrix} -1 & 0 & 0 & 0 \end{bmatrix}.$$

See more problems especially Systems 1–3 of Appendix B.

8

Control Design for Eigenvalue Assignment

The new design principle of this book is divided into two major steps. The first step concerns the design of the dynamic part of the observer/compensator and is covered in Chapters 5 and 6. The second step, which is covered by Chapters 8–10, deals with the design of the output part of the observer/compensator, or the design of the generalized state feedback control $\underline{K}\underline{C}x(t)$ where parameter $\underline{C}$ (= $[T' \; C']'$) is already designed in the first major step.

According to Table 6.2, the generalized state feedback control $\underline{K}\underline{C}x(t)$ unifies direct state feedback control $Kx(t)$ (or rank $(\underline{C}) = n$) and static output feedback control $K_yCx(t)$ (rank $(\underline{C} = C) = m \leq n$). In the following chapters, the design methods for both controls will be presented.

The design of $\underline{K}\underline{C}x(t)$-control is aimed at the single overall feedback system state matrix A–$B\underline{K}\underline{C}$. Therefore, if the design of Chapter 6 is aimed at the realization of the feedback system loop transfer function and is based on the understanding of feedback systems of Chapter 3, then the design of $\underline{K}\underline{C}x(t)$-control is based on the understanding of single systems of Chapter 2. Although this design also determines fully the feedback system loop transfer function $\underline{K}\underline{C}(sI–A)^{-1}B$.

Among the existing design results of this control, the eigenvalue and eigenvector assignment will be presented in Chapters 8 and 9, respectively, and linear-quadratic optimal control will be presented in Chapter 10. It should be emphasized again that based on the analysis of Chapter 2, eigenvalues indicate far more accurately system performance than any other system parameters such as bandwidth, and the sensitivity of an eigenvalue is determined by its eigenvectors. Therefore, eigenvalue and eigenvector assignment can improve system performance and robustness far more directly and effectively, than other design objectives.

8.1 Eigenvalue (Pole) Selection

Although system poles most directly determine system performance, there are no uniquely and generally optimal rules for feedback system pole selection. This is because plant system conditions are very diverse and very complicated, and because the performance and robustness design requirements are contradictory to each other. Therefore, it is impossible to have general,

DOI: 10.1201/9781003259572-8

explicit, and really optimal pole selection rules for all system conditions and all design requirements, and it is also impossible to have a really good pole selection without trial and error.

Nonetheless, there are still some basic and general understandings about the relationship between the system poles and the system performance and robustness (see Figure 2.1 and Conclusion 2.2). The following six general and explicit pole selection rules are guided by these basic understandings (Truxal, 1955).

1. The more negative the real part of the poles, the faster the system transient response before reaches its steady state.
2. In regulator problems, it is often required that the zero-frequency response of the control system $T(s = 0)$ be a specific constant. For example, if the unit-step response $y(t)$ is required to reach a constant 1 as $t \rightarrow \infty$, then according to the final value theorem that $y(t \rightarrow \infty) = sY(s \rightarrow 0) = sT(s \rightarrow 0)/s = T(s \rightarrow 0) = 1$. This implies that if $T(s) = N(s)/D(s)$, then $N(s = 0) = D(s = 0)$ = product of the poles of $T(s)$.
3. From the understanding of root locus, the further away the feedback system poles from the loop transfer function poles, the higher the gain of the loop transfer function required to reach and place these feedback system poles. On the other hand, the higher the loop gain, the higher the chance of feedback system instability, and the higher the required control power, control disturbance, and the chance of control component failure.

   To summarize the first three rules. If rule (1) is concerned with system performance, then rules (2) and (3) are concerned with the fact that higher performance can cause higher sensitivity and lower robustness.
4. If the eigenvalues of a matrix differ too much in magnitude, then the difference between the largest and smallest singular values of that matrix will also differ too much (A.28). This implies the bad condition number and the high sensitivity of the eigenvalues, of that matrix.
5. Multiple eigenvalues can cause defective eigenvectors (5.15d), which are very sensitive to matrix parameter variation (see Golub and Wilkinson, 1976b) and which generally result in rough responses (see Example 2.1 and Figure 2.1). Therefore, multiple poles, even clustered poles, should generally be avoided.
6. For some optimal control systems in the sense of minimal "integral of time multiplied by absolute errors (ITAE)" (Graham and Lathrop, 1953) or in the sense of minimal "integral of quadratic errors (ISE)" (Chang, 1961), the feedback system poles are required to have a similar magnitude and have evenly distributed phase angles between 90° and 270°.

   To summarize, the last three rules (4)–(6) are about the relative positions of the eigenvalues, and these rules conform to each other.

Overall, the pole selection rules are neither exhaustive nor generally optimal. This is perhaps a true and reasonable reflection of the reality of practical engineering systems and imposes perhaps a challenge to real-world control engineers. These six rules are concerned more with the effectiveness and limitations of practical analog plant systems. In contrast, the selections of feedback compensator poles (see the beginning of Section 5.2) are more specifically and explicitly guided. The feedback compensators are usually digital and can therefore be made ideal and precise, while the analog systems usually cannot be ideal and precise.

8.2 Eigenvalue Assignment by State Feedback Control

The eigenvalue assignment by direct state feedback control $Kx(t)$ is presented in this section, while the eigenvalue assignment by generalized state feedback control $\underline{K}\underline{C}x(t)$ (rank $(\underline{C}) \leq n$) will be presented in the next section. The eigenvalue assignment will generally precede the eigenvector assignment (of Chapter 9), although the formulation of the eigenvector assignment and its modifications will also be presented in this chapter.

Let Λ be the Jordan form matrix that is formed by the eigenvalues selected in Section 8.1. Then the eigenstructure assignment problem can be formulated as (1.10)

$$(A - B\boldsymbol{K})V = V\Lambda \tag{8.1a}$$

$$\text{Or } AV - V\Lambda = BK(K = \boldsymbol{K}V) \tag{8.1b}$$

Let matrix F of (4.1) be the same Jordan form matrix Λ of (8.1) (in transpose) and be set to have the same dimension n, then Eq. (4.1), $TA - \Lambda T = LC$, becomes the dual of (8.1b). For example, in the transpose of (4.1), $A'T' - T'\Lambda = C'L'$, parameters A', T', C' and L' match parameters A, V, B, and K of (8.1b), respectively.

Therefore, we can use the dual version of Algorithm 5.3 to compute directly the solution (V, K) of (8.1b) as (T', L') of (4.1). Incidentally, Algorithm 5.3 and its dual version were published formally in the same year, by Tsui (1985) and Kautsky et al. (1985), respectively.

The only difference between these two solutions comes afterwards: solution K of (8.1b) must be post multiplied by V^{-1} to fit the solution $\boldsymbol{K}$ (= KV^{-1}) for the original problem (8.1a). However, the solution L of (4.1) does not need to be adjusted because the observer state matrix is F itself of (4.1) (instead of matrix A–LC).

This dual version of Algorithm 5.3 is presented in the following in a concise and top-down fashion, not as explicitly and as detailed, as the presentation of Algorithm 5.3 itself in Chapter 5.

Let Λ_i be an n_i-dimensional Jordan block of matrix Λ, and let the corresponding

$$V_i \equiv \left[\mathbf{v}_{i1} : \ \dots \ : \mathbf{v}_{i,ni}\right] \text{and } K_i \equiv \left[\mathbf{k}_{i1} : \ \dots \ : \mathbf{k}_{i,ni}\right]$$

be the blocks of V and K respectively corresponding to Λ_i, such that

$$AV_i - V_i\Lambda_i = BK_i \tag{8.2}$$

Using Kronecker product operator $\otimes$, (8.2) can be rewritten as

$$\left[I_{ni} \otimes A - \Lambda_i' \otimes I : -I_{ni} \otimes B\right]\mathbf{w}_i = 0 \tag{8.3a}$$

$$\text{where } \mathbf{w}_i = \left[\mathbf{v}_{i1}' : \ \dots \ : \mathbf{v}_{i,ni}' : \mathbf{k}_{i1}' : \ \dots \ : \mathbf{k}_{i,ni}'\right]' \tag{8.3b}$$

The dimension of matrix of (8.3a) is $(n_i n) \times (n_i (n + p))$. Also, because (system (A, B) is controllable, nonzero solution $\mathbf{w}_i$ of (8.13) always exists with $n_i p$ basis vectors.

For example, if the Jordan block size $n_i = 1$ (real distinct eigenvalue λ_i), then (8.3) becomes

$$[A - \lambda_i I : -B]\begin{bmatrix} \mathbf{v}_i \\ \mathbf{k}_i \end{bmatrix} = 0 \tag{8.4}$$

and the solution vector has p basis vectors. The free linear combination of these p basis vectors forms the eigenvector assignment freedom. Obviously, this freedom exists if $p > 1$ and is shown below

$$\mathbf{v}_i = \left[\boldsymbol{d}_{i1} : \ \dots \ : \boldsymbol{d}_{i,p}\right]\mathbf{c}_i \equiv D_i\mathbf{c}_i \tag{8.5}$$

where the basis vector matrix D_i is $n \times p$ dimensional, and the p-dimensional vector $\mathbf{c}_i$ is free. Vector $\mathbf{k}_i$ will be the same linear combination (same vector $\mathbf{c}_i$) of its own basis vectors.

In this book, like vector $\mathbf{t}_i$ of (4.1), vector $\mathbf{v}_i$ of (8.5) is decoupled from other $\mathbf{v}_j$ vectors ($j \neq i$), and the form of its freedom $\mathbf{c}_i$ is most simple and explicit. Only this form of eigenvectors and their freedom enabled the general, systematic, and effective eigenvector assignment algorithms of Kautsky et al. (1985). These algorithms will be introduced in Chapter 9.

Like the computation of Algorithm 5.3 is based on the block observable Hessenberg form of (5.5), the computation of (8.3) and (8.4) can also be greatly simplified if based on the block controllable Hessenger form:

$$[A : B] = \begin{bmatrix} A_{11} & A_{12} & \dots & \dots & A_{1\mu} & : & B_1 \\ B_2 & A_{22} & \dots & \dots & \vdots & : & 0 \\ 0 & B_3 & \dots & \dots & \vdots & : & 0 \\ \vdots & \ddots & \ddots & & \vdots & : & \vdots \\ 0 & \dots & 0 & B_\mu & A_{\mu\mu} & : & 0 \end{bmatrix} \tag{8.6}$$

where matrix blocks B_i ($i = 1, \ldots, \mu$) are in the upper echelon form, and μ is the largest controllability index of the system.

Definition 8.1

As the dual of the m observability indices of Definition 5.1, there are p controllability indices μ_j ($j = 1, \ldots, p$) of system (A, B). Each index μ_j is corresponding to one (the j-th) input, and

$$\mu_1 + \mu_2 + \cdots + \mu_p = n \tag{8.7}$$

The p basis vectors $\mathbf{d}_{ij}$ ($j = 1, \ldots, p$) of $\mathbf{v}_i$ corresponding to a single λ_i can be computed by back substitution, from each of the p linearly dependent columns corresponding to each of the j-th input of matrix A of (8.6).

In addition, as the dual of Conclusion 5.2, for a fixed value of j, any set of μ_j basis vectors among the total of n $\mathbf{d}_{ij}$ vectors (corresponding to a combination of μ_j eigenvalues out of a total of n eigenvalues) are also linearly independent (Theorem 9.1).

If matrix V is computed based on the form of (8.6), which is an orthonormal similarity transformation of the original system model (A, B), then (8.1b) becomes

$$HAH'V - V\Lambda = HBK \tag{8.8}$$

A comparison of (8.1b) and (8.8) indicates that solution matrix V of (8.8) must be adjusted to $H'V$ in order to be the solution V of the original (8.1b).

As stated following (8.1b), after the above adjustment of $V = H'V$, the new matrix V need to be used, to adjust the original feedback control gain $K = KV^{-1}$.

8.3 Eigenvalue Assignment by Generalized State Feedback Control

The generalized state feedback control gain is $\underline{K}\underline{C}$, where $\underline{K}$ is free to be designed and rank (given $\underline{C}$) $\equiv q \le n$, see Definition 4.2. The case for $q = n$ is equivalent to direct state feedback control and is presented already in the previous section. This section covers the case for $q < n$.

Condition rank $(\underline{C}) = q < n$ means that the state feedback control gain K is now restricted by $K = \underline{K}\underline{C}$, or K must be a linear combination of only q rows of matrix $\underline{C}$. Hence the design of this section is much more difficult than that of the previous section.

Let Λ be a Jordan form matrix with n desired eigenvalues of matrix $A–B\underline{K}\underline{C}$. Then from (1.10), the eigenstructure assignment problem can be expressed in the following dual equations

$$T(A-B\underline{K}\underline{C})=\Lambda T \tag{8.9a}$$

$$\text{and } (A-B\underline{K}\underline{C})V=V\Lambda \tag{8.9b}$$

where T and V ($TV=I$) are the respective left and right eigenvector matrices of $A–B\underline{K}\underline{C}$ corresponding to Λ.

This problem which can be formed by the above two equations, has a remarkable property that is not shared by the state feedback design problem (8.1b) or the observer design problem (4.1) – the duality. This is because the dual system matrix pair B and $\underline{C}$ both appeared in the formulation (8.9), while they do not both appear in either (8.1b) or (4.1).

The following algorithm (Tsui, 1999a) computes the solution to (8.9) with a unique and critical advantage, that the eigenvector assignment is enabled and is formulated in the simplest possible form as (8.5).

Algorithm 8.1: Eigenstructure Assignment by Generalized State Feedback Control (Tsui, 1999a, 2001)

This design algorithm computes the partial solution to (8.9a) and then (8.9b). These two partial solutions joined together can satisfy the whole (8.9a) and (8.9b), because (8.9a) and (8.9b) are redundant, as will be proved in Step 2 of this algorithm.

Before the two steps to satisfy (8.9a) and (8.9b), we need to partition matrix Λ into

$$\Lambda=\text{diag}\left\{\Lambda_{n-q},\Lambda_q\right\}$$

where Λ_{n-q} and Λ_q must contain only eigenvalues which are either real or in complex conjugate pairs. The dimensions of Λ_{n-q} and Λ_q are $n–q$ and q, respectively.

Step 1: Compute solution matrix T_{n-q} of the following equation

$$T_{n-q}A-\Lambda_{n-q}T_{n-q}=L\underline{C} \tag{8.10a}$$

$$\text{and rank}\left(\left[T'_{n-q}:\underline{C}'\right]'\right)=n \tag{8.10b}$$

This problem is the same as the reduced-order state observer design (Example 4.2), with q rows in matrix $\underline{C}$ instead of only m rows in matrix C. Hence, solution T_{n-q} to this problem of (8.10) always exists. Algorithm 5.3 can be used to execute this step.

In addition, T_{n-q} is the left eigenvector matrix of matrix $A - B\underline{KC}$ corresponding to Λ_{n-q}. These eigenvectors have assignment freedom if $q > 1$ and can be used to maximize the angles between the rows of matrix $[T_{n-q}{}':C']'$. The actual eigenvector assignment methods will be presented in Chapter 9.

In the special case of $n = q$ (direct state feedback control), $T_{n-q} = 0$ and Step 1 is avoided.

Step 2: Compute solution matrices V_q and K_q of the following equation pair

$$AV_q - V_q\Lambda_q = BK_q \tag{8.11a}$$

$$\text{and}\, T_{n-q}V_q = 0 \tag{8.11b}$$

For distinct and real eigenvalue λ_i of matrix Λ_q, (8.11a) and (8.11b) together become

$$\begin{bmatrix} A-\lambda_i I & -B \\ T_{n-q} & 0 \end{bmatrix}\begin{bmatrix} \mathbf{v}_i \\ \mathbf{k}_i \end{bmatrix} = 0 \tag{8.12}$$

where $\mathbf{v}_i$ and $\mathbf{k}_i$ are the column of matrices V_q and K_q respectively corresponding to λ_i. V_q is obviously the right eigenvector matrix of A–$B\underline{KC}$ corresponding to Λ_q.

Equation (8.11a and b) is a dual of (4.1) and (4.3), because matrix T_{n-q} of (8.11b) can be considered a new system matrix C and is dual to matrix B, and matrix V_q is dual to matrix T of (4.1) and (4.3) (Section A.5). Therefore, the solution of (4.1) and (4.3) is the dual of the solution of (8.11a and b), as will be revealed in the following.

The dimension of the matrix of (8.12) is $(2n-q) \times (n+p)$. Therefore, nonzero solution of (8.12) exists if $n-q < p$, or $n < p + q$. This is the sufficient condition for arbitrary pole assignment (Kimura, 1975) and partial $(n-q)$ eigenvector assignment at Step 1 (Tsui, 1999a) of the generalized state feedback control.

For systems with $n \geq p + q$, another sufficient condition for the solution of (8.12) to exist is that λ_i equals a transmission zero of system (A, B, T_{n-q}), see (1.18) and Conclusion 6.1.

Two modifications of this design algorithm are raised from this analysis, and for systems with $n \geq p + q$. The first is to design eigenvectors of T_{n-q} in Step 1 to make system (A, B, T_{n-q}) to have q transmission zeros equal the q eigenvalues of Λ_q, or to make eigenvector assignment of T_{n-q} to make an eigenvalue assignment of Λ_q possible. The second is to choose the eigenvalues of Λ_q to be transmission zeros of system (A, B, T_{n-q}) or to make these eigenvalues assignable. These two modifications are discussed in more detail in Section 8.4.

For systems with $n < p + q-1$, there will be $p + q - n$ basis vectors for the right eigenvector $\mathbf{v}_i$ at Step 2. This eigenvector assignment freedom is generally recommended to make matrix $\underline{C}V_q$ as well-conditioned as possible, because the inverse of this matrix will be used in Step 3.

In the special case of $n = q$, $n < p + q$ and arbitrary pole assignment are guaranteed. In addition, $n < p + q - 1$ and eigenvector assignment freedom are also guaranteed if $p > 1$. These conclusions all conform to the existing conclusions of direct state feedback control of Subsection 3.2.1 and Section 8.2.

Step 3: Finally, comparing (8.9b) and (8.11a),

$$\underline{K} = K_q\left(\underline{C}V_q\right)^{-1} \tag{8.13}$$

Here, the nonsingularity of matrix $\underline{C}V_q$ is guaranteed by rank $([T_{n-q}{}' \vdots \underline{C}']') = n$ and $T_{n-q}V_q = 0$. See proof in Example A.7.

Because problem (8.9) is a dual pair, the dual of Algorithm 8.1 is presented below.

Algorithm 8.2: The Dual Version of Algorithm 8.1 (Tsui, 1999a)

Before the two steps to satisfy (8.9b) and (8.9a), we need to partition matrix Λ into

$$\Lambda = \text{diag}\left\{\Lambda_{n-p}, \Lambda_p\right\} \tag{8.14}$$

where Λ_{n-p} or Λ_p contains only eigenvalues that are either real or in complex conjugate pairs.

Step 1: Compute solution matrix V_{n-p} of the following equation pair

$$AV_{n-p} - V_{n-p}\Lambda_{n-p} = BK_p \tag{8.15a}$$

$$\text{and}\,\text{rank}\left(\left[V_{n-p} : B\right]\right) = n \tag{8.15b}$$

The solution to (8.15) always exists because (8.15) is dual to (8.10) of Algorithm 8.1. Matrix V_{n-p} is the right eigenvector matrix of matrix $A-B\underline{KC}$ corresponding to Λ_{n-p}. These eigenvectors have assignment freedom if $p > 1$. The actual eigenvector assignment methods will be presented in Chapter 9, to maximize the angles between the columns of matrix $[V_{n-p}{:}B]$.

In the special case of $n = p$, $V_{n-p} = 0$ and Step 1 is avoided.

Step 2: Compute solution matrices T_p and L_p of the following equation pair

$$T_pA - \Lambda_pT_p = L_p\underline{C} \tag{8.16a}$$

$$\text{and}\,T_pV_{n-p} = 0 \tag{8.16b}$$

Matrix T_p is obviously the left eigenvector matrix of $A-B\underline{KC}$ corresponding to Λ_p.

Matrix V_{n-p} of (8.15) and (8.16b) can be considered a new system matrix B. The computation algorithms for (4.1) and (4.3) can be used to compute the solution of (8.16).

The nonzero solution of (8.16) exists if $n–p < q$ or $n < p + q$. This is the sufficient condition for arbitrary pole assignment (Kimura, 1975) and partial $(n–p)$ eigenvector assignment at Step 1 (Tsui, 1999a) of the generalized state feedback control.

For systems with $n \geq p + q$, another sufficient condition for the nonzero solution of (8.16) to exist is that λ_i equals a transmission zero of system $(A, V_{n-p}, \underline{C})$, see (1.18) and Conclusion 6.1.

For systems with $n < p + q–1$, there will be $p + q - n$ basis vectors for each left eigenvector of T_p. This eigenvector assignment freedom is generally recommended to make matrix T_pB as well-conditioned as possible, because the inverse of this matrix will be used in Step 3.

Step 3: Finally, comparing (8.9a) and (8.16a),

$$\underline{K} = (T_pB)^{-1} L_p \tag{8.17}$$

Here, the nonsingularity of matrix T_pB is guaranteed by rank $([V_{n-p} \vdots B]) = n$ and $T_pV_{n-p} = 0$. □

Algorithms 8.1 and 8.2 are not redundant at all. Because parameters q and p may differ from each other, these two algorithms may complement each other to resolve some otherwise unsolvable problems.

For example, if $p + q = n + 1$, then reducing either p or q (to make either p or q even) will make the key condition $p + q > n$ unsatisfied. Now if all assigned eigenvalues are complex conjugates (n is even), then either the pre-step of Algorithm 8.1 cannot be implemented if q is odd, or the pre-step of Algorithm 8.2 cannot be implemented if p is odd, because complex conjugate pairs cannot break apart. This difficulty has been studied for years since the study of Kimura (1975) without a simple solution (Fletcher and Magni, 1987; Magni, 1987; Rosenthal and Wang, 1992).

However, in this situation of $p + q = n + 1$ and n is even, either one of the two parameters p and q will be even! Thus, this difficulty can be directly resolved by using Algorithm 8.1 if q is even, or by using Algorithm 8.2 if p is even.

Overall, if parameter p is fixed and usually is, then the increase of parameter $q = \text{rank}\,(\underline{C})$ can much improve the power of generalized state feedback control such as its pole assignment capability. The synthesized design principle of this book increases rank $(\underline{C})$ from m to $m + r$, while guaranteeing the full realization of the loop transfer function and robustness of this $\underline{KC}\mathbf{x}(t)$ control.

Example 8.1: Using Algorithm 8.1 to Assign Eigenvalues –1, –2, and –3 to a System of Chu (1993a)

$$\text{Let } (A, B, \underline{C}) = \left(\begin{bmatrix} -4 & 0 & -2 \\ 0 & 0 & 1 \\ 1 & -1 & -2 \end{bmatrix}, \begin{bmatrix} 4 & 2 \\ 0 & -2 \\ 0 & 1 \end{bmatrix}, \begin{bmatrix} 0 & 1 & 0 \\ 0 & 0 & 1 \end{bmatrix} \right)$$

First set $\Lambda_{n-q} = -3$ and $\Lambda_q = \text{diag}\{-2, -1\}$.

Step 1: The q (=2) basis vectors of T_{n-q} are computed from the first column of (8.10a)

$$D_1 = \begin{bmatrix} 1 & 0 & 1 \\ 0 & 1 & 0 \end{bmatrix}$$

Any linear combination of these two rows of D_1 satisfies the first column of (8.10a), and the other two columns can be satisfied by the free parameter L. Thus (8.10a) is satisfied.

The linear combination of rows of D_1 is recommended to maximize the angles between T_{n-q} and the rows of $\underline{C}$. We however simply let $T_{n-q} = [1\ 0]\ D_1 = [1\ 0\ 1]$. It can be verified that (8.10b) is also satisfied.

Step 2: Because $p + q = n + 1$, the solution of this step is unique. By substituting each of the two remaining eigenvalues –2 and –1 into (8.11a), the corresponding two columns of V_q and K_q are computed as

$$V_q = \begin{bmatrix} 0 & 1 \\ -1 & 3 \\ 0 & -1 \end{bmatrix} \quad \text{and} \quad K_q = \begin{bmatrix} -1/2 & 1/4 \\ 1 & -1 \end{bmatrix}$$

It can be verified that both (8.11a) and (8.11b) are satisfied.

Step 3: From (8.13), $\underline{K} = K_q\left(\underline{C}V_q\right)^{-1} = \begin{bmatrix} 1/2 & 5/4 \\ -1 & -2 \end{bmatrix}$.

The corresponding matrix A–$B\underline{KC}$ has the desired eigenvalues –3, –2, –1.

Example 8.2: Repeat Example 8.1 Using Algorithm 8.2

$$\left(A, B, \underline{C}\right) = \left(\begin{bmatrix} -4 & 0 & -2 \\ 0 & 0 & 1 \\ 1 & -1 & -2 \end{bmatrix}, \begin{bmatrix} 4 & 2 \\ 0 & -2 \\ 0 & 1 \end{bmatrix}, \begin{bmatrix} 0 & 1 & 0 \\ 0 & 0 & 1 \end{bmatrix} \right)$$

First set $\Lambda_{n-p} = -3$ and $\Lambda_p = \text{diag}\{-2, -1\}$.

Step 1: The p (= 2) basis column vectors of V_{n-p} and K_{n-p} are computed from (8.15a)

$$D_1 = \begin{bmatrix} 4 & -3 \\ 1 & 0 \\ -3 & 2 \\ \cdots & \cdots \\ -1/2 & -1/4 \\ 0 & 1 \end{bmatrix}$$

Any linear combination of these two columns of D_1 can satisfy (8.15a). This linear combination is recommended to maximize the angles between V_{n-p} and the columns of B. We however simply let $V_{n-q} = D_1\,[1\ 1]' = [1\ 1\ {-1}]'$. It can be verified that (8.15b) is also satisfied.

Step 2: Because $p + q = n + 1$, the solution of this step is unique. By substituting each of the two remaining eigenvalues –2 and –1 into (8.16a), the corresponding two rows of T_p and L_p are computed as

$$T_p = \begin{bmatrix} 1 & 1 & 2 \\ 1 & 2 & 3 \end{bmatrix} \quad \text{and} \quad L_p = \begin{bmatrix} 0 & 1 \\ 1 & 3 \end{bmatrix}$$

It can be verified that both (8.16a) and (8.16b) are satisfied.

Step 3: From (8.17), $\underline{K} = \left(T_p B\right)^{-1} L_p = \begin{bmatrix} 1/2 & 5/4 \\ -1 & -2 \end{bmatrix}$.

This is the same correct answer of Algorithm 8.1.

Example 8.3: Complete Design of an Output Feedback Compensator (Chapters 5–8)

Let a third order, two-input, and one-output system matrix be

$$(A,B,C) = \left(\begin{bmatrix} 0 & 1 & 0 \\ 0 & 0 & 1 \\ 0 & 0 & 0 \end{bmatrix}, \begin{bmatrix} 1 & 0 \\ 3 & 1 \\ 2 & 1 \end{bmatrix}, \begin{bmatrix} 1 & 0 & 0 \end{bmatrix} \right)$$

From (1.9), the transfer function of this system is $G(s) = s^{-3}\,[s^2 + 3s + 2 \quad s + 1]$.

This is a very difficult system to design. Its 2×2 loop transfer function is very difficult even to analyze using the classical control theory. Because $m < p$, this system cannot have an unknown input observer (see Subsection 4.3.5) and thus cannot fully realize the loop transfer function of direct state feedback control based on the existing modern control theory and separation principle.

However, this system has one stable transmission zero –1. Hence, based on Conclusion 6.1 of this book, an output feedback compensator with one pole equal to this –1 can be designed with $F = -1$.

A) Using Algorithm 5.3 to design the dynamic part of the compensator (3.16a)

To satisfy the right n–m columns of (4.1), $T = [1\ {-1}\ 1]$.

To satisfy the left m columns of (4.1), $L = 1$.

It can be verified that (4.1) and (4.3) are satisfied. Hence, an output feedback compensator that can fully realize a $\underline{K}\underline{C}\mathbf{x}(t)$ control is designed, where

$$\underline{C} = \begin{bmatrix} T \\ C \end{bmatrix} = \begin{bmatrix} 1 & -1 & 1 \\ 1 & 0 & 0 \end{bmatrix} \text{ and } q = \text{rank}\,(\underline{C}) = 2.$$

B) Using Algorithm 8.1 to assign eigenvalues −2 and $-1 \pm j\sqrt{3}$ to matrix $A-B\underline{KC}$. Although assigning complex conjugate eigenvalues is more difficult than assigning distinct and real eigenvalues, this kind of eigenvalue is much more common in real-world practice. See Section 8.1.

Step 1: Compute solution T_{n-q} of (8.10a) for $\Lambda_{n-q} = -2$.

First, use a similarity transformation matrix $Q = \begin{bmatrix} 1 & 0 & 0 \\ 0 & 1 & 1 \\ 0 & 0 & 1 \end{bmatrix}$ to make the third column of matrix $\underline{C}Q$ equal zero. Because Λ_{n-q} is a scalar, (8.10a) is now the third column of equation $[T_{n-q}(A-(-2)I)Q - L\underline{C}Q] = 0$.

Solution T_{n-q} of the above equation has q (= 2) basis vectors [−1 −1 2] and [2 0 −1]. We let $T_{n-q} = [-1\ -1\ 2]$.

This solution also satisfies (8.10b) (rank $\begin{bmatrix} T_{n-q} \\ \underline{C} \end{bmatrix} = n$).

Step 2: $\Lambda_q = \begin{bmatrix} -1 & -\sqrt{3} \\ \sqrt{3} & -1 \end{bmatrix}$.

First, use similarity transformation matrix $P = \begin{bmatrix} 1 & 0 & 0 \\ 0 & 1 & 0 \\ -1 & 1 & -1 \end{bmatrix} = P^{-1}$ to make the third row of PB equal zero. Complete similarity transformation on (A, T_{n-q}) as

$$\left(PAP^{-1}, T_{n-q}P^{-1}\right)\left(\begin{bmatrix} 0 & 1 & 0 \\ -1 & 1 & -1 \\ -1 & 0 & -1 \end{bmatrix}, \begin{bmatrix} -3 & 1 & -2 \end{bmatrix}\right)$$

Now, use (8.3a) to satisfy the third row of (8.11a) or $PAP^{-1}V_q - V_q\,\Lambda_q = PBK_q$, or to satisfy

$$\left[I_2 \otimes \left(\text{the third row of } PAP^{-1}\right) - \Lambda_q' \otimes \begin{bmatrix} 0 & 0 & 1 \end{bmatrix}\right]\left[\mathbf{v}_{q1}' : \mathbf{v}_{q2}'\right]' = 0$$

Or

$$\left(\begin{bmatrix} -1 & 0 & -1 & \vdots & 0 & 0 & 0 \\ 0 & 0 & 0 & \vdots & -1 & 0 & -1 \end{bmatrix} - \begin{bmatrix} 0 & 0 & -1 & \vdots & 0 & 0 & \sqrt{3} \\ 0 & 0 & -\sqrt{3} & \vdots & 0 & 0 & -1 \end{bmatrix}\right)$$

$$\begin{bmatrix} v_{q1} \\ v_{q2} \end{bmatrix} = 0$$

The q (=2) columns of solution V_q of the above equation and matrix V_2 for (A, T_{n-q}) before similarity transformation (by matrix P) are

$$V_q = \left[\mathbf{v}_{q1} : \mathbf{v}_{q2}\right] = \begin{bmatrix} \sqrt{3} & : & \sqrt{3} \\ 2+3\sqrt{3} & : & -2+3\sqrt{3} \\ 1 & : & -1 \end{bmatrix},$$

$$V_2 = P^{-1}V_q = \begin{bmatrix} \sqrt{3} & : & \sqrt{3} \\ 2+3\sqrt{3} & : & -2+3\sqrt{3} \\ 1+2\sqrt{3} & : & -1+2\sqrt{3} \end{bmatrix}$$

It can be verified that both (8.11a) (the third row) and (8.11b) are satisfied.

<u>Second</u>, use the first two rows of (8.11a) or
the first two rows of $[(PAP^{-1}V_q) - V_q \Lambda_q = PB\, K_q]$ to compute

$$K_q = \begin{bmatrix} -1+4\sqrt{3} & : & 1+4\sqrt{3} \\ -3-5\sqrt{3} & : & 3-5\sqrt{3} \end{bmatrix}$$

Step 3: Based on (8.13), $\underline{K} = K_q\,(\underline{C}V_2)^{-1} = \begin{bmatrix} 1 & : & 4 \\ 3 & : & -5 \end{bmatrix}$.

It can be verified that the eigenvalues of matrix $A - B\underline{K}\underline{C}$ are indeed -2 and $-1 \pm j\sqrt{3}$.

The whole design result (a complete output feedback compensator) of Example 8.3 is depicted in Figure 8.1, which has the same structure as Figure 4.3.

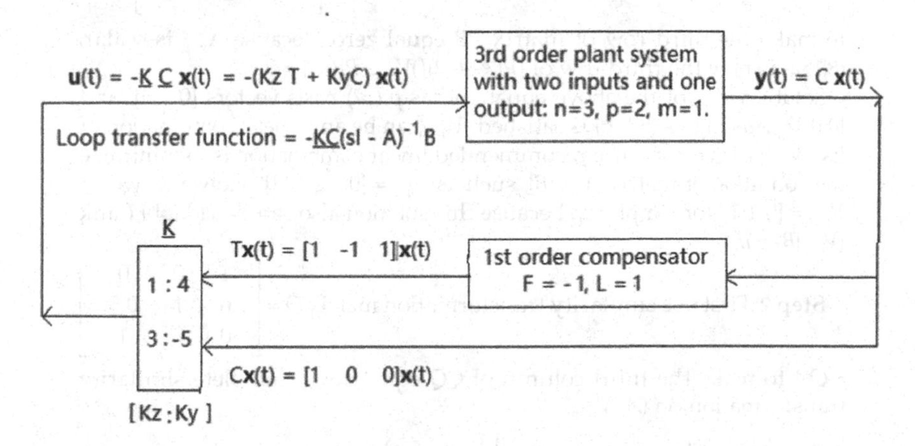

FIGURE 8.1
Block diagram of a complete output feedback compensator that can fully realize a control that is powerful enough to assign all three poles and the sensitivity of one pole.

Example 8.4: Another Complete Design of an Output Feedback Compensator

Let a third order, two-input, and one-output system matrix be

$$(A,B,C)=\left(\begin{bmatrix} 0 & 1 & 0 \\ 0 & 0 & 1 \\ 0 & 0 & 0 \end{bmatrix},\begin{bmatrix} 1 & 0 \\ 3 & 1 \\ 2 & 2 \end{bmatrix},\begin{bmatrix} 1 & 0 & 0 \end{bmatrix}\right)$$

From (1.9), the transfer function of this system is $G(s) = s^{-3}[s^2 + 3s + 2 \quad s + 2]$.

This system has one stable transmission zero −2. Hence, based on Conclusion 6.1 of this book, an output feedback compensator with one pole equal to this −2 can be designed ($F = -2$).

A) Using Algorithm 5.3 to design the dynamic part of the compensator (3.16a).

To satisfy the right $n-m$ columns of (4.1), $T = [2\ {-1}\ 0.5]$.

To satisfy the left m columns of (4.1), $L = 4$.

It can be verified that (4.1) and (4.3) are satisfied. Hence, an output feedback compensator that can fully realize a $\underline{K}\underline{C}\mathbf{x}(t)$ control is designed, where

$$\underline{C} = [T':C']' = \begin{bmatrix} 2 & -1 & 0.5 \\ 1 & 0 & 0 \end{bmatrix} \text{ and } q = \text{rank}\,(\underline{C}) = 2.$$

B) Using Algorithm 8.2 to assign eigenvalues −2 and $-1 \pm j\sqrt{3}$ to matrix $A-B\underline{K}\underline{C}$.

Step 1: Compute solution V_{n-p} of (8.15a) for $\Lambda_{n-p} = -2$.

First, use a similarity transformation matrix $P = \begin{bmatrix} 1 & 0 & 0 \\ 0 & 1 & 0 \\ -4 & 2 & -1 \end{bmatrix} = P^{-1}$ to make the third row of matrix PB equal zero. Because Λ_{n-p} is scalar, (8.15a) is now the third row of $[P(A-(-2)I)V_{n-p}-PBK_p] = 0$.

Solution V_{n-p} of the above equation has p (=2) basis vectors $[0\ 1\ 1]'$ and $[0\ 0\ 1]'$. As long as (8.15b) is satisfied, V_{n-p} can be any linear combination of its two basis vectors. The recommended linear combination is to minimize the condition of matrix $[V_{n-p}:B]$, such as $V_{n-p} = [0 \quad 2 \quad -1]'$. However, we let $V_{n-p} = [0\ 1\ 1]'$ for simplicity, because this solution also satisfies (8.15b) (rank $[V_{n-p}:B] = n$).

Step 2: First use similarity transformation matrix $Q = \begin{bmatrix} 1 & 0 & 0 \\ 0 & 1 & 0.5 \\ 0 & 0 & 1 \end{bmatrix} = Q^{-1}$ to make the third column of $\underline{C}Q$ equal zero. Complete similarity transformation on (A, V_{n-p})

$$\left(Q^{-1}AQ, Q^{-1}V_{n-p}\right)=\left(\begin{bmatrix} 0 & 1 & 0.5 \\ 0 & 0 & 1 \\ 0 & 0 & 0 \end{bmatrix},\begin{bmatrix} 0 \\ 1.5 \\ 1 \end{bmatrix}\right). \text{ Also, } \Lambda_p=\begin{bmatrix} -1 & \sqrt{3} \\ -\sqrt{3} & -1 \end{bmatrix}.$$

First, use (5.13c) to satisfy the third column of (8.16a) ($T_pQ^{-1}AQ - \Lambda_qT_p = L_p\underline{C}Q$):
$[\mathbf{t}_{p1} : \mathbf{t}_{p2}]$ $[I_2 \otimes$ (the third column of $Q^{-1}AQ) - \Lambda_p' \otimes [0\ 0\ 1]'] = 0$
Or

$$\begin{bmatrix} -4 & 1 & -0.5 & : & 0 & \sqrt{3} & -0.5\sqrt{3} \end{bmatrix}$$

$$\left(\begin{bmatrix} 0.5 & : & 0 \\ 1 & : & 0 \\ 0 & : & 0 \\ - & - & - \\ 0 & : & 0.5 \\ 0 & : & 1 \\ 0 & : & 0 \end{bmatrix} - \begin{bmatrix} 0 & : & 0 \\ 0 & : & 0 \\ -1 & : & -\sqrt{3} \\ - & - & - \\ 0 & : & 0 \\ 0 & : & 0 \\ \sqrt{3} & : & -1 \end{bmatrix} \right) = 0$$

The p (=2) rows of solution T_p of the above equation, and corresponding matrix T_2 for original (A, V_{n-p}) before similarity transformation (by matrix Q), are

$$T_p = \begin{bmatrix} -4 & 1 & -0.5 \\ 0 & \sqrt{3} & -0.5\sqrt{3} \end{bmatrix}, T_2 = T_pQ^{-1} = \begin{bmatrix} -4 & 1 & -1 \\ 0 & \sqrt{3} & -\sqrt{3} \end{bmatrix}$$

It can be verified that both (8.16a) (the third column) and (8.16b) ($T_2V_{n-p} = 0$) are satisfied.

Second, use the first two columns of $[(T_pQ^{-1}AQ) - \Lambda_p T_p = L_p\underline{C}Q]$ or the first two columns of (8.16a) to compute $L_p = \begin{bmatrix} 6 & -16 \\ -2\sqrt{3} & 0 \end{bmatrix}$.

Step 3: Based on (8.17), $\underline{K} = (T_2B)^{-1}L_p = \begin{bmatrix} -2 & 4 \\ 0 & 4 \end{bmatrix}$.

It can be verified that the eigenvalues of matrix $A - B\underline{KC}$ are indeed -2 and $-1 \pm j\sqrt{3}$.

The whole design result (a complete output feedback compensator) of Example 8.4 is depicted in the following Figure 8.2, which has the same structure as Figures 4.3 and 8.1.

Examples 8.3 and 8.4 took on a very difficult system to design. Its 2×2 loop transfer function is very difficult even to analyze using the classical control theory. Because $m < p$, this system cannot have an unknown input observer (Subsection 4.3.5) and thus cannot fully realize the loop transfer function of direct state feedback control under the existing modern control theory and separation principle.

However, for such a very difficult yet very basic system, the new design principle and its general and simple design Algorithms (5.3, 6.1, 8.1, and 8.2) have guided successful designs in Examples 8.3 and 8.4, with such a powerful and excellent control that can assign all feedback system poles and sensitivity of one of these three poles, and with an output feedback compensator that can truly and fully realize this control.

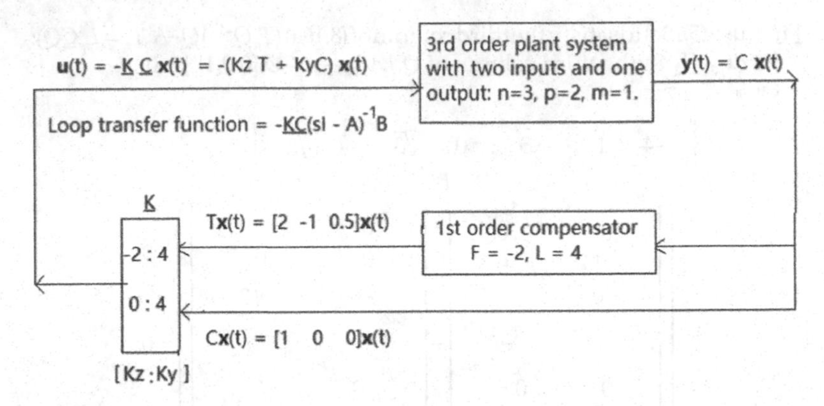

FIGURE 8.2
Block diagram of another complete output feedback compensator that can fully realize a control that is powerful enough to assign all three poles and the sensitivity of one pole.

Ten more problems very similar to that of Examples 8.3 and 8.4 can be found as Exercise problems (8.6–8.12) at the end of this chapter.

The most important guidance for the successful designs of Examples 8.3 and 8.4 is the general and explicit sufficient condition of Conclusion 6.1 of this book, that the plant system has at least one stable transmission zero. This condition is not in other existing literature and is enabled only by the fact that the solution of (4.1) of this book is completely decoupled (Tsui, 2015).

Another sufficient condition of Conclusion 6.1 is that the plant system has more output measurements than control inputs ($m > p$), as demonstrated by five plant system examples in Examples 6.1 and 6.2.

Both sufficient conditions of Conclusion 6.1 together can cover a great majority of plant systems, much more than the existing three conditions of Subsection 4.3.5, which are not satisfied by a great majority of plant systems (see Conclusions 6.4, 6.5 and 6.7).

The designs of this book are not only distinctly powerful and general, but also exceptionally simple. The entire design computation of Examples 8.3 and 8.4 is carried out by hand! This can be proven by the appearance of $\sqrt{3}$ at every intermediate step of design computation. Only the last step of checking and finding the eigenvalues of a 3×3 matrix $A\text{–}B\underline{KC}$ is done by computer.

To summarize, the superiority of the new design principle of this book is sufficiently and convincingly reflected by the exceptional generality, effectiveness, and simplicity of the new design algorithms of this book. This exceptional generality, effectiveness, and simplicity can be no more clearly and convincingly demonstrated by the ample design Examples 8.3, 8.4, and 6.1–6.3.

8.4 Modifications of Generalized State Feedback Control for Eigenstructure Assignment (Tsui, 2004b,c, 2005)

In general, Algorithm 8.1 assigns n arbitrarily given eigenvalues if $n < q + p$, in two groups. A group of n–q eigenvalues are assigned first. Their corresponding n–q left eigenvectors T_{n-q} are also assigned (if $q > 1$), and the recommended purpose of this assignment is to make matrix $[T_{n-q}{}':\underline{C}']$ as well-conditioned as possible (Section 2.2). The second group of q corresponding right eigenvectors V_q have freedom for assignment if $n < q + p-1$. The purpose of this assignment is recommended to make matrix $\underline{C}V_q$ as well-conditioned as possible.

Algorithm 8.2 is simply the dual of Algorithm 8.1. The first group of n–p right eigenvectors V_{n-p} are assigned to make matrix $[V_{n-p}:B]$ as well-conditioned as possible, while the second group of p left eigenvectors T_p are assigned to make matrix T_pB as well-conditioned as possible.

However, the following real situations require corresponding modifications to this standard design algorithm.

1. If $n \geq q + p$ but $n < q \times p$, then Step 2 of Algorithms 8.1 and 8.2 may not be solvable, yet arbitrary pole assignment may still be possible (Wang, 1996). Thus, in this situation, Step 1 of Algorithms 8.1 and 8.2 needs to be modified so that Step 2 is solvable.
2. If $n \geq q \times p$ (Wang, 1996) or even if $n \geq q + p$, then arbitrary pole assignment may be impossible. Thus, in this case, Step 1 of Algorithms 8.1 and 8.2 needs to be modified so that Step 2 may still assign the eigenvalues in a stable region, even though not precisely.
3. Because the first group of poles takes much higher priority than the second group, additional consideration may be needed on the selection of this first group of poles.
4. Unlike static output feedback control where all m rows of matrix C have the same physical meaning, the $m + r$ rows of matrix $\underline{C}$ have very different physical meanings – the m rows of matrix C correspond to the direct system output measurement, while the r rows of matrix T are estimated signals based on the information of direct measurement and the information of the plant system state space model. Thus, the rows of matrix C are more reliable than the rows of matrix T and should have priority over the rows of matrix T during the design.
5. In some practical situations, minimizing the zero-input response is more important than minimizing the sensitivity of the eigenvalues/poles.

The corresponding design modifications for each of these five situations can be as follows.

Modification 1: At Step 1 of Algorithm 8.1, assign matrix T_{n-q} so that system (A, B, T_{n-q}) will have q transmission zeros equal the q given eigenvalues of Λ_q. This is a modified and reversed problem from assigning eigenvectors (of T_{n-q}) to assign transmission zeros and eigenvalues.

This modification will be applied in situation (1) above, because arbitrary pole assignment is not straightforward but still generically possible in this situation (Wang, 1996).

This modification's formulation sounds simple but is actually not easy at all because too many parameters are involved.

Let each row of matrix T_{n-q}, $\mathbf{t}_i = \mathbf{c}_i D_i$, $i = 1, \ldots, n-q$, where the q-dimensional row vector $\mathbf{c}_i$ is the assignment freedom. Then this assignment is to make the matrix

$$\begin{bmatrix} A - \lambda I & : & B \\ \text{diag}\{\mathbf{c}_1, \ \ldots, \ \mathbf{c}_{n-q}\}\left[D_1' : \ \ldots \ : D_{n-q}'\right]' & : & 0 \end{bmatrix} \tag{8.18}$$

have linearly dependent column when λ equals one of the q eigenvalues of Λ_q.

It is obvious that this problem is far more likely to have a solution if $n = q + p$. This is because $n = q + p$ or $n - q = p$ makes the above matrix a square matrix and generically have transmission zeros (Davison and Wang, 1974). Recent studies have gradually moved the condition $n < q + p$ to $n \leq q + p$ (Bachelier et al., 2006; Bachelier and Mehdi, 2008; Konigorski, 2012; Wang and Konigorski, 2013). However, these studies treated this problem strictly as a mathematical problem that requires a precise solution, even though the precise assignment of these q eigenvalues is not necessary from a control system perspective.

Example 8.5: An Example of Modification 1

$$\text{Let}\left(A, B, \underline{C}\right) = \left(\begin{bmatrix} 0 & 1 & 0 & 0 \\ 0 & 0 & 1 & 0 \\ 0 & 0 & 0 & 1 \\ 1 & 0 & 1 & 0 \end{bmatrix}, \begin{bmatrix} 0 & 0 \\ 1 & 0 \\ 0 & 0 \\ 0 & 1 \end{bmatrix}, \begin{bmatrix} 1 & 0 & 0 & 0 \\ 0 & 1 & 0 & 0 \end{bmatrix}\right)$$

and let the assigned eigenvalues be −1, −2, −3, and −4.

Because $n = q + p$, we need Modification (1) on Algorithm 8.1.

Let us first assign $\Lambda_{n-q} = \text{diag}\{-1, -2\}$, and $\Lambda_q = \text{diag}\{-3, -4\}$.

Step 1: Each of the $n-q$ (= 2) rows of T_{n-q} has a q (= 2) × n-dimensional basis vector matrix D_i, and each D_i corresponds to an eigenvalue $-i$, $i = 1$, 2. They are

$$D_1 = \begin{bmatrix} u & 0 & -1 & 1 \\ v & 0 & 0 & 0 \end{bmatrix} \quad \text{and} \quad D_2 = \begin{bmatrix} x & 3 & -2 & 1 \\ y & 0 & 0 & 0 \end{bmatrix}$$

where u, v, x, and y are arbitrary constants.

Based on Modification (1), we will make the linear combinations of these basis vectors such that not only (8.10b) or rank $[T_{n-q}{}' \vdots \underline{C}']' = n$ is satisfied, but also matrix (8.18) has linearly dependent row/column at λ equals −3 and −4. Fortunately, a solution of this linear combination can be found for this problem as $\mathbf{c}_1 = \mathbf{c}_2 = [1\ 0]$ and $u = 60$ and $x = 84$. Thus,

$$T_{n-q} = \begin{bmatrix} \mathbf{c}_1 D_1 \\ \mathbf{c}_1 D_2 \end{bmatrix} = \begin{bmatrix} 60 & 0 & -1 & 1 \\ 84 & 3 & -2 & 1 \end{bmatrix}$$

Step 2: The above specially designed matrix T_{n-q} guaranteed the existence of the unique solution of (8.11) and (8.12) as

$$V_q = \begin{bmatrix} 1 & 1 \\ -3 & -4 \\ 15 & 12 \\ -45 & -48 \end{bmatrix} \quad \text{and} \quad K_q = \begin{bmatrix} -6 & 4 \\ 119 & 179 \end{bmatrix}$$

Step 3: Based on (8.13), $\underline{K} = K_q \left(\underline{C} V_q \right)^{-1} = \begin{bmatrix} 36 & 10 \\ 61 & 60 \end{bmatrix}$

It can be verified that matrix A–$B\underline{KC}$ does have the desired eigenvalues −1, −2, −3, and −4. □

Modification 2: Assign the n–q left eigenvectors T_{n-q} at Step 1 of Algorithm 8.1 so that system (A, B, T_{n-q}) will have q transmission zeros not necessarily equal the q eigenvalues of Λ_q, but instead be inside acceptable regions such as a stable region.

This modification actually raised two very novel ideas worthy of persuasion: (a) Using the freedom of some eigenvector's assignment for some other eigenvalue's assignment is novel regarding the existing design procedure of only using the remaining eigenvalue assignment freedom to assign the corresponding eigenvectors and (b) using the pole selection freedom to find a group of q poles that can be assigned.

Modification 3: Because out of the two groups to be assigned in Algorithms 8.1 and 8.2, the first group of eigenvalues is guaranteed assigned and whose corresponding eigenvectors also have assignment freedom whenever q (or p) is greater than one, while the second group of poles can be precisely assigned only when $n < q + p$, and whose eigenvectors have assignment freedom only if $n < q + p - 1$. Therefore, the first group of poles has much higher priority than the second group.

We will therefore and naturally select only the more important poles, such as the poles closer to the unstable region (see Subsection 2.2.2) into that first group. This idea of differentiating the individual eigenvalues is also used in all eigenvector assignment algorithms of Chapter 9.

Example 8.6: An Example of Modification 3

The system $(A, B, \underline{C})$ and the assigned eigenvalues −3, −2, and −1 are the same as Examples 8.1 and 8.2. However, we select a different eigenvalue −1 instead of −3, into the first group Λ_{n-q}, in this example. After all, −1 is closer to the unstable region than −2 and −3. It is called the "dominant pole" in the classical control theory.

Step 1: For $\Lambda_{n-q} = -1$, Eq. (8.10a) implies

$$\begin{bmatrix} 1 & 0 & 3 & : & -3 & -5 \\ 0 & 1 & 0 & : & 1 & 1 \end{bmatrix} \begin{bmatrix} -3 & 0 & -2 \\ 0 & 1 & 1 \\ 1 & -1 & -1 \\ & & \\ 0 & -1 & 0 \\ 0 & 0 & -1 \end{bmatrix} = 0$$

$$D_1 \quad : \quad E_1$$

Matrix T_{n-q} can be any linear combination of the two rows of D_1, provided that (8.10b) or rank $[T_{n-q}{}':\underline{C}']' = n$. We set

$$T_{n-q} = \mathbf{c}_1 D_1 = \begin{bmatrix} 1 & -2 \end{bmatrix} D_1 = \begin{bmatrix} 1 & -2 & 3 \end{bmatrix}$$

Step 2: The unique solution of (8.11) or (8.12) for assigning eigenvalues −2 and −3 is

$$V_q = \begin{bmatrix} 2 & -9 \\ -5 & 3 \\ -4 & 5 \end{bmatrix} \quad \text{and} \quad K_q = \begin{bmatrix} -5/2 & 13/4 \\ 7 & -7 \end{bmatrix}.$$

Step 3: Based on (8.13), $\underline{K} = K_q\left(\underline{C}V_q\right)^{-1} = \begin{bmatrix} -1/26 & 35/42 \\ -7/13 & -14/13 \end{bmatrix}$.

The eigenvalues are correctly assigned. Compare this $\underline{K}$ to the $\underline{K}$ of Examples 8.1 and 8.2, every entry of this $\underline{K}$ is smaller (and more robust). This is the difference of Modification 3.

Modification 4: This modification is for situation (4) listed above that out of the two different groups of rows in matrix $\underline{C}$, rows of matrix C should have priority over the rows of matrix T during the design.

The main occasion involved with the linear dependency of the rows of matrix $\underline{C}$ is the eigenstructure assignment of matrix $A - B\underline{K}\underline{C} = A - B[K_zT + K_yC]$.

We should pay extra attention to reduce the gain K_z as compared to the gain K_y because matrix T is less reliable than matrix C. In addition, (4.8) shows that BK_z is the only coupling factor (not BK_y) between the observer F and the feedback system $A-B\underline{KC}$. Thus, it certainly makes sense to reduce K_z. This consideration is also made heavily in fault control design. Please see Subsection 7.2.4.

Two more specific occasions of design involve linear dependency of the rows of matrix $\underline{C}$, such as Steps 1 and 2 of Algorithm 8.1 (making matrices $[T_{n-q}' \vdots \underline{C}']'$ and $[\underline{C}V_q]$ as well-conditioned as possible). The numerical eigenvector assignment Algorithm 9.2 enables this differentiation among eigenvectors by putting different weighting factors to each eigenvector.

It should be noticed that matrix T of $\underline{C}$ is not the matrix T_{n-q} of Step 1 of Algorithm 8.1 (nor the matrix T_p of Step 2 of Algorithm 8.2).

Modification 5: If we know a-priori of initial condition $\mathbf{x}(0)$, then the zero-input response $Ve^{\Lambda t}T\,\mathbf{x}(0)$ (see Eq. 2.2) may be reduced by designing matrix T_{n-q} so that $T_{n-q}\mathbf{x}(0) = 0$, instead of making matrix $[T_{n-q}' \vdots \underline{C}']'$ as well-conditioned as possible, in Step 1 of Algorithm 8.1. This is because matrix T_{n-q} is part of matrix T of (2.2).

Example 8.7: An Example of Modification 5

Same plant system and the same assigned eigenvalues of Examples 8.1 and 8.6. But according to Modification 5, the design of matrix T_{n-q} $(= \mathbf{c}_1 D_1)$ at Step 1 will be aimed at making $T_{n-q}\mathbf{x}(0) = 0$ instead.

If we know the initial state $\mathbf{x}(0) = [0\ x\ 0]'$ $(x \neq 0)$, then $D_1\mathbf{x}(0) = [0\ x]'$. Thus, we will design $\mathbf{c}_1 = [1\ 0]$ such that $T_{n-q}\mathbf{x}(0) = \mathbf{c}_1 D_1 \mathbf{x}(0) = 0$. This result of $T_{n-q} = \mathbf{c}_1 D_1 = [1\ 0\ 1]$ is the same as that of Example 8.1.

If we know the initial state $\mathbf{x}(0) = [x\ 2x\ x]'$ $(x \neq 0)$, then $D_1\mathbf{x}(0) = [4x\ 2x]'$. Thus, we will design $\mathbf{c}_1 = [1\ {-2}]$ such that $T_{n-q}\mathbf{x}(0) = \mathbf{c}_1 D_1 \mathbf{x}(0) = 0$. This result of $T_{n-q} = \mathbf{c}_1 D_1 = [1\ {-2}\ \ 3]$ is the same as that of Example 8.6.

8.5 Summary of Eigenstructure Assignment Designs

There is no general and optimal pole selection rule, and there is no optimal pole selection for all real and practical applications without trial and error and adjustment. This situation should be the same as the selection of weighting matrices of the LQ optimal control criteria (see Chapter 10) and should be a true reflection of the reality. Nonetheless, the poles are far more directly related to system performance and stability, than any other system or criteria parameters, and as a result, there is far more rational and explicit guidance on the selection of the poles in Section 8.1.

Consider assigning n arbitrarily given eigenvalues to matrix $A-B\underline{KC}$. This can be achieved if rank $(\underline{C}) \equiv q = n$ (Theorem 3.1) and can be achieved by Algorithms 8.1 and 8.2 for systems with $q < n$ if $q + p > n$. Some recent research has gradually relaxed this condition to $q+p \geq n$. This assignment is generically possible even if $q \times p > n$ (Wang, 1996). The new design principle of this book has greatly increased the value of q from m to $m+r$ while guaranteeing the full realization of this control (see Conclusions 6.2, 6.7, and Table 6.2).

For situations of $q+p \leq n$ when direct pole assignment cannot be guaranteed, Modifications 1 and 2 of Step 1 of Algorithms 8.1 and 8.2 try to make this assignment possible. These two modifications raised the idea of making eigenvector assignment for the purpose of making (another group of) eigenvalue assignments possible, and the idea of approximately assigning poles into a region instead of the current pole assignment into a precise position. A very interesting result of this kind of approximate pole assignment may be found in the work of Chu (1993a).

The new design principle of this book allows the state feedback control gain K to be constrained by $K = \underline{KC}$, and thus makes this control fully realizable for a great majority of plant systems, as improved from the existing separation principle whose unconstrained $K\mathbf{x}(t)$ control cannot be fully realized for a great majority of plant systems. Likewise, the idea of selecting the second group of poles to be based on the result of the assignment of the first group of eigenvectors, and the idea of selecting and assigning the second group of poles only approximately, should improve greatly the corresponding assignability from the existing unconstrained pole selection.

Therefore, the author believe that approximate and synthesized pole selection and assignment can be a very promising and very practical research direction. It is also in the full spirit of the new synthesized design principle of this book.

Because the left and right eigenvectors determine the sensitivity of the corresponding eigenvalue (Theorem 2.1), eigenvector assignment can determine and improve the condition of the eigenvalue assignment problem. Thus, eigenvector assignment is as important as eigenvalue assignment, just like system robustness is as important as performance.

There have been many eigenvalue assignment algorithms, either for the case of $q=n$ (Miminis and Paige, 1982; Petkov et al., 1984; Duan, 1993a; Kautsky et al., 1985) or for the case of $q<n$ (Misra and Patel, 1989; Syrmos and Lewis, 1993; Syrmos, 1994: Wang, 1996). But none except Kautsky et al. (1985) considered eigenvector assignment and thus tried to improve the condition of the eigenvalue assignment problem itself (see Section 2.2). A most basic principle of numerical linear algebra is that the result of an ill-conditioned problem is unreliable, and that unreliability cannot be reduced by the improvement of the reliability of the computational algorithm for that problem.

Consider the eigenvector assignment freedom revealed by Algorithms 8.1 and 8.2. For the case $q=n$, all eigenvectors have the assignment freedom if $p>1$. For the case of $q<n$ but $q+p>n$, only one group of $n-q$ (or $n-p$) eigenvectors have assignment freedom if q (or p) >1, and the remaining eigenvectors have assignment freedom if $q+p>n+1$. Therefore, there usually is a large amount of freedom for eigenvector assignment. This makes eigenvector assignment not only important but also potentially very effective.

The most important advantage of Algorithms 8.1 and 8.2 of this book is that its eigenvector assignment is in the simplest possible form – as a linear combination of its basis vectors. Only this form makes possible the really general and effective numerical eigenvector assignment algorithms of Kautsky et al. (1985) and makes possible the general and effective analytical eigenvector assignment rules, which are presented in Chapter 9.

Finally, five novel modifications on the purpose of eigenvector assignment part of Algorithm 8.1 are proposed in this book. This kind of substantial and solid progress is possible, only because of the substantial and solid mathematical progress such as the general and decoupled solution of matrix equation (8.1b) or (4.1) and such as the closed-form solution to the matrix equation pair (8.11a and b) or (4.1 and 4.3). See Figure 5.1 and Section A.4.

Exercises

8.1 Suppose $n = 10$ and all assigned eigenvalues must be complex conjugates.

a. If $q = 7$ and $p = 4$, can Algorithm 8.1 be used directly? Why? Can Algorithm 8.2 be used directly? Why?

b. If $q = 6$ and $p = 5$, can Algorithm 8.1 be used directly? Why? Can Algorithm 8.2 be used directly? Why?

c. If $q = 7$ and $p = 5$, how can Algorithm 8.1 be used directly?

d. If $q = 7$ and $p = 5$, how can Algorithm 8.2 be used directly?

8.2 Make Modification 3 on Examples 8.1 and 8.2 by assigning eigenvalue –2 at Step 1 and eigenvalues –1 and –3 at Step 2. Compare the design result $\underline{K}$ with that of Examples 8.1 and 8.6.

8.3 Use Algorithms 8.1 and 8.2 to assign eigenvalues {–1, –2, –3, –4} to the following two systems (Chu, 1993). Notice that the provided answer $\underline{K}$ is not unique.

(a)

$$(A, B, \underline{C}, \underline{K}) = \left(\begin{bmatrix} 0 & 1 & 0 & 0 \\ 1 & 1 & 0 & 0 \\ -1 & 0 & 0 & 0 \\ 0 & 0 & 0 & 0 \end{bmatrix}, \begin{bmatrix} 0 & 0 \\ 1 & 0 \\ 0 & 0 \\ 0 & 1 \end{bmatrix}, \begin{bmatrix} 1 & 0 & 0 & 0 \\ 0 & 0 & 1 & 0 \\ 0 & 0 & 0 & 1 \end{bmatrix}, \begin{bmatrix} -47 & 34 & 10 \\ 49 & -35 & -11 \end{bmatrix} \right)$$

(b)

$$(A,B,\underline{C},\underline{K}) = \left(\begin{bmatrix} 0 & 1 & 0 & 0 \\ 0 & 0 & 1 & 0 \\ 0 & 0 & 0 & 1 \\ -1 & 0 & 0 & 0 \end{bmatrix}, \begin{bmatrix} 1 & 0 & 0 \\ 0 & 0 & 0 \\ 0 & 1 & 0 \\ 0 & 0 & 1 \end{bmatrix}, \begin{bmatrix} 1 & 0 & 0 & 0 \\ 0 & 1 & 0 & 0 \end{bmatrix}, \begin{bmatrix} -10 & 4.32 \\ 62 & -35 \\ 52.58 & -29/84 \end{bmatrix} \right)$$

8.4 Use Algorithms 8.1 and 8.2 and Modification 1 to assign eigenvalues {–1, –2, –3, –4} to the following system (Chu, 1993). Notice that the provided answer $\underline{K}$ is not unique.

$$(A,B,\underline{C},\underline{K}) = \left(\begin{bmatrix} 0 & 1 & 0 & 0 \\ 0 & 0 & 0 & 0 \\ 0 & 0 & 0 & 1 \\ 0 & 0 & -1 & 0 \end{bmatrix}, \begin{bmatrix} 0 & 0 \\ 1 & 0 \\ 0 & 0 \\ 0 & 1 \end{bmatrix}, \begin{bmatrix} 0 & 1 & 1 & 0 \\ 1 & 0 & 0 & 1 \end{bmatrix}, \begin{bmatrix} -50 & -49.47 \\ 40.49 & 40 \end{bmatrix} \right)$$

8.5 Use Algorithms 8.1 and 8.2 and Modification 1 to assign eigenvalues {–1, –1, –2, –2} to the following system (Kwon and Youn, 1987).

$$(A,B,\bar{C},\bar{K}) = \left(\begin{bmatrix} 0 & 1 & 0 & 0 \\ 0 & 0 & 1 & 0 \\ 0 & 0 & 0 & 1 \\ 1 & 0 & 1 & 0 \end{bmatrix}, \begin{bmatrix} 0 & 0 \\ 1 & 0 \\ 0 & 0 \\ 0 & 1 \end{bmatrix}, \begin{bmatrix} 1 & 0 & 0 & 0 \\ 0 & 1 & 0 & 0 \end{bmatrix}, \begin{bmatrix} 14 & 6 \\ 19 & 18 \end{bmatrix} \right)$$

8.6 Let the plant system state space model be

$$(A,B,C) = \left(\begin{bmatrix} 0 & 1 & 0 \\ 0 & 0 & 1 \\ 0 & 0 & 0 \end{bmatrix}, \begin{bmatrix} 1 & 0 \\ 4 & 1 \\ 3 & 3 \end{bmatrix}, \begin{bmatrix} 1 & 0 & 0 \end{bmatrix} \right)$$

a. Repeat Example 8.3 by setting the observer pole equal –3.

b. Repeat Example 8.4 by setting the observer pole equal –3.

c. Repeat part (a), assigning eigenvalues –1 and $-1 \pm j$ to matrix A–$B\underline{K}\underline{C}$.

Partial answer: $T = [9\ -3\ \ 1]$, $L = 27$

d. Repeat part (b), assigning eigenvalues –1 and $-1 \pm j$ to matrix A–$B\underline{K}\underline{C}$.

Partial answer: $T = [9\ -3\ \ 1]$, $L = 27$.

8.7 Repeat part (b) of Example 8.3, using Algorithm 8.2 instead.

8.8 Repeat part (b) of Example 8.3, assigning eigenvalues –1 and $-1 \pm j$ instead.

8.9 Repeat part (b) of Example 8.3, using Algorithm 8.2 and assign eigenvalues –1 and $-1 \pm j$.

8.10 Repeat part (b) of Example 8.4, using Algorithm 8.1 instead.

8.11 Repeat part (b) of Example 8.4, assigning eigenvalues –1 and $-1 \pm j$ instead.

8.12 Repeat part (b) of Example 8.4, using Algorithm 8.1 and assign eigenvalues –1 and $-1 \pm j$.

8.13 For a plant system with $n = 5$ and $p = m = 2$. This kind of system has generically $n–m$ (= 3) transmission zeros (Davison and Wang, 1974).

The static output feedback control of this system cannot assign all poles because $m \times p < n$ (Wang, 1996). However, the compensator designed by the new design principle of this book can fully realize a generalized state feedback control that is equivalent to $r + m$ ($\equiv q$) outputs and that is much more powerful, where r equals the number of stable transmission zeros out of the three zeros. The probability of r stable zeros out of three zeros can be found in Exercises 4.2 and 4.6, for Pzstable = ½ and ¾, respectively, where Pzstable is the probability for a zero to be stable.

Find the probabilities P_1 and P_2 using these two assumptions of Pzsable of achieving each of the following four design objectives or better, based on the results of Exercises 4.2 and 4.6.

a. Direct state feedback control for full pole assignment and full eigenvector assignment.

   Answer: $q = r + m = n = 5$ implies $r = n–m = 3$. For $r \geq 3$, the two probabilities are $P_1 = 0.125$ and $P_2 = 0.422$. The probabilities are quite small.

b. Assign all poles and at least part of ($n–q$) or ($n–p$) eigenvectors or better.

   Answer: $q + p > n$ implies $r \geq n–p–m + 1 = 2$. For $r \geq 2$, $P_1 = 0.5$ and $P_2 = 0.844$. Hint: The probability for static output feedback control to achieve this objective is 0.

c. Assign generically all poles or better.

   Answer: $q\ x\ p > n$ implies $r \geq n/p–m + 1 = 1$. For $r \geq 1$, $P_1 = 0.875$ and $P_2 = 0.98$. The probability for static output feedback control to achieve this objective is 0.

d. Same control as static output feedback control ($r = 0$): $P_1 = 0.125$ and $P_2 = 0.0156$. The probability of our design to have this low power control is very low.

8.14 Repeat the first three parts of Exercise 8.13 with new system parameters $n = 4, p = m = 2$ (and $p \times m$ is still not greater than n, as required for generic pole assignment).

Answer:

	(a) $q = n, r = 2$	(b) $q \geq 3, r \geq 1$	(c) $q \geq 3, r \geq 1$
$P_1 =$	0.25	0.75	0.75
$P_2 =$	0.56	0.94	0.94

The probability of the last two columns (our new design principle) is much higher than that of the first column (the probability of exact LTR), and is much higher than the probability (= 0) of static output feedback control ($r \equiv 0$).

8.15 Repeat the first three parts of Exercise 8.13 with new system parameters $n = 9, p = m = 3$ (and $p \times m$ is still not greater than n, as required for generic pole assignment).

Answer:

	(a) $q = n, r = 6$	(b) $q \geq 7, r \geq 4$	(c) $q \geq 4, r \geq 1$
$P_1 =$	0.0156	0.344	0.98
$P_2 =$	0.178	0.83	0.999

The probability of the last two columns (our new design principle) is much higher than that of the first column (probability of exact LTR under separation principle), and is much higher than the probability (= 0) of static output feedback control ($r \equiv 0$).

9

Control Design for Eigenvector Assignment

This chapter describes exclusively eigenvector assignment techniques.

The normal design sequence of eigenstructure assignment is assigning eigenvalues first and then using the remaining freedom to assign the eigenvectors (see Section 8.5). This is why eigenvalue assignment and its associate eigenvector assignment formulation are presented in the previous chapter.

More importantly, all eigenvector assignment of this book is formulated in a unified form – to assign the linear combination of the basis vectors of each eigenvector. This form is the simplest possible, which has enabled the general and effective eigenvector methods of this chapter.

The purpose of this assignment is also unified in this chapter as to maximize the angles between all n eigenvectors (or to minimize the condition number of the eigenvector matrix V, or to minimize eigenvalue sensitivity $s(\Lambda)$ (2.16)), even though some other purposes of eigenvector assignment were raised in Section 8.4.

From Section 2.2, a small condition number of V, $k(V)=\|V\|\|V^{-1}\|$, can improve directly the sensitivity of system poles (or system robust performance) and robust stability margins M_1 to M_3. It also has the following additional important technical advantages.

1. Smoother response (2.2):

$$\mathbf{x}(t)=Ve^{\Lambda t}V^{-1}\mathbf{x}(0)+\int_0^t Ve^{\Lambda(t-\tau)}V^{-1}B\mathbf{u}(\tau)d\tau \tag{9.1}$$

   It is obvious that after eigenvalues of Λ are assigned, the most important system parameters of (9.1) to be assigned are V and V^{-1}.

2. Smaller control gain (8.1b) and (8.6):

$$\boldsymbol{K}=KV^{-1}=\left[B_1^{-1}:0\ \right]\left(A-V\Lambda V^{-1}\right) \tag{9.2}$$

   It is obvious that the smaller the control gain $\boldsymbol{K}$, the smaller the control power and the lower the chance of disturbance and control component failure.

DOI: 10.1201/9781003259572-9

In addition, from (4.8), the state matrix of the observer feedback system is

$$\underline{A_c} = \begin{bmatrix} A - B\underline{\underline{K}}C & -BK_z \\ 0 & F \end{bmatrix} \tag{9.3}$$

The only coupling factor between the plant feedback system state matrix $A–B\underline{K}C$ and the observer state matrix F is BK_z, where K_z is a part of control gain $\underline{K}$ (= $[K_z{:}K_y]$).

The research of eigenvector assignment started in the middle of the 1970s (Moore, 1976; Klein and Moore, 1977; Fahmy and O'Reilly, 1982; Van Dooren, 1981; Van Loan, 1984, Duan 1993b). But only until 1985 did people use the formulation of assigning the linear combination of the basis vector matrix, for eigenvector assignment, and only this simplest possible formulation enabled really general and effective eigenvector assignment algorithms of Kautsky et al. (1985). This simplest possible formulation enabled the full use of its freedom in several other applications, as presented in Figure 5.1 and Section A.4.

The eigenvector assignment of this chapter is also fully based on this form. Two numerical iterative methods of Kautsky et al. (1985) are presented in Section 9.1. An analytical decoupling method is presented in Section 9.2.

All design methods are under a unified formulation, as to assign n right eigenvectors $D_i\mathbf{c}_i$, $i=1, \ldots, n$. There are quite a few variations of this form in this book from various other applications, such as (6.1), (6.6), (8.10), (8.11), (8.15), and (8.16). It is quite easy to adapt and apply Algorithm 9.1 to these various applications and formulations.

9.1 Numerical Iterative Methods (Kautsky et al., 1985)

The single purpose of numerical iterative eigenvector assignment methods is to maximize the angles between the eigenvectors or to minimize the condition number of eigenvector matrix V.

For simplicity and unified formulation, all two numerical iterative methods of this section require that all p basis vectors $\mathbf{d}_{ij}$, $j=1, \ldots, p$ within each matrix D_i be orthonormal. Meaning that $\mathbf{d}_{ij}'\mathbf{d}_{ik}=\delta_{jk}$ and $\|\mathbf{d}_{ij}\|=1$ for all $i=1, \ldots, n$, $j=1, \ldots, p$, and $k=1, \ldots, p$. This requirement can be met in the following two different ways.

Way one is to satisfy this requirement during the computation of D_i itself. For example, in the computation of (8.4), we first make the QR decomposition of the matrix

$$[A - \lambda_i I : -B] = [R_1 : 0]Q_i', \ i = 1,\ldots,n \tag{9.4a}$$

where the dimension of matrix Q_i is $(n+p)\times(n+p)$.

Then,

$$D_i = [\, I_n : 0]Q_i\big[0 : I_p\big]',\ i = 1,\ldots,n \qquad (9.4b)$$

Or D_i equals the top n rows and right p columns of the orthonormal matrix Q_i.

Way two is to compute the basis vector matrix D_i first by other means (such as back substitution on the block controllable Hessenberg form (8.6)) and then making the QR decomposition of matrix D_i:

$$D_i = Q_i R_i,\ i = 1,\ldots,n \qquad (9.5a)$$

Matrix D_i will be updated as the left p columns of matrix Q_i (9.5b)

Because Way two takes one more step of computation than Way one, its result may be less reliable than Way one. However, the much higher dimension makes the QR decomposition computation of Way one much more difficult than that of Way two.

The following Algorithms 9.1 and 9.2 are called "Rank one method" and "Rank two method", respectively. The Rank one method updates one eigenvector at each iteration to maximize the angle between this eigenvector and other n–1 eigenvectors. The Rank two method updates two vectors out of a separate set (say S) of n orthonormal vectors at each iteration, to minimize the angle between these two vectors and their corresponding pair of D_i matrices while maintaining the orthonormality of set S.

Algorithm 9.1: Rank One Eigenvector Assignment

Step 1: Let j=0. Set arbitrarily an initial set of n eigenvectors,

$$\mathbf{v}_i = D_i \mathbf{c}_i,\ i = 1,\ldots,n$$

Step 2: Let j=j +1. Select a vector $\mathbf{v}_j$ for updating. Then set the $n\times(n-1)$ dimensional corresponding matrix V_j, which contains the other $n-1$ eigenvectors.

Step 3: Make QR upper triangularization of matrix V_j

$$V_j = Q_j R_j = \big[\bar{Q}_j\ :\ \mathbf{q}_j\big] \begin{bmatrix} \bar{R}_j \\ 0 \end{bmatrix} \begin{matrix} \\ n-1 \end{matrix}$$

where Q_j is n-dimensional unitary, and $\bar{R}_j$ is (n–1)-dimensional and upper triangular. Hence, $\mathbf{q}_j$ is orthogonal to $R(V_j)$ because $\mathbf{q}_j'V_j$=0 (see (2.7) and (A.3)).

Step 4: Compute the normalized least-square solution $\mathbf{c}_j$ of $\mathbf{q}_j = D_j\mathbf{c}_j$, or compute the projection of $\mathbf{q}_j$ on D_j. (see Example A.8 or Golub and Van Loan (1989)):

$$\mathbf{c}_j = \frac{D_j'\mathbf{q}_j}{\|D_j'\mathbf{q}_j\|} \tag{9.6}$$

Step 5: Update eigenvector vector

$$\mathbf{v}_j = D_j\mathbf{c}_j = \frac{D_jD_j'\mathbf{q}_j}{\|D_j'\mathbf{q}_j\|} \tag{9.7}$$

Steps 4 and 5 can be computed together in actual computation.

Step 6: Check the condition number of matrix V, which contains all n eigenvectors. Stop the iteration if satisfactory. Otherwise, go to Step 2 and update one more eigenvector.

It is a normal practice to stop when all n eigenvectors are updated once.

The main computation of this algorithm, the QR decomposition of Step 3, may not need to perform from start to finish at every iteration. According to Kautsky et al. (1985), this decomposition can be updated from the result of the QR decomposition of the previous iteration, and thus reduce the order of computation from $2n^3/3$ to n^2. The actual procedure of this updating seems not yet published.

Based on experience, Kautsky et al. concluded that Algorithm 9.1 is very effective in lowering $k(V)$, at the first sweep of all n eigenvectors, but cannot guarantee the convergence to a minimum $k(V)$. This is because maximizing the angle between *one* eigenvector and the others cannot guarantee the maximization of angles between *all* eigenvectors.

Tits and Yang (1996) claim that each iteration of Algorithm 9.1 can increase the determinant of matrix V, $|V|$, and the algorithm can converge to a locally maximum $|V|$ depending on the initial value of V.

Tits and Yang also extended Algorithm 9.1 to complex conjugate eigenvalue cases, using complex number arithmetic operations. To extend this algorithm using real number arithmetic operation only, the results of Step 1 of Algorithm 5.3 and of (8.3) may be used.

Algorithm 9.2: Rank Two Eigenvector Assignment

Step 1: Select a set of n orthonormal vectors $\mathbf{x}_i$, $i=1, \ldots, n$. An example of this set is $[\mathbf{x}_1 : \ldots : \mathbf{x}_n] = I$.

Compute the complement basis vector matrix $\underline{D}_i$ of D_i, $i=1, \ldots, n$.

$$\underline{D}_i = Q_i\begin{bmatrix} 0 \\ I_{n-p} \end{bmatrix} \quad \text{where } D_i = Q_i\begin{bmatrix} R_i \\ 0 \end{bmatrix}$$

In other words, $\underline{D}_i$ equals the right $n-p$ columns of Q_i, where Q_i is the QR decomposition of D_i.

Step 2: Select two vectors $\mathbf{x}_j$ and $\mathbf{x}_{j+1}$ for updating. Rotate these two vectors by an angle θ such that

$$\left[\underline{\mathbf{x}}_j \ : \ \underline{\mathbf{x}}_{j+1}\right] = \left[\mathbf{x}_j \ : \ \mathbf{x}_{j+1}\right]\begin{bmatrix} \cos(\theta) & \sin(\theta) \\ -\sin(\theta) & \cos(\theta) \end{bmatrix} \tag{9.8}$$

and such that both the angle ψ_j between $\underline{\mathbf{x}}_j$ and $\underline{D}_j$ and the angle ψ_{j+1} between $\underline{\mathbf{x}}_{j+1}$ and $\underline{D}_{j+1}$ are maximized. This is expressed in terms of θ as

$$\min\left\{r_j^2\cos^2(\psi_j) + r_{j+1}^2\cos^2(\psi_{j+1})\right\} = \min\left\{r_j^2\left\|\underline{D}_j'\underline{\mathbf{x}}_j\right\| + r_{j+1}^2\left\|\underline{D}_{j+1}'\underline{\mathbf{x}}_{j+1}\right\|\right\}$$

$$= \min\left\{c_1\sin^2(\theta) + c_2\cos^2(\theta) + c_3\sin(\theta)\cos(\theta)\right\} \tag{9.9}$$

$$\equiv \min\left\{f(\theta)\right\} \tag{9.10}$$

where in (9.9), r_j and r_{j+1} are weighting factors for $\mathbf{v}_j$ and $\mathbf{v}_{j+1}$, respectively. For example, the maximization of a far more generally accurate robust stability measure M_3 (2.25) requires that $r_i = |\text{Re}(\lambda_i)|^{-1}$, $i=j$, and $j+1$. Also, in (9.9),

$$c_1 = r_j^2\mathbf{x}_j'\underline{D}_{j+1}\underline{D}_{j+1}'\mathbf{x}_j + r_{j+1}^2\mathbf{x}_{j+1}'\underline{D}_j\underline{D}_j'\mathbf{x}_{j+1}$$

$$c_2 = r_j^2\mathbf{x}_j'\underline{D}_j\underline{D}_j'\mathbf{x}_j + r_{j+1}^2\mathbf{x}_{j+1}'\underline{D}_{j+1}\underline{D}_{j+1}'\mathbf{x}_{j+1} \tag{9.11}$$

$$c_3 = 2\mathbf{x}_j'\left(r_{j+1}^2\underline{D}_{j+1}\underline{D}_{j+1}' - r_j^2\underline{D}_j\underline{D}_j'\right)\mathbf{x}_{j+1}$$

The function $f(\theta)$ of (9.9 and 9.11) is positive, continuous, periodic, and has a global minimum. The examination of $f(\theta)$ shows that if $c_1=c_2$ or if $c_3=0$, then $f(\theta)$ of (9.9) is at the minimum when $\theta=0$, or the updating of this pair of vectors is no longer needed.

For $c_1 \neq c_2$ and $c_3 \neq 0$, the nonzero value of θ is determined by solving $df(\theta)/d\theta=0$:

$$df(\theta)/d\theta = (c_1 - c_2)\sin(2\theta) + c_3\cos(2\theta) = 0$$

$$\text{or} \quad \frac{c_3}{c_2 - c_1} = \tan(2\theta) = \frac{2\tan\theta}{1-\tan^2\theta} \tag{9.12}$$

From the first equation of (9.12),

$$\theta = \tan^{-1}\left[c_3/(c_2 - c_1)\right]/2 + k\pi \tag{9.13}$$

where k is an integer such that $d^2f(\theta)/d\theta^2 > 0$:

$$d^2 f(\theta)/d\theta^2 = 2\left[(c_1 - c_2)\cos(2\theta) - c_3\sin(2\theta)\right] > 0 \text{ or } \tan(2\theta) < (c_1 - c_2)/c_3 \tag{9.14}$$

From the second equation of (9.12), which is a quadratic equation, solution $\tan(\theta)$ and θ is

$$\theta = \tan^{-1}\left[-c_4 + \left(c_4^{\,2} + 1\right)^{1/2}\right] + k\pi \tag{9.15a}$$

where

$$c_4 = c_3/(c_2 - c_1) \tag{9.15b}$$

After θ is determined either from (9.13) or from (9.15), with (9.14) guaranteed, we can proceed with the rotation and updating of this iteration (9.8).

Step 3: If θ is close enough to 0 or $k\pi$ (k is any integer), then $\mathbf{x}_j$ and $\mathbf{x}_{j+1}$ are almost linear combination of D_j and D_{j+1}, and they no longer need updating.

If this is not true for all vectors, go back to Step 2 for more updating. Otherwise, compute the projections of all vectors $\mathbf{x}_i$ to $R(D_i)$ or $\mathbf{v}_i = D_i(D_i'\mathbf{x}_i)/\|D_i'\mathbf{x}_i\|$, $i = 1, \ldots, n$.

The critical step of this algorithm is Step 2, which has not appeared in the literature either. This version is based on and updated from Chu (1993b).

According to Kautsky et al. (1985), the order of computation of each iteration of this algorithm is similar to that of Algorithm 9.1. More specifically, the order of computation of (9.11) is $4pn$ (4 pairs of $\mathbf{x}_i'\underline{D}_k$), while the order of computation of Algorithm 9.1 is n^2.

Also, according to Kautsky et al. (1985), Algorithm 9.2 requires less iteration and is more efficient than Algorithm 9.1 for well-conditioned problems. This is understandable because Algorithm 9.2 starts from an ideal orthonormal set and then makes it approach the actual solution, while Algorithm 9.1 starts from a random solution and then improves its condition.

However, both Algorithms 9.1 and 9.2 cannot yield reliable results for ill-conditioned problems (Kautsky et al., 1985). In such cases, we can use the simplest analytical eigenvector assignment (Section 9.2) or the very complicated "Method 1" of Kautsky et al. (1985). It turns out people still preferred Algorithms 9.1 and 9.2 over Method 1 because of simplicity and have made Algorithms 9.1 and 9.2 standard procedures in MATLAB (Grace et al., 1990).

An advantage of Algorithm 9.2 over Algorithm 9.1 is the use of weighting factors r_i. Thus, a sensible improvement of Algorithm 9.1 is not to treat or weigh every eigenvector indifferently – some eigenvectors corresponding to more important eigenvalues should be updated first and more times, than the less important eigenvectors. The analytical eigenvector assignment of the next section can fully consider analytical parameters and properties such as eigenvalues, controllability indices, and decoupling.

Algorithm 9.2 can also be improved, by adding more deliberation on the initial pairing between $\mathbf{x}_i$ and D_i ($i=1, \ldots, n$) at Step 1. For example, in the initial random pairing, the angle between $\mathbf{x}_i$ and D_i may be large while the angle between $\mathbf{x}_i$ and D_j ($i \neq j$) may be small. Thus, it is certainly more reasonable to pair $\mathbf{x}_i$ with D_j instead of D_i. Algorithm 9.1 also can improve its initial computation of the n eigenvectors. The analytical eigenvalue assignment of the next section can help provide a good set of initial eigenvectors.

9.2 Analytical Decoupling Method

The single purpose of numerical eigenvector assignment algorithms is minimizing the condition number of eigenvector matrix V, $k(V)$.

However, $k(V)$ is still not accurate enough in indicating individual eigenvalue sensitivity and the system's robust stability (see Section 2.2). In addition, numerical methods overlook some important and analytical system parameters and properties such as eigenvalues, controllability indices, and decoupling. From Examples 2.4 and 2.5 and their analysis, decoupling is exceptionally important in eigenvalue/pole sensitivity and robust stability.

The analytical eigenvector assignment method of this section is based on decoupling. It also takes full account of the eigenvalues and controllability indices. However, it cannot claim any optimality such as the minimized $k(V)$, as the numerical eigenvector assignment methods.

The analytical eigenvector assignment is also based on the block controllable Hessenberg form of system matrices (8.6). Like Algorithm 5.3, the basis vector matrix D_i corresponding to eigenvalue λ_i, $i=1, \ldots, n$, is computed first. Each of the p columns of D_i, $\mathbf{d}_{ij}$, $j=1, \ldots, p$, is computed by back substitution on one of the p linearly dependent columns of matrix $A-\lambda_i I$. As the dual of Conclusions 5.1 and 5.2, each eigenvector $\mathbf{v}_i$ has p linearly independent basis vectors in D_i. In addition, for each input say the j-th, μ_j $\mathbf{d}_{ij}$ vectors (for μ_j different values of i and λ_i) are also linearly independent.

We will analyze this linear dependency of the $\mathbf{d}_{ij}$ vectors completely and explicitly in the following. For simplicity of presentation and without loss of generality, we assume that the p controllability indices satisfy $\mu=\mu_1 \geq \mu_2 \geq \ldots \geq \mu_p$.

With this assumption, the $\mathbf{d}_{ij}$ basis vectors can be expressed as Tsui (1987a, b, 1993a)

$$\mathbf{d}_{ij} = \left[\begin{array}{ccccccc}
x & \vdots & \cdots & \vdots & x & \vdots & x \\
\vdots & \vdots & \cdots & \vdots & \vdots & \vdots & \vdots \\
\vdots & \vdots & \cdots & \vdots & \vdots & \vdots & x \\
\vdots & \vdots & \cdots & \vdots & \vdots & \vdots & * \\
\vdots & \vdots & \cdots & \vdots & \vdots & \vdots & 0 \\
\vdots & \vdots & \cdots & \vdots & \vdots & \vdots & \vdots \\
x & \vdots & \cdots & \vdots & x & \vdots & 0 \\
x & \vdots & \cdots & \vdots & x & \vdots & \\
\vdots & \vdots & \cdots & \vdots & \vdots & \vdots & \\
\vdots & \vdots & \cdots & \vdots & x & \vdots & \\
\vdots & \vdots & \cdots & \vdots & * & \vdots & \\
\vdots & \vdots & \cdots & \vdots & 0 & \vdots & \\
\vdots & \vdots & \cdots & \vdots & \vdots & \vdots & \\
x & \vdots & \cdots & \vdots & 0 & \vdots & \\
 & \vdots & & & & & \\
x & \vdots & & & & & \\
\vdots & \vdots & & & & & \\
x & \vdots & & & & & \\
* & \vdots & & & & & \\
0 & \vdots & & & & & \\
\vdots & \vdots & & & & & \\
0 & \vdots & & & & & \\
0 & & & & & & \\
\vdots & & & & & & \\
0 & & & & & &
\end{array}\right]
\begin{array}{l}
\\ \\ \\ \}p_1 \\ \\ \\ \\ \\ \\ \\ \}p_2 \\ \\ \\ \\ \\ \\ \\ \\ \}p_{\mu j} \\ \\ \\ \\ \\ \\
\end{array}
\begin{bmatrix} 1 \\ \lambda_i \\ \vdots \\ \lambda_i^{\mu j-1} \end{bmatrix} \underline{\Delta U_j \mathbf{v}_{ij}} \tag{9.16}$$

In the above $n \times \mu_j$ dimensional matrix U_j, the x elements are arbitrary, and the * elements are nonzero and are at the j-th position of their vector block and all blank entries are 0. Parameter p_k ($k=1, \ldots, \mu$) indicates the number of controllability indices that are $\geq k$ (see Definition 5.1 for the dual case).

The above partition of $\mathbf{d}_{ij}$ vectors is presented here for analysis purposes only. The actual $\mathbf{d}_{ij}$ vectors can be computed by direct back substitution from (8.6) without U_j and $\mathbf{v}_{ij}$.

Example 9.1

Let $\mu_1=4$, $\mu_2=\mu_3=2$, and $\mu_4=1$. Then from (9.16),

$$\mathbf{d}_{i1} = \underbrace{\begin{bmatrix}
x & x & x & * \\
x & x & x & 0 \\
x & x & x & 0 \\
x & x & x & 0 \\
x & x & * & 0 \\
x & x & 0 & 0 \\
x & x & 0 & 0 \\
x & * & 0 & 0 \\
* & 0 & 0 & 0
\end{bmatrix}
\begin{bmatrix} 1 \\ \lambda_i \\ \lambda_i^2 \\ \lambda_i^3 \end{bmatrix}}_{\underline{\Delta U_1 \mathbf{v}_{i1}}}, \quad
\mathbf{d}_{i2} = \underbrace{\begin{bmatrix}
x & x \\
x & * \\
x & 0 \\
x & 0 \\
x & 0 \\
* & 0 \\
0 & 0 \\
0 & 0 \\
0 & 0
\end{bmatrix}
\begin{bmatrix} 1 \\ \lambda_i \end{bmatrix}}_{\underline{\Delta U_2 \mathbf{v}_{i2}}}$$

$$\mathbf{d}_{i3} = \underbrace{\begin{bmatrix} x & x \\ x & x \\ x & * \\ x & 0 \\ x & 0 \\ x & 0 \\ * & 0 \\ 0 & 0 \\ 0 & 0 \end{bmatrix}\begin{bmatrix} 1 \\ \lambda_i \end{bmatrix}}_{\Delta U_3 \mathbf{v}_{i3}}, \quad \mathbf{d}_{i4} = \underbrace{\begin{bmatrix} x \\ x \\ x \\ * \\ 0 \\ 0 \\ 0 \\ 0 \\ 0 \end{bmatrix}}_{\Delta U_4 \mathbf{v}_{i4}}$$

Theorem 9.1: Dual of Conclusions 5.1 and 5.2

(A) For each fixed value of i (or λ_i), $i=1, \ldots, n$, its corresponding eigenvector $\mathbf{v}_i$'s p basis vectors $\mathbf{d}_{ij}$ ($j=1, \ldots, p$, corresponding to each of the p inputs) are linearly independent.

(B) For each fixed value of j (or j-th input or matrix U_j), $j=1, \ldots, p$, there are up to μ_j linearly independent $\mathbf{d}_{ij}$ vectors (for up to μ_j different values of i or λ_i) that are linearly independent.

Proof

A. For each value of i, it is clear from (9.16) and Examples 5.5 and 9.1 that the corresponding p different $U_j\mathbf{v}_{ij}$ vectors ($j=1, \ldots, p$) are linearly independent.

B. For each value of j, up to μ_j $\mathbf{d}_{ij}$ vectors (say $i=1, \ldots, \mu_j$) are, from (9.16),

$$\left[\mathbf{d}_{1j} : \ \ldots \ : \mathbf{d}_{\mu j, j}\right] = U_j\left[\mathbf{v}_{1j} : \ \ldots \ : \mathbf{v}_{\mu j, j}\right] \equiv U_j V_j \tag{9.17}$$

Because the μ_j columns of matrix U_j of (9.17) are linearly independent, and because up to u_j of $\mathbf{v}_{ij}$ vectors of (9.17) are also linearly independent since they form a Vandermonde matrix V_j of size μ_j), up to μ_j $\mathbf{d}_{ij}$ vectors are linearly independent.

The above part (B) can be extended to general eigenvalue cases. Equations (9.16) and (9.17) indicate that all eigenvalues and their exponentials are associated only with the Vandermonde matrix V_j of (9.17) which is the right eigenvector matrix of a companion form matrix, which is the transpose of (1.14) (Brand, 1968). Because the general eigenvalues cannot change the linear dependency of the corresponding eigenvectors of V_j, the general eigenvalues cannot change the linear dependency of the columns of U_jV_j.

General Rule of Decoupling Eigenvector Assignment

The analytical decoupling eigenvector assignment is to decouple the n eigenvectors (and the corresponding n eigenvalues) into p separate groups, each with μ_j eigenvectors U_jV_j, or each of the p groups is associated with only one of the p inputs:

$$V = \left[U_1V_1 : \ \ldots \ : U_pV_p\right]$$
$$= \left[U_1 : \ \ldots \ : U_p\right]\text{diag}\left\{V_1,\ldots,V_p\right\} \equiv U\ \text{diag}\left\{V_1,\ldots,V_p\right\} \qquad (9.18)$$

In general, each ith eigenvector $\mathbf{v}_i = D_i\mathbf{c}_i = [U_1 : \ \ldots \ : U_p]\text{diag}\{\mathbf{v}_{i1}, \ldots, \mathbf{v}_{ip}\}\ \mathbf{c}_i$ (see (9.16)) or

$$V = [\mathbf{v}_1 : \ \ldots \ : \mathbf{v}_n]$$
$$= \left[U_1 : \ \ldots \ : U_p\right]\left[\text{diag}\left\{\mathbf{v}_{11},\ldots,\mathbf{v}_{1p}\right\}\mathbf{c}_1 : \ \ldots \ : \text{diag}\left\{\mathbf{v}_{n1},\ldots,\mathbf{v}_{np}\right\}\mathbf{c}_n\right] \qquad (9.19)$$

Compare the two matrices of each of (9.18) and (9.19), the first or the left matrix is the same U, only the second matrix (or the right matrix) is different. Our analytical eigenvector assignment (9.18) has its right and second matrix decoupled into p diagonal and decoupled blocks V_j, $j=1, \ldots, p$.

The remaining freedom of this decoupling eigenvector assignment is how to distribute the n eigenvalues (and their eigenvectors) into these p groups (or blocks) of sizes μ_j, $j=1, \ldots, p$. We have the following three specific rules, based on the general understanding of Section 8.1.

Rule 1

Distribute multiple eigenvalues into different V_j groups. This is because only at the same group or only for the same value of j, can the eigenvectors corresponding to multiple eigenvalues be defective and be bad conditioned (see (5.15d), rule (5) of Section 8.1, and Golub and Wilkinson (1976b)).

Rule 2

Distribute relatively more important eigenvalues (such as the ones that are closer to the unstable region, see Subsection 2.2.2) into relatively smaller groups. This is because the smaller matrix block V_j is usually better conditioned than the larger blocks. On the other hand, the condition of a larger block with a larger controllability index μ_j has more dominant effect on the condition of the overall eigenvector matrix V.

Rule 3

Distribute within each of the p blocks the eigenvalues with similar magnitude and evenly distributed phase angles between 90° and 270°. See (4) and (6) of Section 8.1.

The above three rules fully considered the important analytical parameters such as the eigenvalues (poles) and controllability indices. These important analytical parameters are currently not considered at all in numerical eigenvector assignment methods.

It should be mentioned again that in the actual design computation, after the eigenvalue λ_i is distributed into the j-th input group, then the corresponding eigenvector $\mathbf{v}_i$ can be computed directly, by back substitution operation on the dependent column corresponding to the j-th input, of matrix $A - \lambda_i I$, where the system matrices are in the block controllable Hessenberg form (8.6). This direct computation does not need the computation of the intermediate results U_j and $\mathbf{v}_{ij}$ of (9.16).

Unlike numerical eigenvector assignment methods, which have no intermediate steps – only a one-step run to the final numerical answer with a single goal of either minimized value of $k(V)$ or minimized value of $\|K\|$, the analytical eigenvector assignment method has multiple and intermediate steps, considerations or goals such as rules 1–3. Thus, this method can be illustrated by examples with several steps.

Example 9.2: Let the System Matrices Be

$$[A : B] = \begin{bmatrix} -20 & 0 & & 0 & 0 & 0 & : & 20 & 0 \\ 0 & -20 & & 0 & 0 & 0 & : & 0 & 20 \\ \ldots\ldots & \ldots\ldots & \ldots & \ldots\ldots & \ldots\ldots & \ldots & : & \ldots\ldots \\ -0.08 & -0.59 & : & -0.174 & 1 & 0 & : & 0 & 0 \\ -18.95 & -3.6 & : & -13.41 & -1.99 & 0 & : & 0 & 0 \\ 2.07 & 15.3 & : & 44.79 & 0 & 0 & : & 0 & 0 \end{bmatrix} \tag{9.20}$$

This is the state space model of a fighter plane at flight condition of 3,048-meter height and Mach 0.77 (Sobel et al., 1984; Spurgeon, 1988). The five system states are elevator angle, flap angle, incidence angle, pitch rate, and normal acceleration integrated, respectively, while the two inputs are elevator angle demand and flap angle demand, respectively.

The problem is to design a state feedback control law K so that eigenvalues $\lambda_1=-20$, $\lambda_{2,3}=-5.6\pm j4.2$, and $\lambda_{4,5}=-10\pm j10\sqrt{3}$ are assigned to matrix $A - BK$. Because $p=2>1$, there is eigenvector assignment freedom to make and design.

We first compute the block controllable Hessenberg form using the dual of Algorithm 5.2. Because matrix B is already in the desired form, only one triangularization (at $j=2$ in Algorithm 5.2) of the lower-left 3×2 block of matrix A of (9.20) is needed.

Hence the unitary operator matrix H is

$$H = \begin{bmatrix} 1 & 0 & 0 & 0 & 0 \\ 0 & 1 & 0 & 0 & 0 \\ 0 & 0 & & & \\ 0 & 0 & & H_2 & \\ 0 & 0 & & & \end{bmatrix} = \begin{bmatrix} 1 & 0 & 0 & 0 & 0 \\ 0 & 1 & 0 & 0 & 0 \\ 0 & 0 & -0.0042 & -0.9941 & 0.1086 \\ 0 & 0 & 0.0379 & -0.1086 & 0.9931 \\ 0 & 0 & 0.9991 & -0 & 0.0386 \end{bmatrix}$$

The resulting block controllable Hessenberg form of system matrices is

$$[H'AH : H'B]$$

$$= \begin{bmatrix} -20 & 0 & 0 & 0 & 0 & : & 20 & 0 \\ 0 & -20 & 0 & 0 & 0 & : & 0 & 20 \\ \cdots & \cdots & \cdots & \cdots & \cdots & \cdots & \cdots & \cdots \\ 19.0628 & 5.2425 & : & -2.0387 & 0.4751 & 18.1843 & : & 0 & 0 \\ 0 & -14.8258 & : & -0.0718 & -1.6601 & -43.0498 & : & 0 & 0 \\ \cdots & \cdots & : & \cdots & \cdots & \cdots & \cdots & \cdots & \cdots \\ 0.0002 & 0.0005 & : & -0.9931 & -0.109 & : & -0.0114 & : & 0 & 0 \end{bmatrix} \quad (9.21)$$

which is still denoted as $[A : B]$ in the rest of this example for simplicity of presentation, even though the rest of the design computation is based on (9.21) instead of (9.20). The two nonzero elements [0.0002 0.0005] at the lower-left corner are computational errors (should be 0 and will be considered zero). From this result, the controllability indices are $\mu_1 = 3$ and $\mu_2 = 2$.

We take the back substitution operation starting at the fifth (for U_1) and fourth (for U_2) columns and the fifth row, of matrix $A - \lambda_i I$, and derive the U_1 and U_2 matrices of the basis vector matrices of (9.16–9.17):

$$U = [U_1 : U_2]$$

$$= \begin{bmatrix} 0.67923 & -0.114 & -0.05395 & : & 0.9648 & 0.012415 \\ -2.889 & 0.015167 & 0 & : & -0.108166 & -0.06743 \\ -0.0043141 & -1.02945 & 0 & : & -0.11683 & 0 \\ 0 & 0 & 0 & : & 1 & 0 \\ 1 & 0 & 0 & : & 0 & 0 \end{bmatrix}$$

Because among the five assigned eigenvalues there are two complex conjugate pairs, we only have two possible distributions of these n eigenvalues in V_1 and V_2:

Assignment 1:

$$\operatorname{diag}\{V_1,V_2\}=\operatorname{diag}\left\{\begin{bmatrix}1&1&1\\ \lambda_1&\lambda_2&\lambda_3\\ \lambda_1^2&\lambda_2^2&\lambda_3^2\end{bmatrix},\begin{bmatrix}1&1\\ \lambda_4&\lambda_5\end{bmatrix}\right\} \tag{9.22}$$

$$\Lambda_1=\operatorname{diag}\left\{\operatorname{diag}\{\lambda_1,\lambda_2,\lambda_3\},\operatorname{diag}\{\lambda_4,\lambda_5\}\right\}$$

Assignment 2:

$$\operatorname{diag}\{V_1,V_2\}=\operatorname{diag}\left\{\begin{bmatrix}1&1&1\\ \lambda_1&\lambda_4&\lambda_5\\ \lambda_1^2&\lambda_4^2&\lambda_5^2\end{bmatrix},\begin{bmatrix}1&1\\ \lambda_2&\lambda_3\end{bmatrix}\right\} \tag{9.23}$$

$$\Lambda_2=\operatorname{diag}\left\{\operatorname{diag}\{\lambda_1,\lambda_4,\lambda_5\},\operatorname{diag}\{\lambda_2,\lambda_3\}\right\}$$

The final eigenvector matrix V is $[U_1V_1 \vdots U_2V_2]$ according to (9.18). For the above two different assignments of V_1 and V_2, the corresponding final eigenvector matrices are denoted as V^1 and V^2.

To compute V^1 and V^2 directly from the matrix A of (9.21), we compute the $\mathbf{d}_{ij}$ vectors such that the lower $n-p$ (=3) rows of $(A-\lambda_i I)\mathbf{d}_{ij}=0$. The first group ($j=1$) of three columns of V^1 and V^2 will be computed by back substitution on the fifth column of matrix $A-\lambda_i I$ (using the first three eigenvalues of (9.22) and (9.23) for V^1 and V^2, respectively), starting at the element $a_{55}-\lambda_i$. The second ($j=2$) group of two columns of V^1 and V^2 will be computed by back substitution on the fourth column of matrix $A-\lambda_i I$ (using the fourth and fifth eigenvalues of (9.22) and (9.23) for V^1 and V^2, respectively), starting at the element a_{54}.

To broaden the comparison, in addition to V^1 and V^2, a third eigenvector matrix V^3 is computed as

$$V^3=QV_1$$

$$=\begin{bmatrix}-0.0528&0.0128&-0.1096&-0.0060&-0.1555\\ 0&-0.06745&0.0049&-0.1114&-2.904\\ 0&0&-1.007&-0.1098&-0.01148\\ 0&0&0&1&0\\ 0&0&0&0&1\end{bmatrix}\begin{bmatrix}\lambda_1^4&\lambda_2^4&\lambda_3^4&\lambda_4^4&\lambda_5^4\\ \lambda_1&\lambda_2&\lambda_3&\lambda_4&\lambda_5\\ \lambda_1^3&\lambda_2^3&\lambda_3^3&\lambda_4^3&\lambda_5^3\\ 1&1&1&1&1\\ \lambda_1^2&\lambda_2^2&\lambda_3^2&\lambda_4^2&\lambda_5^2\end{bmatrix}$$

(9.24a)

and $\Lambda_3 = \text{diag}\{\lambda_1, \lambda_2, \lambda_3, \lambda_4, \lambda_5\}$ (9.24b)

Here, the nonsingular matrix Q is the similarity transformation matrix that makes $(Q^{-1}AQ, Q^{-1}B)$ into a controllable canonical form (a dual of (1.16)) (Wang and Chen, 1982). There is no decoupling of this eigenvector assignment because component matrix V_1 has no partition at all.

Although eigenvector matrices V^i, $i=1, 2, 3$ are computed from (9.21), which is a similarity transformation of the original system state space model (9.20), the conditions of the eigenvector matrices of these two state space models are the same, because the similarity transformation matrix H between these two systems is unitary.

The state feedback control gain K^i, $i=1, 2, 3$ for the three eigenvector matrices can be computed first by the formula (9.2)

$$\underline{K}^i = K^i\left(V^i\right)^{-1} = \left[B_1^{-1} : 0\right]\left(A - V^i\Lambda_i\left(V^i\right)^{-1}\right) \quad (9.25)$$

where K^i and V^i are the solution matrices of (8.1b) with (A, B) in the block controllable Hessenberg form (9.21). Therefore, the actual state feedback control gain $\boldsymbol{K}$ for the original system state space model (9.20), $\boldsymbol{K}_i$, should be

$$\boldsymbol{K}_i = \underline{K}^i H, i = 1, 2, 3 \quad (9.26)$$

These three $\boldsymbol{K}_i$ values are:

$$\boldsymbol{K}_1 = \begin{bmatrix} 0.4511 & 0.7991 & -1.3619 & -0.4877 & 1.0057 \\ 0.0140 & -0.0776 & 2.6043 & 0.2662 & 1.357 \end{bmatrix}$$

$$\boldsymbol{K}_2 = \begin{bmatrix} 0.8944 & 0.9611 & -20.1466 & -1.7643 & 0.8923 \\ 0.0140 & -0.5176 & 1.3773 & 0.1494 & 0.1800 \end{bmatrix}$$

and

$$\boldsymbol{K}_3 = \begin{bmatrix} 1 & 5{,}044 & 14{,}565 & 23 & 1{,}033 \\ 0 & -1 & 0 & 0 & 0 \end{bmatrix}$$

It can be verified that the eigenvalues of matrices $A-B\boldsymbol{K}_i$ are correctly placed for $i=1, 2$, but eigenvalues of matrix $A-B\boldsymbol{K}_3$ differ noticeably from the desired values. This difference is caused by the computational error but not by the methods. However, this computational error does demonstrate that a good eigenvector assignment such as V^1 and V^2 is very important on the numerical accuracy aspect of this eigenvalue assignment problem (see Section 8.5).

Table 9.1 compares the feedback control gain $\|\boldsymbol{K}\|$ and eigenvector matrix condition number $k(V)$ of these three eigenstructure assignment designs. As introduced at the beginning of this chapter, these two aspects of eigenvector assignment are perhaps most technically important. This should be why the comparisons between the three $\|K_i\|$ and between the three $k(V^i)$ all conform fully, with the comparison of the zero-input response simulation for a common initial state $\mathbf{x}(0)=[1\ 1\ 1\ 1\ 1]'$ of these three assignments, in Figure 9.1.

TABLE 9.1

Comparison of Three Eigenvector Assignments of Example 9.2

Eigenvector Assignment i	i=1:Eq. (9.22)	i=2:Eq. (9.23)	i=3:Eq. (9.24)
$\|K_i\|_F$	3.556	20.34	15,448
$k(V^i)$	71.446	344.86	385,320

It is clear that the smaller the technical indicators $\|K\|_F$ and $k(V)$ in Table 9.1, the smoother the corresponding zero-input response of Figure 9.1. This is predicted at the beginning of this chapter. An even more important technical indicator about the overall system performance and robustness – the robust stability criteria (2.25), should also conform with this comparison.

This comparison indicates that eigenvector assignment makes a great difference in the technical aspects (performance and robustness) of the feedback system.

Unlike the numerical eigenvector assignment methods of Section 9.1 and optimal control design methods of Chapter 10, there is a very clear analytical understanding of the formulation, intermediate eigenvalue distribution decision, and final technical indicators and simulations, of the analytical eigenvector assignment method of this chapter. Only this kind of understanding can guide the reversed adjustment based on the final result data like Table 9.1 and simulation like Figure 9.1, on different design formulations of these three eigenvector assignments. The guide on such design aspects as whether to decouple and how to distribute eigenvalue distribution into different Jordan blocks.

For example, the final results show that assignment 3, which has no decoupling, has far worse (rougher) response simulation and far higher sensitivity $k(V)$ and control gain$\|K\|$. This final result reversely indicates the great effect of decoupling.

Another example is the comparison between the two decoupling eigenvector assignment of V^1 and V^2. This comparison reversely indicates the dominant effect of larger blocks such as V_1 (size $\mu_1=3$) over the smaller block such as V_2 (size $\mu_2=2$), on the overall condition number of V^1 and V^2. It can be calculated that $k(V_2)$ is greater in V^1 than in V^2 (23.6 vs. 11), yet the overall $k(V^1)$ is five times smaller than $k(V^2)$ in Table 9.1. This is because $k(V_1)$ is smaller in V^1 than in V^2 (653.7 vs. 896.5). This understanding reversely implies that Rule 2 (emphasize more on larger blocks) is more important than Rule 3 (emphasize more on relative eigenvalue positions within each block) of our analytical eigenvector assignment.

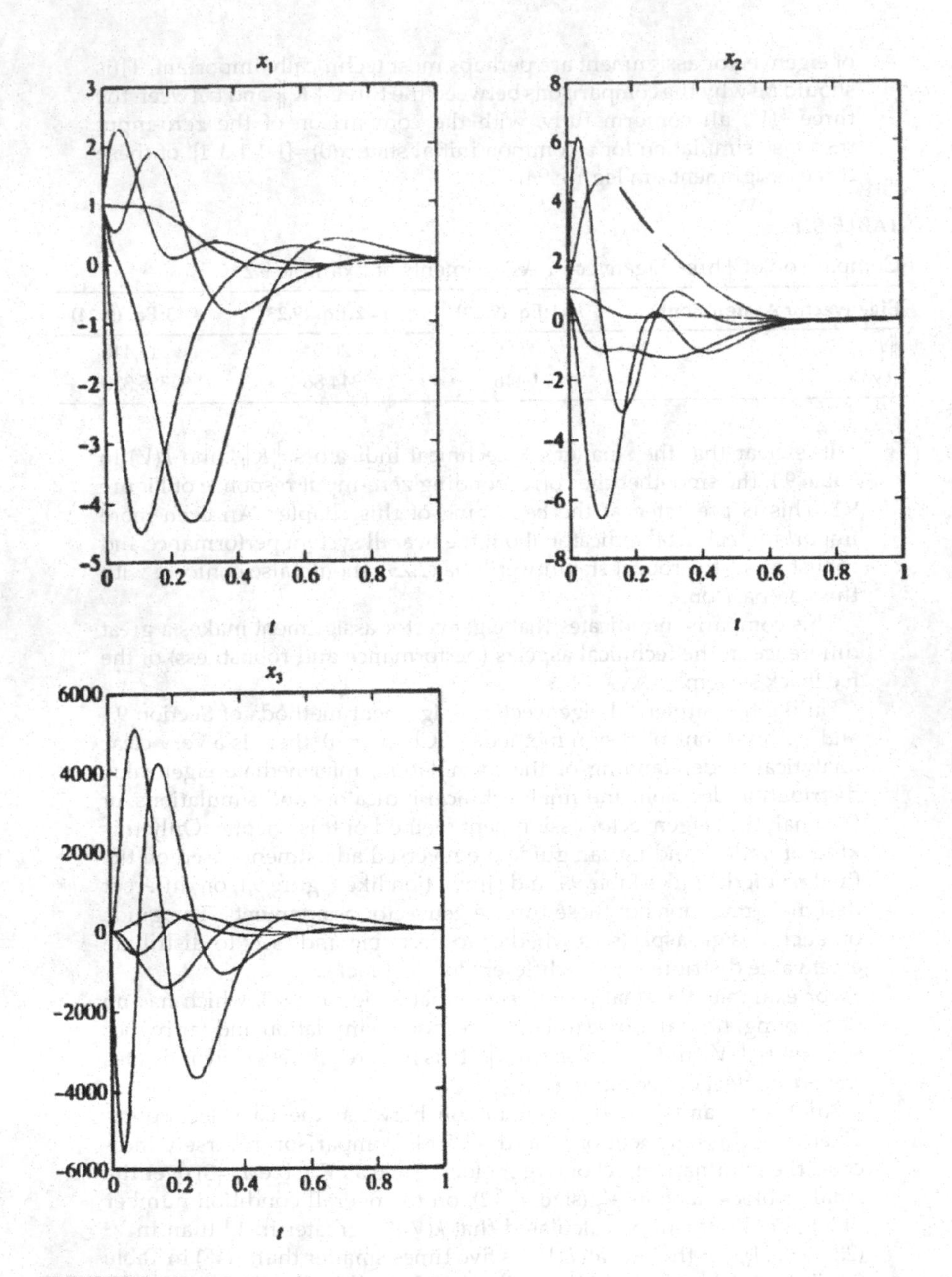

FIGURE 9.1
Comparison of zero-state responses of three eigenvector assignments.

9.3 Summary of Eigenstructure Assignment of Chapters 8 and 9

Chapter 8 presented the eigenvalue assignment design Algorithms 8.1 and 8.2 for generalized state feedback control (a special case of which is direct state feedback control). This design also computed all basis vectors for the associated eigenvectors, for the first time. Chapter 9 presented two numerical and one analytical eigenvector assignment methods. These three methods can compute the linear combinations of the basis vectors of the eigenvectors generally, explicitly, and effectively, to make these eigenvectors as linearly independent as possible.

Based on the analysis of Chapter 2, system poles (eigenvalues) can determine system response and performance most directly and accurately, and far more directly and accurately than any other system parameters such as bandwidth. Furthermore, eigenvectors can determine the sensitivity and robustness of the corresponding eigenvalues and therefore determine (together with eigenvalues) system robust stability far more directly and accurately than any other criteria such as gain margin and phase margin. Therefore, eigenstructure assignment can improve system performance and robustness far more directly and effectively, than other designs such as optimal control designs.

There are two more distinct advantages of the eigenstructure assignment design results of this book over other optimal control designs such as the linear quadratic control design of Chapter 10.

First, this design can consider fully some key technical parameters and properties, such as system poles, controllability indices, and decoupling. This advantage is enabled by the very clear, simple, and sharp analytical understandings of Chapter 2. The selection of the LQ optimal criterion cannot consider these technical parameters and properties.

Second, this design enables the rational adjustment of design parameters and procedures, such as the distribution of system poles into different decoupled groups, the purpose of eigenvector assignment (Section 8.4), weighting factors of eigenvectors (Algorithm 9.2), etc., based on the final design results (see the end of Example 9.2). The optimal designs cannot have this adjustment which is essential in practice. This advantage is enabled not only by the clear and sharp understanding of system parameters and properties but also by much simpler and more explicit design computation procedures – even the numerical Algorithms 9.1 and 9.2 are much simpler than other numerical design computations (see Chapter 10). This computational simplicity itself is also enabled by the clear and sharp advances in theoretical analysis.

To summarize, the superiority of eigenstructure assignment formulation is supported by the full and basic analytical results of the control systems theory. This design has become practical and useful because of the development of general, simple, and effective design algorithms and methods of Chapters 8 and 9.

The eigenstructure assignment, especially the eigenvector assignment, is achievable *only* by $Kx(t)$-control or $\underline{KC}x(t)$-control, and the new design principle of Chapters 4–6 of this book enabled *for the first time the full realization* of this control for the great majority of plant systems. These two developments have major implications on the whole control design theory: If the superiority of the modern control theory over the classical control theory used to be reflected in the control system *modeling and analysis*, then this superiority is now reflected in the *design* of control systems. This monumental development will enable far more general and successful practical applications of the modern control theory, even though this brilliant prospect has not yet been fully realized by the mainstream control community.

Exercises

9.1 Assign eigenvalues $\lambda_1=-1$, $\lambda_{2,3}=-2\pm j$, and $\lambda_{4,5}=-1\pm j2$ to the following system

$$A=\begin{bmatrix} 0 & 1 & 0 & : & 0 & 0 \\ 0 & 0 & 1 & : & 0 & 0 \\ 2 & 0 & 0 & : & 1 & 1 \\ 0 & 0 & 0 & : & 0 & 1 \\ 0 & 0 & 0 & : & -1 & -2 \end{bmatrix} \text{ and } B=\begin{bmatrix} 0 & 0 \\ 0 & 0 \\ 1 & 0 \\ 0 & 0 \\ 0 & 1 \end{bmatrix}.$$

Repeat Example 9.2 for the following three eigenstructure assignments (1) – (3). Compare $\|K\|$, $k(V)$, robust stability M_3 of (2.25), and the waveforms of zero-input responses for the common initial condition $\mathbf{x}(0)=\begin{bmatrix} 2 & 1 & 0 & -1 & -2 \end{bmatrix}'$ of these three assignments.

Partial answers:

a. $(\Lambda,K)=\left(\operatorname{diag}\{\lambda_{1,2,3},\ \lambda_{4,5}\},\ \begin{bmatrix} 7 & 9 & 5 & 1 & 1 \\ 0 & 0 & 0 & 4 & 0 \end{bmatrix}\right)$

b. $(\Lambda,K)=\left(\operatorname{diag}\{\lambda_{1,4,5},\ \lambda_{2,3}\},\ \begin{bmatrix} 7 & 7 & 3 & 1 & 1 \\ 0 & 0 & 0 & 4 & 2 \end{bmatrix}\right)$

c. $(\Lambda,K)=\left(\operatorname{diag}\{\lambda_{1,2,3,4,5}\},\ \begin{bmatrix} 2 & 0 & 0 & 0 & 1 \\ 25 & 55 & 48 & 23 & 5 \end{bmatrix}\right)$

Hint: System matrices (A, B) is a sequence-permutation similarity transformation of the block controllable Hessenberg (or canonical) form (8.6), on which Example 9.2 is based.

To derive form (8.6) from this (A, B), change the sequence of rows of (A, B) from (1, 2, 3, 4, 5) to (3, 5, 2, 4, 1). Then make the same change of the sequence of columns of matrix A.

9.2 Assign the same eigenvalues of Exercise 9.1 to the following system

$$A=\begin{bmatrix} 0 & 1 & 0 & : & 0 & 0 \\ 0 & 0 & 1 & : & 0 & 0 \\ 3 & 1 & 0 & : & 1 & 2 \\ 0 & 0 & 0 & : & 0 & 1 \\ 4 & 3 & 1 & : & -1 & -4 \end{bmatrix} \text{ and } B=\begin{bmatrix} 0 & 0 \\ 0 & 0 \\ 1 & 0 \\ 0 & 0 \\ 0 & 1 \end{bmatrix}$$

Repeat Example 9.2 for the following three eigenstructure assignments (1)–(3). Compare $\|K\|$, $k(V)$, robust stability M_3 of (2.25), and the waveforms of zero-input responses for the common initial condition $\mathbf{x}(0)=\begin{bmatrix} 2 & 1 & 0 & -1 & -2 \end{bmatrix}'$ of these three assignments. Partial answers:

a. $$(\Lambda, K)=\left(\text{diag}\{\lambda_{1,2,3},\ \lambda_{4,5}\},\ \begin{bmatrix} 8 & 10 & 5 & 1 & 2 \\ 4 & 3 & 1 & 4 & -2 \end{bmatrix}\right)$$

b. $$(\Lambda, K)=\left(\text{diag}\{\lambda_{1,4,5},\ \lambda_{2,3}\},\ \begin{bmatrix} 8 & 8 & 3 & 1 & 2 \\ 4 & 3 & 1 & 4 & 0 \end{bmatrix}\right)$$

c. $$(\Lambda, K)=\left(\text{diag}\{\lambda_{1,2,3,4,5}\},\ \begin{bmatrix} 3 & 1 & 0 & 0 & 2 \\ 29 & 58 & 49 & 23 & 3 \end{bmatrix}\right)$$

9.3 Assign the same eigenvalues of Exercise 9.1 to the following system

$$A=\begin{bmatrix} 0 & 1 & 0 & : & 0 & 0 \\ 0 & 0 & 1 & : & 0 & 0 \\ -10 & -16 & -7 & : & 1 & 2 \\ 0 & 0 & 0 & : & 0 & 1 \\ 4 & 3 & 1 & : & -2 & -2 \end{bmatrix} \text{ and } B=\begin{bmatrix} 0 & 0 \\ 0 & 0 \\ 1 & 0 \\ 0 & 0 \\ 0 & 1 \end{bmatrix}$$

Repeat Example 9.2 for the following three eigenstructure assignments (1)–(3). Compare $\|K\|$, $k(V)$, robust stability M_3 of (2.25), and the waveforms of zero-input responses for the common initial condition $x(0)=[2\ 1\ 0{-}1{-}2]'$, of these three assignments. Partial answers:

a. $$(\Lambda, K)=\left(\text{diag}\{\lambda_{1,2,3},\ \lambda_{4,5}\},\ \begin{bmatrix} -5 & -7 & -2 & 1 & 2 \\ 4 & 3 & 1 & 3 & 0 \end{bmatrix}\right)$$

b. $(\Lambda, K) = \left(\text{diag}\{\lambda_{1,4,5},\ \lambda_{2,3}\},\ \begin{bmatrix} -5 & -9 & -4 & 1 & 2 \\ 4 & 3 & 1 & 3 & 2 \end{bmatrix} \right)$

c. $(\Lambda, K) = \left(\text{diag}\{\lambda_{1,4,5,2,3}\},\ \begin{bmatrix} -10 & -16 & -7 & 0 & 2 \\ 29 & 58 & 49 & 22 & 5 \end{bmatrix} \right)$

Also see the last three design projects of Appendix B, for numerical eigenvector assignment algorithms.

10

Control Design for LQ Optimal Control

Optimal control is a main result of the whole control systems theory. Before the 1980s, optimal control is limited to linear-quadratic (LQ) optimal control. This control is formulated with state space models and is within the state space control theory. It is usually covered in a course next to the first course of the state space control theory.

Unlike the eigenstructure assignment control which is developed from the basic results of control theory and control system analysis of Chapter 2, the LQ optimal control is aimed entirely at minimizing an abstract and mathematical criterion called quadratic criterion:

$$J = (1/2)\int_0^\infty [\mathbf{x}(t)'Q\mathbf{x}(t) + \mathbf{u}(t)'R\mathbf{u}(t)]dt \tag{10.1}$$

where Q is a symmetrical, positive semidefinite matrix and R is a symmetrical positive definite matrix. The LQ optimal design is aimed at minimizing J under the constraint (1.1a)

$$d\mathbf{x}(t)/dt = A\mathbf{x}(t) + B\mathbf{u}(t), \text{ and a given } \mathbf{x}(0)$$

Inspection of (10.1) shows that to minimize or to have a finite value of J, $\mathbf{x}(t \to \infty)$ must be zero or the system must be stable (see Definition 2.1). In addition, among the two terms of J, the first term reflects the quickness and smoothness of $\mathbf{x}(t)$ before it converges to zero, which is closely related to system performance, while the second term reflects the control energy, which is closely related to control gain and system robustness (see Example 9.2).

It is well known that performance and robustness are contradictory to each other. For example, the faster the $\mathbf{x}(t)$ converges to 0, the higher the control power needed to steer $\mathbf{x}(t)$.

The magnitude of Q and R can reflect the relative importance of these two properties. A relatively large Q (compared to R) indicates the higher priority for performance over control energy and cost. When $R=0$, the corresponding LQ problem is called the "minimal (response) time problem" (Friedland, 1962, 1989). Anti-air missile control can be such a problem. On the other hand, a relatively small Q (compared to R) indicates a higher priority on saving control energy over performance. When $Q=0$, the corresponding LQ problem is called the "minimal fuel problem" (Athanassiades, 1963). Remote orbit space craft control can be considered as such a problem.

DOI: 10.1201/9781003259572-10

However, besides the magnitude, there are no more general, analytical, and explicit considerations of system performance and robustness, made on the more than $n^2/2$ parameters of Q and $p^2/2$ parameters of R. Hence, the LQ criterion/problem itself is still very abstract, and still reflects very grossly and vaguely the overall system performance and robustness.

To summarize, matrices Q and R (or J) are not really the direct and accurate reflection of actual system performance and robustness. The result of minimizing the value of J is optimal only in a mathematical sense, but not really in an actual control system sense, or not really at optimizing control system performance and robustness. For example, it is not even possible to compare the overall performance and robustness levels corresponding to two different J's.

The state feedback control that minimizes the criterion J of (10.1) is unique (see Theorem 10.1), or the J of more than $n^2/2+p^2/2$ almost randomly selected parameters fully determines the feedback system! Isn't it more important to optimize the selection of this many parameters of J themselves? How can such a design problem/formulation be really rational, since that criterion J cannot accurately measure the overall system performance and robustness level?

This critical deficiency of LQ optimal control is further compounded by the fact that unlike eigenstructure assignment, it is very hard to adjust generally, systematically, and wisely the criterion J, based on the finally computed design solution (K). This is because only a numerically approximate solution can be computed that minimizes J. This situation is like the numerical eigenvector assignment algorithms but much worse (see the summary of Chapter 9).

The computational difficulty and complexity of all control design algorithms of Chapters 8–10 will be summarized and compared in Section 10.3.

Furthermore, and even more serious, that only $K\mathbf{x}(t)$-control with a unique gain K can minimize a given J, yet such a control cannot be fully realized by a great majority of the plant systems (see Chapters 4 and 6)! In other words, the LQ optimal design is rarely realizable and realistic.

These three critical drawbacks of the LQ optimal design listed above are at least shared by all other optimal control designs, if not more severe. For example, some other optimal control designs are based on the system's frequency response, which is even less generally accurate in reflecting system performance and robustness (see Chapter 2). Furthermore, most of the other optimal control designs require even much more difficult computation than that of the LQ optimal control design. For example, the H_∞ optimal control design requires the solving of multiple algebraic Riccati equations (the LQ optimal control design requires the solving of only one such equation).

The LQ optimal control problem has been extensively studied and covered in the literature since 1960. This book intends to introduce only the essential and practical design computational algorithms as well as the basic and practical physical meanings, of this design problem.

As with the presentation of eigenstructure assignment design, the LQ design in this chapter is divided into state feedback control $K\mathbf{x}(t)$ in

Section 10.1 and then generalized state feedback control $\underline{K}\underline{C}\mathbf{x}(t)$ in Section 10.2. This $\underline{K}\underline{C}\mathbf{x}(t)$-control unifies the existing $K\mathbf{x}(t)$-control and the existing static output feedback control (corresponding to $\underline{C}=C$), see Conclusion 6.6.

10.1 Direct State Feedback Control Design

The direct state feedback control design for LQ optimal control has been extensively covered in the literature (Kalman, 1960; Chang, 1961; Pontryagin et al., 1962; Athans and Falb, 1966; Bryson and Ho, 1969; Anderson and Moore, 1971; Kwakernaak and Sivan, 1972; Sage and White, 1977). The following solution of this problem can be formulated using Calculus of Variation with Lagrange multipliers.

Theorem 10.1

The unique solution that minimizes J of (10.1) and that is subject (1.1a) is

$$u^*(t) = -K^*x(t) \quad \text{where}\, K^* = R^{-1}B'P \tag{10.2}$$

and where P is the symmetric and positive definite solution matrix of the following algebraic Riccati equation (ARE)

$$PA + A'P + Q - PBR^{-1}B'P = 0 \tag{10.3}$$

Based on this optimal control $\mathbf{u}^*(t)$, there is an optimal state trajectory $\mathbf{x}^*(t)$, which is the solution of

$$d\mathbf{x}^*(t)/dt = A\mathbf{x}^*(t) + B\mathbf{u}^*(t), \quad \text{and}\, \mathbf{x}^*(0) = \text{the given } \mathbf{x}(0) \tag{10.4}$$

and the minimized LQ criterion is

$$J^* = (1/2)\mathbf{x}(0)'P\mathbf{x}(0) \tag{10.5}$$

Theorem 10.1 indicates that the main difficulty of the LQ optimal design is at solving ARE (10.3). There are a number of numerical methods available for solving (10.3) such as the method of eigenstructure decomposition of the Hamiltonian matrix

$$H = \begin{bmatrix} A & -BR^{-1}B' \\ -Q & -A' \end{bmatrix} \tag{10.6}$$

Van Loan (1984), Byers (1983, 1990), Xu (1991), the method of matrix sign function (Byers, 1987), etc. The basic version of computing the Hamiltonian decomposition using Schur triangularization (Laub, 1979) is described in the following.

Algorithm 10.1: Solving Algebraic Riccati Equation

Step 1: Compute the Hamiltonian matrix H of (10.6).

Step 2: Make Schur triangularization (Francis, 1961, 1962) of matrix H such that

$$U'HU = S = \begin{bmatrix} S_{11} & S_{12} \\ 0 & S_{22} \end{bmatrix}, \quad U'U = I \tag{10.7}$$

where matrix S is an essentially upper triangular (called the Schur triangular form) matrix, whose diagonal elements equal the eigenvalues of H (2×2 diagonal block for complex conjugate eigenvalues), and the eigenvalues of S_{11} are stable.

Step 2a: Let $k=1$ and $H_1=H$.

Step 2b: Compute the unitary matrix Q_k such that

$$Q_k' H_k = R_k \tag{10.8}$$

where R_k is upper triangular (see Section A.2).

Step 2c: Compute

$$H_{k+1} = R_k Q_k \tag{10.9}$$

Step 2d: If H_{k+1} is already in the Schur triangular form (10.7), then go to Step 3. Otherwise let $k=k+1$. Then go to Step 2b for another iteration.

Step 3: Based on (10.8) and (10.9),

$$H_{k+1} = Q_k' H_k Q_k = Q_k' \ldots Q_1' H_1 Q_1 \ldots Q_k$$

Therefore, if the solution matrix U of (10.7) is

$$U = Q_1 \ldots Q_k \triangleq \begin{bmatrix} U_{11} & U_{12} \\ U_{21} & U_{22} \end{bmatrix} \tag{10.10},$$

then by substituting (10.6) and (10.10) into (10.7) and then comparing with (10.3), we have

$$P = U_{21} U_{11}^{-1} \tag{10.11}$$

To accelerate the convergence or reduce the number of iterations of Step 2, Step 2b can be adjusted such that it becomes

$$Q_k'(H_k - s_k I) = R_k$$

and Step 2c should be adjusted back correspondingly such that

$$H_{k+1} = R_k Q_k + s_k I$$

This adjusted version of Step 2 is called the "shifted version", where the shift s_k is determined by the two eigenvalues of the bottom right 2×2 corner block of H_k (Wilkinson, 1965).

The main computation of Algorithm 10.1 is at the iterative Step 2. The main computation of Step 2 is at Step 2b (10.8). The computation of (10.8) using the Householder method is of order $2(2n)^3/3$ (see Section A.2 and notice that the dimension of matrix H is $2n$ instead of n).

Because of some special properties of Hamiltonian matrix H, it is possible to half the size of H from $2n$ to n, before the computation of Step 2. The general procedure is outlined by Xu (1991) as follows.

First, compute H^2, which is skew symmetrical ($H^2 = -(H^2)'$). The Schur triangularization will be made on part of H^2 instead of H.

Second, make elementary symplectic transformation on H^2 (Paige and Van Loan, 1981) such that

$$V'H^2V = \begin{bmatrix} H_1 & X \\ 0 & H_1' \end{bmatrix}, (V'V = I) \tag{10.12}$$

where matrix block H_1 is in the upper Hessenberg form (5.1).

Third, make the Schur triangularization on H_1 instead of H. This way not only the computation of each iteration is reduced by eight times ($2n^3/3$ (instead of $2(2n)^3/3$), but also the number of iterations needed before convergence is reduced even much more.

Finally, compute the square root of the result of Schur triangularization of H_1 (Bjorck and Hammaling, 1983), to recover this result to that of the original Hamiltonian matrix H.

It should be noticed again that for the state feedback control designed in this section, its loop-transfer function and robustness properties *cannot* be actually realized for a great majority of plant systems (A, B, C) (see Chapters 4 and 6).

10.2 Design of Generalized State Feedback Control

Generalized state feedback control $\underline{KC}x(t)$ is a state feedback control with or without constraint $K = \underline{KC}$ (for rank $(\underline{C}) < n$ or $= n$, respectively). Therefore, this control can be weaker and more difficult to design if rank $(\underline{C})<n$. The decisive advantage of this control is that at the great majority of plant system conditions where the unconstrained $K\mathbf{x}(t)$ control cannot be fully realized, the full realization of the loop transfer function and robust properties of $\underline{KC}\mathbf{x}(t)$ control is still guaranteed.

Among the existing methods of this design (Levine and Athans, 1970; Choi and Sirisena, 1974; Horisberger and Belanger, 1974; Toivonen, 1985; Zheng, 1989; Yan et al., 1993), that of Yan et al. (1993) is briefly described in the following. This method is called the gradient method and is presented here because it can generally and clearly unify the cases of rank $(\underline{C})<n$ and $=n$.

This method is based on the partial derivative of J of (10.1) with respect to $\underline{K}$:

$$\partial J/\partial \underline{K} = [R\underline{KC} - B'P]L\underline{C}' \tag{10.13}$$

where P and L are positive semidefinite solution matrices of the following two consecutive Lyapunov equations:

$$P(A - B\underline{KC}) + (A - B\underline{KC})'P = -\underline{C}'\underline{K}'R\underline{KC} - Q \tag{10.14}$$

$$\text{and } L(A - B\underline{KC})' + (A - B\underline{KC})L = -P \tag{10.15}$$

Based on this result, the gradient flow of $\underline{K}$ with respect to J is the homogeneous differential equation

$$\Delta K/\Delta t = [B'P - R\underline{KC}]L\underline{C}' \tag{10.16}$$

whose solution $\underline{K}$ can be computed by a number of numerical methods. The simplest is the first order "Euler method":

$$\underline{K}_{i+1} = \underline{K}_i + \Delta\underline{K}_i\Delta t = \underline{K}_i + \left([B'P_i - R\underline{K}_i\underline{C}]\ L_i\underline{C}'\right)\Delta t \tag{10.17}$$

where $\Delta\underline{K}_i$ or P_i and L_i at each i-th iteration must satisfy the Lyapunov equations (10.14) and (10.15) for the current value of $\underline{K}_i$. The initial condition of this iteration $\underline{K}_0$, and a small enough interval Δt, must guarantee the convergence (Helmke and Moore, 1992). However, a larger interval Δt may accelerate the convergence.

Theorem 10.2

Define **J** as a set of finite $J(\underline{K})$ of (10.1). This set **J** should include the global and local minima of $J(\underline{K})$. Then under the assumption that $J(\underline{K}_0)$ is finite, the gradient method (10.14)–(10.16) has the following four properties:

1. The gradient flow of (10.16) has a unique solution $\underline{K}$ such that $J(\underline{K})$ is within the set **J**;
2. The value of $J(\underline{K}_i)$ is nonincreasing with each increment of i;
3. $\Delta\underline{K}_i$ approaches 0 as i approaches infinity.
4. There is a convergent sequence $\underline{K}_i$ to an equilibrium of (10.16) whose corresponding matrix $A-B\underline{KC}$ is stable and whose corresponding J is at minimum (can be a local minimum).

Proof: Can Be Found in the study of Yan et al. (1993)

In addition, an inspection of (10.13) or (10.16) or (10.17) shows that when rank $(\underline{C})=n$ and $\underline{C}^{-1}$ exists, then the convergent value of $\underline{K}=R^{-1}B'P\underline{C}^{-1}$. This result conforms with that of (10.2) when $\underline{C}=I$.

Like the main computation of Section 10.1 is at solving of ARE, the main computation of (10.16) and (10.17) is at solving of Lyapunov equations (10.14) and (10.15). There are a set of numerical methods (Rothschild and Jameson, 1970; Davison, 1975). The following is a method using Schur triangularization (Golub et al., 1979).

Algorithm 10.2: Solving Lyapunov Equation $AP+PA'=-Q$

Step 1: Make a Schur triangularization of matrix A

$$U'AU=\begin{bmatrix} A_{11} & A_{12} & \cdots & A_{1r} \\ 0 & A_{22} & \cdots & A_{2r} \\ \vdots & \ddots & \ddots & \vdots \\ 0 & \cdots & 0 & A_{rr} \end{bmatrix} \quad (U'U=I) \tag{10.18}$$

where A_{ii}, $i=1, \ldots, r$, are either scalars or 2×2 real blocks for complex conjugate eigenvalues. The computation of (10.18) is discussed in Step 2 of Algorithm 10.1.

Step 2: Compute

$$U'QU=\begin{bmatrix} Q_{11} & \cdots & Q_{1r} \\ \vdots & & \vdots \\ Q_{r1} & \cdots & Q_{rr} \end{bmatrix}$$

where Q_{ij} has the same dimension as A_{ij} of (10.18), for all i and j.

Step 3: Replace A, P, and Q of the Lyapunov equation by $U'AU$, $U'PU$, and $U'QU$, respectively:

$$\begin{bmatrix} A_{11} & \cdots & A_{1r} \\ & \ddots & \vdots \\ 0 & & A_{rr} \end{bmatrix} \begin{bmatrix} P_{11} & \cdots & P_{1r} \\ \vdots & & \vdots \\ P_{r1} & \cdots & P_{rr} \end{bmatrix}$$

$$+ \begin{bmatrix} P_{11} & \cdots & P_{1r} \\ \vdots & & \vdots \\ P_{r1} & \cdots & P_{rr} \end{bmatrix} \begin{bmatrix} A_{11}' & & 0 \\ \vdots & \ddots & \\ A_{1r}' & \cdots & A_{rr}' \end{bmatrix} = -\begin{bmatrix} Q_{11} & \cdots & Q_{1r} \\ \vdots & & \vdots \\ Q_{r1} & \cdots & Q_{rr} \end{bmatrix} \tag{10.19}$$

Solve (10.19) by back substitution, we have for $i=r, r-1, \ldots, 1$ and $j=r, r-1, \ldots, 1$:

$$P_{ij} = \begin{cases} P_{ji}', & \text{if } i<j \\ -\left(A_{ii}+A_{jj}\right)^{-1}\left[Q_{ij} + \sum_{k=i+1}^{r} A_{ik}P_{kj} + \sum_{k=j+1}^{r} P_{ik}A_{jk}'\right] \text{(for scalar } A_{jj}) & \text{if } i \geq j \\ -\left[Q_{ij} + \sum_{k=i+1}^{r} A_{ik}P_{kj} + \sum_{k=j+1}^{r} P_{ik}A_{jk}'\right]\left(A_{ii}+A_{jj}\right)^{-1} \text{(for scalar } A_{ii}) & \end{cases} \tag{10.20}$$

There are two formulas for the case of $i \geq j$, because blocks A_{ii} and A_{jj} can have different dimensions (one is a 1×1 scalar and the other is a 2×2 block). In such cases, the 1×1 scalar must multiply I_2 before being added to the other 2×2 block.

If both blocks are 2×2, then the corresponding P_{ij} will be the solution of a 2×2 Lyapunov equation

$$A_{ii}P_{ij} + P_{ij}A_{jj}' = -\underline{Q}_{ij} \tag{10.21}$$

where $\underline{Q}_{ij}$ equals the matrix inside the square bracket [ ] of (10.20).

Let $P_{ij}=[\mathbf{p}_1 : \mathbf{p}_2]$ and $\underline{Q}_{ij}=[\mathbf{q}_1 : \mathbf{q}_2]$, the Eq. (10.21) can be expressed as

$$\left[I_2 \otimes A_{ii} + A_{jj} \otimes I_2\right]\left[\mathbf{p}_1' : \mathbf{p}_2'\right]' = -\left[\mathbf{q}_1' : \mathbf{q}_2'\right]'$$

People can solve this set of linear equations for P_{ij}.

The main computation of Algorithm 10.2 is still at Step 1 of Schur triangularization (10.18). The dimension of (10.18) is n which is only half of the dimension of Hamiltonian matrix H of (10.6), indicating that the Lyapunov equation is easier to solve than the Riccati equation. However, Algorithm 10.2 will need to run repeatedly in an iterative loop of (10.17) until convergence is achieved. Hence, the LQ generalized state feedback control design is much more difficult to compute than the LQ direct state feedback control design.

It should be pointed out that the optimal state feedback control gain solution K^* of (10.2) is unique. Therefore, the control gain $\underline{K}$ designed in this section generically cannot converge to that K^* (= $\underline{KC}$), because that K^* is generically not a linear combination of the rows of matrix $\underline{C}$.

10.3 Comparison and Conclusion of Feedback Control Designs

The order of computation of design methods of Chapters 8–10 is summarized in Table 10.1.

TABLE 10.1

Computation of Feedback Control Design Methods

Design Methods for K of A–$B\underline{KC}$	Number of Loops of Iterations	Order of Computation of the Main Step of Each Iteration
Pole Assignment		
Computation of (8.6)	0	$4n^3/3$
Computation of (8.4)	0	$np(n-q)^2/2$
Computation of (9.4a)	0	$2n^4/3$
Eigenvector Assignment		
Algorithm 9.1	1	$n^2/2$ to $2n^3/3$
Algorithm 9.2	1	$4pn$
Analytical methods	0	$n^2/2$
LQ Optimal Control Design		
Algorithm 10.1	1	$2n^3/3$ to $2(2n)^3/3$
Algorithm 10.2	2	$2n^3/3$

Because similar orthonormal matrix operations are used in the main step of each of the above design computation methods, the comparison of Table 10.1 is fair. This is also the reason that the orders of computation of the main step of each iteration of all design methods, as listed in the third column of Table 10.1, are about the same.

In each numerical iteration loop, the actual number of iterations needed, before convergence to a reasonably accurate value, differs from problem to problem, if convergent at all (otherwise the number of iterations can be infinite). The actual number of iterations is usually very large – more than n^3. That means the number of iterations needed is at least 10^3 if n =10, and at least 10^6 if n=100, and at least 10^{12} if two loops of iterations are needed. This is why applied mathematicians make a great effort just to half the size of a Hamiltonian matrix before letting it go through the iteration for Schur triangularization (see the end of Section 10.1).

Hence, the computational difficulty is determined by the number of loops of iteration, as listed in the middle column of Table 10.1.

This middle column of Table 10.1 shows that the design computation of eigenstructure assignment generally requires one loop of iteration less than that of LQ optimal design and is, therefore, much easier. The computation of direct state feedback control design is also one loop of iteration less than that of the generalized state feedback control design.

Compare the feedback control design of Chapters 8–10 of this book to that of the other books and literature. Chapters 8–10 deal with the design of $\underline{KC}\mathbf{x}(t)$-control gain only or the design of static output part (3.16b) of the observer/controller only. The dynamic part (3.16a) has already been designed in Chapters 5 and 6. The new design principle of this book pre-guarantees that the design of the $\underline{KC}\mathbf{x}(t)$-control also determines fully the feedback system loop transfer function $\underline{KC}(sI–A)^{-1}B$. On the other hand, many of the other designs such as H_∞ or H_2 or L_1 are formulated for the design of the entire observer or controller (Zhou, 1992; Yeh et al., 1992; Dahleh and Pearson, 1987; Dullerud and Francis, 1992). Therefore, the control design of this book is far easier.

There are some existing results on the H_∞ design using state feedback control (Khargoneker et al., 1988) or using static output feedback control (Geromel et al., 1993; Stoustrup and Niemann, 1993). But such a state feedback control cannot be fully realized by a state observer for a great majority of plant systems, and no effort was made to increase the number of rows of C (= m) of the existing static output feedback control. The new design principle of this book unified these two exiting controls as its two extremes under the guarantee of full realization and thus completely overcomes the critical disadvantages of these two exiting controls (see Conclusion 6.5).

Of so many optimal design objectives and formulations and their combinations, for the $\underline{KC}\mathbf{x}(t)$-control design, each of them claimed exclusive optimality (the unique solution of one optimal design formulation is not the solution of other optimal design formulations, or each of them claimed the other optimal design results not optimal), none of them can compare with the objective of eigenstructure assignment because of the following two overwhelming reasons.

The first overwhelming advantage of eigenstructure assignment is the effectiveness of this design formulation. Eigenvalues determine system performance far more directly and generally accurately (see Section 2.1). The sensitivity of eigenvalues is determined by their corresponding eigenvectors! See Subsection 2.2.1. Hence, their assignment should improve feedback system performance and robustness far more directly and therefore effectively (see Subsection 2.2.2).

In sharp contrast, the selection of up to n^2+p^2 parameters of the LQ criterion has essentially no general guidance from system analysis results at all. Therefore, the existing optimal designs have far less generally accurate reflection and far less guarantee of combined overall control system performance and robustness, as compared to the robust stability measure M_3 (2.25) of this book. The existing performance measure is bandwidth in the classical control theory, which is far less generally accurate (see Example 2.2). The existing sensitivity function is too gross and too conservative, yet still ignored the all-important phase angle of the sensitivity/loop-transfer function (see the comments of Figure 3.5) and still ignored performance. This general indirectness and inaccuracy are also reflected by the fact of so many existing and exclusive optimal design objectives such as H_∞ or H_2 or L_1, each is essentially claiming the other is not really optimal!

To summarize, a really effective and practical control system design objective and design requirement must guarantee the control system performance and robustness properties and must be based on the solid understanding of system analysis such as the roles of system poles and their sensitivities, as the eigenstructure assignment. In sharp contrast, the optimal design formulations are essentially abstract mathematical problem formulations that reflect far less accurately the actual and overall system performance and robustness.

The second overwhelming advantage of eigenstructure assignment is the ability of the adjustment of the original design formulation and parameter, based on the final design results and their simulation. This adjustment is required for all practical design formulation, especially the sophisticated and general design formulations of the control theory.

The eigenstructure assignment Algorithms (8.1 and 8.2) are very simple and explicit and thus can be fully adjusted and modified (see Section 8.4) even back to the original eigenvalue selection (Modification 2). Each step of the analytical decoupling eigenvector assignment (Section 9.2) can also be fully understood by the final design results and simulations (see Example 9.2). In the design of the dynamic part of the observers (Chapters 5 and 6), the most important observer/controller parameter – order, is also fully adjustable (see the whole Section 6.4). This observer order adjustment and this general and systematic eigenstructure assignment are enabled only by the mathematical breakthrough on a general and decoupled solution of (4.1) and (8.1b). This solution of (4.1) and (8.1b) of 1985 also enabled the new design principle of this book (Tsui, 2015).

In sharp contrast, Table 10.1 shows that numerical iteration (to reach finally a numerical equilibrium) operation is needed to compute the solution of LQ optimal design. Other optimal designs can require even much heavier computation. For example, LQ designs require solving only one ARE, while H_∞ design requires solving several AREs (Zhou, 1992; Zhou et al., 1995). It is almost unimaginable to adjust generally and analytically the original design formulation from this kind of numerical computation design results. Even the very existence of the numerical solution is often questionable. We should be reminded from the first overwhelming disadvantage of optimal designs that its original formulation does not reflect generally accurately the actual and overall control system performance and robustness in the first place.

To summarize, optimal control designs, which formed the main state-of-art results of the modern control theory since the start of this theory in the 1960s, are in practical reality *not* optimal in improving system performance and robustness as their name suggested.

These overwhelming advantages of eigenstructure assignment should be a major reflection of the advantage of the modern control theory over the classical control theory, because eigenvector assignment can only be achieved by (generalized) state feedback control and only based on the state space model.

We all know that the modern control theory surpassed the classical control theory in the 1960s (when observers/Kalman filters extended the application of state feedback control of the late 1950s to most plant systems with only terminal measurements), because of its superiority to the classical control theory in mathematical modeling information and analysis. However, the modern control theory in other existing literature and over six decades cannot solve the following two critical design problems: how to fully realize the critical loop transfer function and robust properties of (generalized) state feedback control for the great majority of plant systems, and how to assign eigenvectors using the generalized state feedback control.

The new design principle of this book (Chapters 4–6) solved decisively the first problem, while the algorithms of Chapters 8 and 9 solved the second problem. With these two momentous developments, the superiority of the modern control theory to the classical control theory can now be extended from modeling and analysis to design. Thus, the modern control theory (and the whole control theory) has reached a new stage – able to achieve high control system performance and robustness really and generally. This momentous development opened two bright prospects of the control theory: far more successful practical applications of the control theory and a far more solid and realistic foundation for future research and development of the control theory.

Exercises

10.1 Let the plant system be

$$A = \begin{bmatrix} 0 & 1 \\ 0 & 0 \end{bmatrix} \quad \text{and} \quad B = \begin{bmatrix} 0 \\ 1 \end{bmatrix}$$

a. Design a state feedback control law K to minimize the LQ criterion (10.1) with

$$Q = \begin{bmatrix} 4 & 0 \\ 0 & 0 \end{bmatrix} \quad \text{and} \quad R = 1$$

Answer: $K = [\ -2 \quad -2\ ]$

b. Given $C = [1 \quad x]$ $(x \neq 1)$. Can J of (10.1) be minimized by static output feedback control?

c. Repeat part b if given $C = [1 \quad 1]$.

10.2 Repeat Exercise 10.1 with the following change of the LQ criterion (10.1):

$$J = \int_0^\infty \left[2x_1(t)^2 + 2x_1(t)x_2(t) + x_2(t)^2 + \mathbf{u}(t)^2 \right] dt$$

Hint:

$$Q = \begin{bmatrix} 2 & 1 \\ 1 & 1 \end{bmatrix} \quad \text{and} \quad R = 1$$

10.3 Repeat Exercise 10.1 with the following change of the LQ criterion (10.1):

$$J = \int_0^\infty \left[y(t)^2 + 2\mathbf{u}(t)^2 \right] dt \quad \left(\text{Given } \mathbf{y}(t) = C\mathbf{x}(t) = \begin{bmatrix} 1 & 2 \end{bmatrix} \mathbf{x}(t) \right)$$

Hint: $Q = C'C$ and $R = 2$.

10.4 For the same plant system of Exercises 10.1 and 10.3, (using the same system matrix C of Exercise 10.3), design a static output feedback control for the three different LQ criteria of Exercises 10.1–10.3.

10.5 Test the computational time and difficulty of LQ optimal design.

a. Randomly select five 10×10 $(n=5)$ Hamiltonian matrices (10.6). Program and calculate the Schur triangularization (Step 2 of Algorithm 10.1) and notice the average computational time.

b. Repeat part (a) for five 12×12 $(n=6)$ Hamiltonian matrices.

c. Repeat part (a) for five 20×20 $(n=10)$ Hamiltonian matrices.

Appendix A: Linear Algebra & Numerical Linear Algebra

This appendix introduces the relevant mathematical background to this book. In addition, an attempt is made to use the simplest possible language, even though such a presentation may sacrifice certain degree of mathematical rigor.

The appendix is divided into four sections.

Section A.1 introduces some basic results of linear algebra, especially the geometrical meanings and numerical importance of orthogonal linear transformation.

Section A.2 describes and analyzes some basic matrix operations, which transform a given matrix into echelon form. A special case of the echelon form is triangular form. This operation is the one used most often in this book.

Section A.3 introduces a basic result of numerical linear algebra – the singular value decomposition (SVD). Several applications of SVD are also introduced.

Section A.4 describes matrix equation pair $TA-FT=LC$ (4.1) and $TB=0$ (4.3), its five important applications, and the computation of its solution. These applications made this matrix equation pair by far the most important mathematical requirement of the state space control systems design theory.

A.1 Linear Space and Linear Operators

A.1.1 Linear Dependence, Linear Independence, and Linear Space

Definition A.1

A set of n vectors $\{\mathbf{x}_1, \ldots, \mathbf{x}_n\}$ is linearly dependent if there exists an n-dimensional nonzero vector $\mathbf{c} \underline{\Delta} [c_1,\ldots,c_n]' \neq 0$ such that

$$[\mathbf{x}_1 : \ldots : \mathbf{x}_n]\mathbf{c} = \mathbf{x}_1 c_1 + \cdots + \mathbf{x}_n c_n = 0 \tag{A.1}$$

Otherwise, this set of vectors is linearly independent. At least one vector of a set of linear dependent vectors is a linear combination of other vectors in that set. For example, if a set of vectors satisfies (A.1), then

$$\mathbf{x}_i = -\frac{\left(\sum_{i \neq j} \mathbf{x}_j c_j\right)}{c_i}, \quad \text{if } c_i \neq 0 \tag{A.2}$$

Example A.1

Let a set of two vectors be

$$[\mathbf{x}_1 : \mathbf{x}_2] = \begin{bmatrix} 1 & -2 \\ -1 & 2 \end{bmatrix}$$

Because there exist a vector $\mathbf{c}=[2 \;\; 1]' \neq 0$ such that $[\mathbf{x}_1 \colon \mathbf{x}_2]\mathbf{c}=0$, $\mathbf{x}_1$, and $\mathbf{x}_2$ are linearly dependent of each other.

Example A.2

Let another set of two vectors be

$$[\mathbf{x}_3 : \mathbf{x}_4] = \begin{bmatrix} 2 & 1 \\ 0 & 1 \end{bmatrix}$$

Because only a zero vector $\mathbf{c}=0$ can make $[\mathbf{x}_3 \colon \mathbf{x}_4]\mathbf{c}=0$, $\mathbf{x}_3$ and $\mathbf{x}_4$ are linearly independent of each other.

Similarly, any combination of two vectors, with one from the set $\{\mathbf{x}_3, \mathbf{x}_4\}$ and another from the set $\{\mathbf{x}_1, \mathbf{x}_2\}$ of Example A.1, is linearly independent. However, any set of three vectors out of $\mathbf{x}_i$ ($i=1, \ldots, 4$) is linearly dependent.

Examples A.1 and A.2 can be interpreted geometrically from Figure A.1, which can be interpreted to have the following three points.

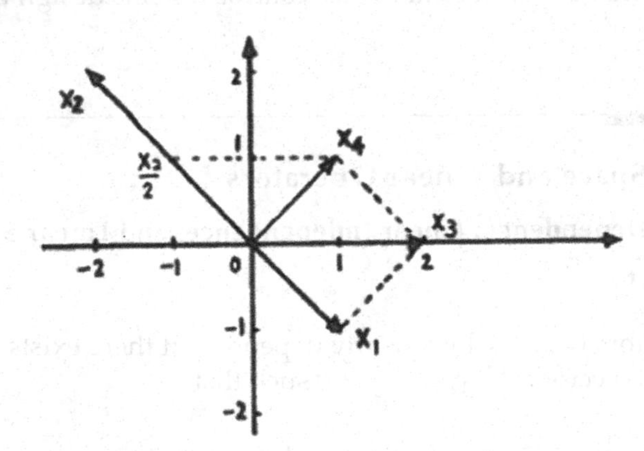

FIGURE A.1
Four two-dimensional vectors.

1. Because $\mathbf{x}_1$ and $\mathbf{x}_2$ vectors are parallel in Figure A.1, or because the angle between them is 180° (or 0°), $\mathbf{x}_1$ and $\mathbf{x}_2$ are linearly dependent, or $\mathbf{x}_1$ differs from $\mathbf{x}_2$ only by a scalar factor. Because the angles between all other vector pairs in Figure A.1 are not equal to 0° or 180°, all other pairs of vectors are linearly independent.

2. From analytical geometry, the angle θ between two vectors $\mathbf{x}_i$ and $\mathbf{x}_j$ satisfies the relation

$$\mathbf{x}_i'\mathbf{x}_j = \mathbf{x}_j'\mathbf{x}_i = \|\mathbf{x}_i\|\|\mathbf{x}_j\|\cos\theta \tag{A.3}$$

where the vector norm $||\mathbf{x}||$ is defined in Definition 2.3. For example,

$$\begin{aligned}\mathbf{x}_1'\mathbf{x}_2 &= [1\ \ -1][-2\ \ 2]' = -4\\ &= \|[1\ \ -1]\|\|[-2\ \ 2]\|\cos 180° = \left(\sqrt{2}\right)\left(2\sqrt{2}\right)(-1)\end{aligned}$$

$$\begin{aligned}\mathbf{x}_1'\mathbf{x}_3 &= [1\ \ -1][1\ \ 0]' = 1\\ &= \|[1\ \ -1]\|\|[1\ \ 0]\|\cos 45° = \left(\sqrt{2}\right)(1)\left(1/\sqrt{2}\right)\end{aligned}$$

and

$$\begin{aligned}\mathbf{x}_1'\mathbf{x}_4 &= [1\ \ -1][1\ \ 1]' = 0\\ &= \|\mathbf{x}_1\|\|\mathbf{x}_4\|\cos 90° = \|\mathbf{x}_1\|\|\mathbf{x}_4\|(0)\end{aligned}$$

We define two vectors as "orthogonal" if the angle between them is ±90°. For example, $\{\mathbf{x}_1, \mathbf{x}_4\}$ and $\{\mathbf{x}_2, \mathbf{x}_4\}$ are orthogonal pairs, while other vector pairs of Figure A.1 are not. Because cos 0° and cos 180° have the largest magnitude (1) among cosine functions and cos(±90°)=0 has the smallest, orthogonal vectors are considered "most linearly independent."
We also define $||\mathbf{x}_i||\cos\theta$ as the "projection" of $\mathbf{x}_i$ on $\mathbf{x}_j$, if θ is the angle between $\mathbf{x}_i$ and $\mathbf{x}_j$. Obviously, a projection of $\mathbf{x}_i$ is always less than or equal to $||\mathbf{x}_i||$ and is equal to 0 if $\theta = \pm 90°$.

3. Any two-dimensional vector is a linear combination of any two linearly independent vectors on the same plane. For example, $\mathbf{x}_3 = \mathbf{x}_1 + \mathbf{x}_4$ and $\mathbf{x}_4 = (1/2)\mathbf{x}_2 + \mathbf{x}_3$. These two relations are shown in Figure A.1 by the dotted lines. Therefore, the vectors in any set of three two-dimensional vectors are linearly dependent of each other. For example, if $\mathbf{x} = [\mathbf{y}:\mathbf{z}]\mathbf{c}$, then $[\mathbf{x}:\mathbf{y}:\mathbf{z}][1: -\mathbf{c}']' = 0$.
If two vectors $(\mathbf{y}, \mathbf{z})$ are orthogonal to each other, and if $[\mathbf{y}: \mathbf{z}]\mathbf{c}$ equals a third vector $\mathbf{x}$, then the two coefficients of $\mathbf{c}$ equal the projections of $\mathbf{x}$ on $\mathbf{y}$ and $\mathbf{z}$ respectively, after dividing these two projections by their respective $||\mathbf{y}||$ and $||\mathbf{z}||$. For example, for $\mathbf{x}_3$ of Figure A.1, the linear combination coefficients (1 and 1) of the orthogonal vectors $\mathbf{x}_1$ and $\mathbf{x}_4$ equal the projections ($\sqrt{2}$ and $\sqrt{2}$) of $\mathbf{x}_3$ on $\mathbf{x}_1$ and $\mathbf{x}_4$, divided by the norms ($\sqrt{2}, \sqrt{2}$) of $\mathbf{x}_1$ and $\mathbf{x}_4$.

Definition A.2

A linear space **S** can be formed by a set of vectors such that any vector within this set (defined as ∈ **S**) can be represented as a linear combination of some other vectors $X=[\mathbf{x}_1 : \ldots : \mathbf{x}_n]$ within this set (defined as the span of X). The largest number of linearly independent vectors needed to represent the vectors in this space is defined as the dimension dim(**S**) of that space.

For example, vectors $\mathbf{x}_1$ and $\mathbf{x}_2$ of Example A.1 can span only a straight-line space, which is parallel to $\mathbf{x}_1$ and $\mathbf{x}_2$. Any of these parallel vectors is a linear combination of another parallel vector only. Hence, the dimension of this straight line space is 1.

In Examples A.1 and A.2, each of the vector pair $\{\mathbf{x}_1, \mathbf{x}_3\}$, $\{\mathbf{x}_1, \mathbf{x}_4\}$, and $\{\mathbf{x}_3, \mathbf{x}_4\}$ can span a plane space, because any vector on this plane can be represented as a linear combination of one of these three vector pairs. In fact, because any one vector in this plane is a linear combination of two linearly independent vectors on this plane, the dimension of this plane space is 2.

Example A.3

The above result can be extended to higher dimensional vectors. Let a set of three-dimensional vectors be

$$\left[\mathbf{y}_1 : \mathbf{y}_2 : \mathbf{y}_3 : \mathbf{y}_4 : \mathbf{y}_5 : \mathbf{y}_6 : \mathbf{y}_7\right] = \begin{bmatrix} 2 & 1 & 1 & 0 & 0 & -1 & -2 \\ 0 & 1 & 1 & 1 & 0 & 0 & 1 \\ 1 & 0 & 1 & 1 & 2 & 1 & 0 \end{bmatrix}$$

which are plotted in Figure A.2.

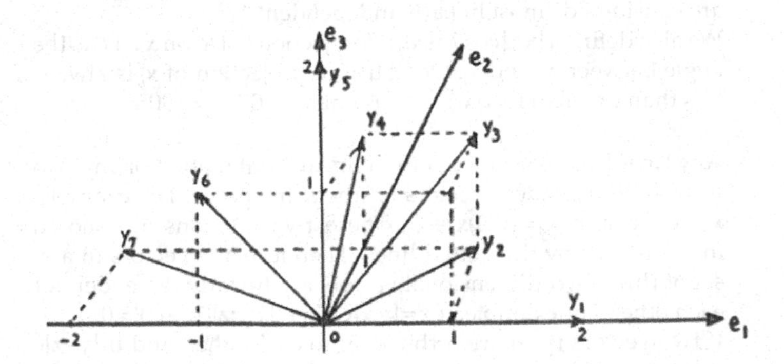

FIGURE A.2
Seven three-dimensional vectors.

From Figure A.2, vectors $\mathbf{y}_1$ and $\mathbf{y}_2$ span a horizontal two-dimensional plane space. Any three-dimensional vector with form $[x \;\; x \;\; 0]'$ ("x" stands for an arbitrary entry) or with 0 at the third (vertical) direction equals a linear combination of $\mathbf{y}_1$ and $\mathbf{y}_2$, and therefore lies within this horizontal plane space. For example, $\mathbf{y}_7=[-2\ \ 1\ \ 0]'=[\mathbf{y}_1{:}\mathbf{y}_2][-3/2\ \ 1]'$ belongs to this space. However, all other vectors $\mathbf{y}_3$ to $\mathbf{y}_6$ which stretch on the vertical direction are linearly independent of the vectors $\{\mathbf{y}_1, \mathbf{y}_2, \mathbf{y}_7\}$ of this horizontal plane space, and hence do not belong to this horizontal plane space.

Although $\mathbf{y}_3$ to $\mathbf{y}_6$ are linearly independent of the vectors $\{\mathbf{y}_1, \mathbf{y}_2, \mathbf{y}_7\}$ on the horizontal plane space, only $\mathbf{y}_5$ $(=[0\ \ 0\ \ x]')$ is orthogonal to all vectors of this horizontal space (also called orthogonal to that space). Finally, any one of the vectors $\mathbf{y}_3$ to $\mathbf{y}_6$, together with two linearly independent vectors of this horizontal plane space, forms a three-dimensional cubic space.

Similarly, vectors $\mathbf{y}_1$ and $\mathbf{y}_6$ span a two-dimensional plane space which is parallel to this page flat. Any three-dimensional vector with form $[x\ 0\ x]$ or with 0 at the second (depth) direction equals a linear combination of $\mathbf{y}_1$ and $\mathbf{y}_6$. For example, $\mathbf{y}_5=[0\ 0\ 2]'=[\mathbf{y}_1 : \mathbf{y}_6][2/3\ \ 4/3]'$ belongs to this space. However, all other vectors of Figure A.2 have nonzero projection on the depth direction. Therefore, these vectors are linearly independent of the vectors $\{\mathbf{y}_1, \mathbf{y}_5, \mathbf{y}_6\}$ and do not belong to this space. Among these vectors, none is orthogonal to this two-dimensional space because none has the form $[0\ \ x\ \ 0]'$, even though within each pair of $\{\mathbf{y}_4, \mathbf{y}_1\}$, $\{\mathbf{y}_2, \mathbf{y}_5\}$, and $\{\mathbf{y}_3, \mathbf{y}_6\}$, the two vectors are orthogonal to each other.

In the literature, there is a more rigorous definition than Definition A.2 for the linear space **S** (Gantmacher, 1959). For example, if we generalize the vectors of a linear space **S** as "elements" of that space, then **S** must also have "0" and "1" elements (Gantmacher, 1959).

Example A.4

We define the space formed by all n-dimensional vectors $\mathbf{b}$ satisfying the equation $\mathbf{b}=A\mathbf{x}$ (matrix A is given and x is arbitrary) as $\mathbf{R}(A)$, or as the "range space of A." We also define the number of linearly independent columns/rows of A as the "column rank/row rank" of A. It is clear that the necessary and sufficient condition for $\dim[\mathbf{R}(A)]=n$ [or for $\mathbf{R}(A)$ to include any possible nonzero $\mathbf{b}$] is that the column rank of A equals n.

If the column/row rank of a matrix equals the number of columns/rows of that matrix, then we call this matrix "full-column rank"/ "full-row rank."

We also define the space formed by all vectors $\mathbf{x}$ satisfying $A\mathbf{x}=0$ as $\mathbf{N}(A)$ or the "null space of A." It is clear that if matrix A is full-column rank, then the only vector in $\mathbf{N}(A)$ is $\mathbf{x}=0$.

However, the set of all vectors $\mathbf{x}$ satisfying $A\mathbf{x}=\mathbf{b}$ ($\mathbf{b}\neq 0$ is given) cannot form a linear space, because this set lacks a "0" element (or 0 vector) such that $A0=\mathbf{b}\neq 0$.

A.1.2 Basis, Linear Transformation, and Orthogonal Linear Transformation

Definition A.3

If any vector $\mathbf{x}$ of a linear space **S** is a linear combination of a set of linearly independent vectors of **S**, then this set of linear independent vectors is defined as the "basis" of **S**. The linear combination coefficient is defined as the "representation" of $\mathbf{x}$ with respect to this set of basis vectors.

Because any set of n linearly independent n-dimensional vectors can span an n-dimensional linear space **S**, by Definition A.3 any of these sets can be considered as a basis of **S**.

Definition A.4

An n-dimensional linear space can have many different sets of basis vectors. The operation which transforms the representation of a vector from one basis to another basis is called a "linear transformation."

For example, the simplest and most commonly used basis is a set of orthogonal unit coordinate vectors

$$I \underline{\underline{\Delta}} [\mathbf{e}_1 : \ldots : \mathbf{e}_n] \, \Delta \begin{bmatrix} 1 & 0 & \cdots & & 0 \\ 0 & 1 & \ddots & & \vdots \\ \vdots & & \ddots & \ddots & \\ & & & & 0 \\ 0 & \cdots & & 0 & 1 \end{bmatrix}$$

Because any n-dimensional vector $\mathbf{b} = [b_1, \ldots, b_n]'$ is a linear combination of the vectors $[\mathbf{e}_1 : \ldots : \mathbf{e}_n]$ such that

$$\mathbf{b} = I\mathbf{b} \tag{A.4}$$

and because the representation of **b** on I is **b** itself, I is a basis and is called an "identity matrix."

For another example, if we let the vectors of $A = [\mathbf{a}_1 : \ldots : \mathbf{a}_n]$ be the basis for a vector **b**, then $A\mathbf{x} = \mathbf{b}$ implies that $\mathbf{x} = A^{-1}\mathbf{b}$ is the representation of **b** on A.

Now let another set of vectors $\mathbf{V} = [\mathbf{v}_1 : \ldots : \mathbf{v}_n]$ be the basis for the same b. Then,

$$V\bar{\mathbf{x}} = \mathbf{b} = A\mathbf{x} \tag{A.5a}$$

implies

$$\bar{\mathbf{x}} = V^{-1}\mathbf{b} = V^{-1}A\mathbf{x} \tag{A.5b}$$

is the representation of **b** on V.

Definition A.5

A set of orthogonal basis vectors $\left\{\mathbf{u}_1, \ldots, \mathbf{u}_n, \left(\mathbf{u}_i'\mathbf{u}_j = x\delta_{ij}\right)\right\}$ is called an "orthogonal basis." The linear transformation which transforms to an orthogonal basis is called "orthogonal linear transformation."

Furthermore, if all vectors of this orthogonal basis are "normalized" ($\|\mathbf{u}_i\| = 1, \forall i$), then the basis is called "orthonormal" and the corresponding orthogonal linear transformation becomes an "orthonormal linear transformation." A matrix U, which is formed by a set of orthonormal basis vectors, satisfies $U'U = I$ and is called a "unitary matrix."

Example A.5

Let a vector $\mathbf{x} = \begin{bmatrix} 1 & \sqrt{3} \end{bmatrix}'$. Table A.1 shows some two-dimensional linear transformation examples in which the orthonormal linear transformation can preserve the norms of any vector $\mathbf{x}$ and its representation $\bar{\mathbf{x}}$ on the new orthonormal basis.

TABLE A.1

Some Examples of Two-Dimensional Linear Transformation

Basis Vectors	Representation of x, $\bar{\mathbf{x}}$	$\|\bar{\mathbf{x}}\|$	x and Its New Basis	Transformation Form
$E = \begin{bmatrix} 1 & 0 \\ 0 & 1 \end{bmatrix}$	$\begin{bmatrix} 1 \\ \sqrt{3} \end{bmatrix}$	2	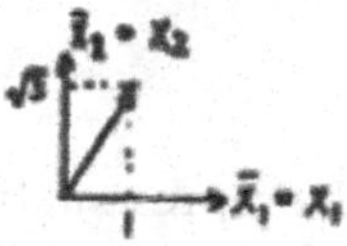	Identity
$O = \begin{bmatrix} 0 & 2 \\ 2 & 0 \end{bmatrix}$	$\begin{bmatrix} \sqrt{3}/2 \\ 1/2 \end{bmatrix}$	1	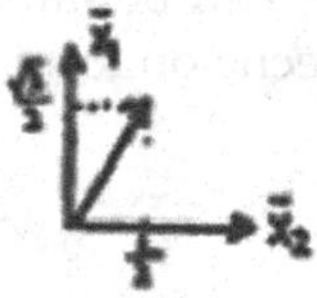	Orthogonal
$U = \begin{bmatrix} 0 & -1 \\ 1 & 0 \end{bmatrix}$	$\begin{bmatrix} \sqrt{3} \\ -1 \end{bmatrix}$	2	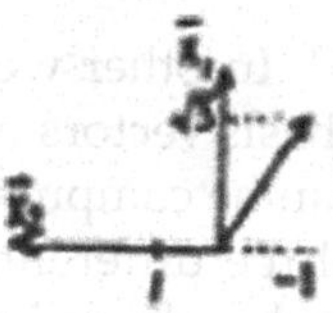	Orthonormal (Givens 90°)
$G_1 = \begin{bmatrix} 1/2 & -\sqrt{3}/2 \\ \sqrt{3}/2 & 1/2 \end{bmatrix}$	$\begin{bmatrix} 2 \\ 0 \end{bmatrix}$	2		Orthonormal (Givens 60°)
$G_2 = \begin{bmatrix} \sqrt{3}/2 & -1/2 \\ 1/2 & \sqrt{3}/2 \end{bmatrix}$	$\begin{bmatrix} \sqrt{3} \\ 1 \end{bmatrix}$	2		Orthonormal (Givens 30°)
$H = \begin{bmatrix} -1/2 & -\sqrt{3}/2 \\ -\sqrt{3}/2 & 1/2 \end{bmatrix}$	$\begin{bmatrix} -2 \\ 0 \end{bmatrix}$	2	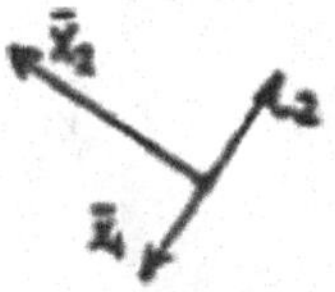	Orthonormal (Householder)
$P = \begin{bmatrix} 1 & -\sqrt{3}/2 \\ 0 & 1/2 \end{bmatrix}$	$\begin{bmatrix} 4 \\ 2\sqrt{3} \end{bmatrix}$	$2\sqrt{7}$		Ordinary

This property can be interpreted geometrically from the fourth column of Table A.1, which shows that every element of $\bar{\mathbf{x}}$ equals the projection of $\mathbf{x}$ on the corresponding axis [see interpretation (3) of Figure A.1]. This property can be proved mathematically that

$$\|\bar{\mathbf{x}}\| = (\bar{\mathbf{x}}'\bar{\mathbf{x}})^{1/2} = \left(\mathbf{x}'(V^{-1})'(V^{-1})\mathbf{x}\right)^{1/2} = (\mathbf{x}'\mathbf{x})^{1/2} \tag{A.6}$$

if V (or V^{-1}) is a unitary matrix. This property implies that the orthonormal matrix operation is numerically stable (Wilkinson, 1965).

A.2 Computation of Matrix Decomposition

In solving a set of linear equations

$$A\mathbf{x} = \mathbf{b} \tag{A.7a}$$

or in computing the representation $\mathbf{x}$ of $\mathbf{b}$ on the column vectors of A, a nonsingular matrix V^{-1} can be multiplied on the left side of A and $\mathbf{b}$ to make matrix $\bar{A} = V^{-1}A$ in echelon form. Then based on the equation

$$\bar{A}\mathbf{x} = V^{-1}\mathbf{b} \,\underline{\underline{\Delta}}\, \bar{\mathbf{b}} \tag{A.7b}$$

$\mathbf{x}$ can be computed. In other words, the representation $\bar{\mathbf{b}}$ of $\mathbf{b}$ can be computed on the new basis vectors of $\mathbf{V}$ such that $\bar{A}$ of (A.7b) is in a decomposed form, and then $\mathbf{x}$ can be computed based on (A.7b).

We will study three different matrices of V^{-1}. All three matrices can be introduced and used in the following unified algorithm.

Algorithm A.1: QR Decomposition (Dongarra et al., 1979)

Let $A = [\mathbf{a}_1 : \ldots : \mathbf{a}_n]$ be an $n \times n$ dimensional square matrix.
Step 1: Compute $n \times n$ dimensional matrix V_1^{-1} such that

$$V_1^{-1}\mathbf{a}_1 = [x,\ 0 \ldots 0]' \tag{A.8a}$$

Step 2: Let

$$V_1^{-1}A = A_1 = \begin{bmatrix} x & : & & & \mathbf{a}_{11} & & \\ .. & .. & .... & .. & .... & .. & .... \\ 0 & : & & : & & : & \\ : & : & \mathbf{a}_{12} & : & .... & : & \mathbf{a}_{1n} \\ 0 & : & & : & & : & \end{bmatrix}$$

Step 3: Compute $(n-1)\times(n-1)$ dimensional matrix $\bar{V}_2^{-1}$ such that

$$\bar{V}_2^{-1}\mathbf{a}_{12} = [x,\ 0\ldots0]' \tag{A.8b}$$

Step 4: Let

$$\begin{bmatrix} 1 : & \ldots & 0 \\ 0 : & & \\ \vdots & \bar{V}_2^{-1} & \\ 0 : & & \end{bmatrix} \left(V_1^{-1}A\right) \underline{\Delta} V_2^{-1}A_1 = \begin{bmatrix} x & : & & & & & \mathbf{a}_{11}' & & \\ .. & .. & .. & .. & \cdots & .. & \cdots & .. & \cdots \\ 0 & : & x & : & & & \mathbf{a}_{22}' & & \\ & : & .. & .. & \cdots & .. & \cdots & .. & \cdots \\ & : & 0 & : & & : & & : & \\ : & : & : & : & \mathbf{a}_{23} & : & \cdots & : & \mathbf{a}_{2n} \\ 0 & : & 0 & : & & : & & : & \end{bmatrix}$$

Continuing in this fashion at most $n-1$ times, we will have

$$\left(V_{n-1}^{-1}\ldots V_2^{-1}V_1^{-1}\, A\ \underline{\Delta} V^{-1}A\right) = \begin{bmatrix} x & -\mathbf{a}_{11}' - & & \\ 0 & x & -\mathbf{a}_{22}' - & \\ \vdots & & \ddots & \\ & & & \ddots \\ 0 & & \cdots & x \end{bmatrix} \tag{A.9}$$

During this basic procedure, if $\mathbf{a}_{i,\,i+1}=0$ is encountered ($i=1, 2, \ldots$), or if

$$V_i^{-1}\ldots V_2^{-1}V_1^{-1}A = \begin{bmatrix} x & X & & : & & & & & & & \\ & \ddots & & : & & & & X & & & \\ & & x & : & & & & & & & \\ .. & \cdots & \cdots & \cdots & \cdots & \cdots & \cdots & \cdots & \cdots & \cdots & \cdots \\ & & & : & 0 & : & & : & & : & \\ & 0 & & : & : & : & a_{i,i+2} & : & .. & : & a_{in} \\ & & & : & 0 & : & & : & & & \end{bmatrix}$$

then the matrix $\bar{V}_{i+1}^{-1}$ will be computed based on the next nonzero vector positioned on the right side of $\mathbf{a}_{i,\ i+1}$ (for example, if $\mathbf{a}_{i,\ i+2} \neq 0$) such that $\bar{V}_{i+1}^{-1}\mathbf{a}_{i,i+2} = [x, 0 \ldots 0]'$. The algorithm will then proceed normally.

The above situation can happen more than once. However, as long as this situation happens at least once, the corresponding result of (A.9) will become a so-called upper-echelon form, such as

$$
V^{-1}A = \left[\begin{array}{ccc:c:c:ccc:c:c:ccc:c}
x & R_1 & & & & & & & & & & & & & \\
 & \ddots & & & & & & & & X & & & & & \\
0 & & x & & & & & & & & & & & & \\
\hdashline
0 & \cdots & 0 & 0 & & x & & R_2 & & & & & & & \\
 & & & \vdots & \vdots & & \ddots & & & & & X & & & \\
 & & & 0 & 0 & & & x & & & & & & & \\
\hdashline
0 & \cdots & 0 & 0 & 0 & \cdots & & 0 & 0\ldots0 & x & & R_3 & & & \\
 & & & & & & & & & & \ddots & & & X & \\
 & & & & & & & & 0\ldots0 & 0 & & x & & & \\
\hdashline
 & & & 0 & & & & & & 0 & \cdots & \cdots & \cdots & 0 & \\
 & & & & & & & & & \vdots & & & & \vdots & \\
 & & & & & & & & & 0 & \cdots & \cdots & \cdots & 0 & \\
 & \underbrace{}_{p} & & & & & & & \underbrace{}_{q} & & & & & &
\end{array}\right] \triangleq R \tag{A.10}
$$

where "x"s are nonzero elements.

In the upper-echelon form, the nonzero elements appear only at the upper right-hand side of the upper triangular blocks [such as R_1, R_2, and R_3 in (A.10)]. These upper triangular blocks appear one after another after shifting one or more columns to the right. For example, in (A.10), R_2 follows R_1 after shifting one column to the right, and R_3 follows R_2 after shifting q columns to the right.

If two upper triangular blocks appear one next to the other without column shifting, then the two blocks can be combined as one upper triangular block. If there is no column shifting at all, then the entire matrix is an upper triangular matrix as in (A.9). Hence the upper triangular form is a special case of the upper-echelon form.

The main feature of an upper-echelon-form matrix is that it reveals clearly the linear dependency among its columns. More explicitly, all columns corresponding to the upper triangular blocks are linearly independent of each other, while all other columns are linear combinations of their respective linearly independent columns at their left.

For example in matrix R of (A.10), the $(p+1)$-th column is linearly dependent on the columns corresponding to R_1, while the q columns between R_2 and R_3 are linearly dependent on the columns corresponding to R_1 and R_2.

The above property of an upper-echelon-form matrix enables the solving of Eq. (A.7a). We first let matrix

$$\tilde{A} = [A \ : \mathbf{b}] \tag{A.11}$$

Then apply Algorithm A.1 to matrix $\tilde{A}$. If after V_r^{-1} is applied,

$$V_r^{-1}...V_1^{-1}\tilde{A} = \begin{bmatrix} A_{11} & : A_{12} & : \mathbf{b}_1 \\ 0 & : A_{22} & : 0 \end{bmatrix} \begin{matrix} \}r \\ \}n-r \end{matrix} \tag{A.12a}$$

then $\mathbf{b}_1$ is already a linear combination of the columns of A_{11}, and the coefficients of this linear combination form the solution $\mathbf{x}$ of (A.7a). In other words, if $A_{11}\mathbf{x}_1 = \mathbf{b}_1$, then the solution of (A.7a) is $\mathbf{x} = [\mathbf{x}_1' : 0...0]'$ with $n-r$ 0's.

In general, we cannot expect the form of (A.12a) for all A and $\mathbf{b}$. Instead, we should expect

$$V_{n-1}^{-1}...V_1^{-1}\mathbf{b} = [x...x]' \tag{A.12b}$$

For (A.12b) to be represented as a linear combination of the columns of $\bar{A} \ \underline{\Delta} V_{n-1}^{-1}...V_1^{-1}A$, $\bar{A}$ must be in the upper triangular form or must have all n columns linearly independent of each other. This is the proof that to have $A\mathbf{x}=\mathbf{b}$ solvable for all $\mathbf{b}$, the square matrix A must have full-column rank (see Example A.4).

In the basic procedure of Algorithm A.1, only matrix $V_i\left(V_i^{-1}\mathbf{a}_i = [x,0...0]'\right)$ can be nonunique. We will introduce three kinds of such matrices in the following. The last two matrices among the three are unitary. We call Algorithm A.1 "QR decomposition" when matrix V is unitary.

For simplicity of presentation, let us express

$$\mathbf{a}_i \ \underline{\Delta} \ \mathbf{a} = [a_1, ..., a_n]'$$

A. **Gaussian Elimination with Partial Pivoting**

$$E = E_2E_1 = \begin{bmatrix} 1 & 0 & & \cdots & & 0 \\ -a_2/a_j & 1 & \ddots & & & \vdots \\ \vdots & & \ddots & & & \vdots \\ -a_{j-1}/a_j & & & \ddots & & \vdots \\ -a_1/a_j & & & & \ddots & \vdots \\ -a_{j+1}/a_j & & & & \ddots & \vdots \\ \vdots & & & & & 0 \\ -a_n/a_j & & & & & 1 \end{bmatrix}$$

(continue to the next line)

$$\begin{bmatrix} 0 & 0 & 0 & 0 & 1 & 0 & \cdots & 0 \\ 0 & 1 & 0 & 0 & 0 & 0 & & \vdots \\ \vdots & & \ddots & & & & & \\ 0 & 0 & 0 & 1 & 0 & 0 & & \\ 1 & 0 & 0 & 0 & 0 & 0 & \cdots & 0 \\ 0 & 0 & 0 & 0 & 0 & 1 & & \vdots \\ \vdots & & & 0 & & & \ddots & 0 \\ 0 & & & \cdots & & & 0 & 1 \end{bmatrix} \begin{matrix} \\ \\ \\ \\ \leftarrow j-\text{th row} \\ \\ \\ \\ \end{matrix} \tag{A.13}$$

$$\uparrow$$
$$\text{the } j-\text{th column}$$

where $|a_j| = \max_i \{|a_i|\}$ is called the "pivotal element."

Because $E_1\mathbf{a} \,\Delta\, \bar{\mathbf{a}} = [a_j, a_2, \ldots, a_{j-1}, a_1, a_{j+1}, \ldots, a_n]'$, it can be easily verified that $E_2E_1\mathbf{a} = E_2\bar{\mathbf{a}} = [a_j, 0 \ldots 0]'$.

Because of (A.13), all unknown parameters of E_2

$$|-a_i/a_j| \le 1, \quad \forall\, i \text{ and } j \tag{A.14}$$

Therefore, the Gaussian elimination with partial pivoting is fairly numerically stable (Wilkinson, 1965).

The order of the computation (multiplications only) of $E\mathbf{x}$ ($\mathbf{x} \ne \mathbf{a}$) is n, excluding the computation of matrix E itself. Hence, the order of computation for Algorithm A.1 using the Gaussian elimination method is $\sum_{i=2 \text{ to } n} i^2 \approx n^3/3$.

B. **Householder Method (Householder, 1958)**

$$H = I - 2\overline{\mathbf{a}\mathbf{a}}' \tag{A.15a}$$

where

$$\bar{\mathbf{a}} = \frac{\mathbf{b}}{\|\mathbf{b}\|} \tag{A.15b}$$

and

$$\mathbf{b} = \begin{cases} \mathbf{a} + \|\mathbf{a}\|\mathbf{e}_1, & \text{if } a_1 \ge 0 \\ \mathbf{a} - \|\mathbf{a}\|\mathbf{e}_1, & \text{if } a_1 < 0 \end{cases} \tag{A.15c}$$

Because

$$\|\mathbf{b}\| = (\mathbf{b}'\mathbf{b})^{1/2} = \left(2\|\mathbf{a}\|^2 \pm 2a_1\|\mathbf{a}\|\right)^{1/2} \tag{A.16}$$

$$\begin{aligned} H\mathbf{a} &= \left(I - 2\mathbf{b}\mathbf{b}'/\|\mathbf{b}\|^2\right)\mathbf{a} \\ \text{(A.15)}: \quad &= \mathbf{a} - 2\mathbf{b}\left(\|\mathbf{a}\|^2 \pm a_1\|\mathbf{a}\|\right)/\|\mathbf{b}\|^2 \\ \text{(A.16)}: \quad &= \mathbf{a} - 2\mathbf{b}/2 \\ \text{(A.15)}: \quad &= \mathbf{a} - \left(\mathbf{a} \pm \|\mathbf{a}\|[1,0\ldots0]'\right) \\ &= \mp\left[\|\mathbf{a}\|,0\ldots0\right]' \end{aligned} \tag{A.17}$$

In addition, because

$$\begin{aligned} H'H &= \left(I - 2\overline{\mathbf{a}}\overline{\mathbf{a}}'\right)\left(I - 2\overline{\mathbf{a}}\overline{\mathbf{a}}'\right) \\ &= I - 4\overline{\mathbf{a}}\overline{\mathbf{a}}' + \overline{\mathbf{a}}\overline{\mathbf{a}}'\overline{\mathbf{a}}\overline{\mathbf{a}}' \\ \text{(A.15b)} &:= I - 4\overline{\mathbf{a}}\overline{\mathbf{a}}' + 4\overline{\mathbf{a}}\left(\mathbf{b}'\mathbf{b}/\|\mathbf{b}\|^2\right)\overline{\mathbf{a}}' \\ &= I - 4\overline{\mathbf{a}}\overline{\mathbf{a}}' + 4\overline{\mathbf{a}}\overline{\mathbf{a}}' \\ &= I \end{aligned}$$

matrix H is unitary. Hence this computation is numerically stable (see Example A.5).

The actual computation of $H\mathbf{x}$ ($\mathbf{x} \neq \mathbf{a}$) does not need to compute the matrix H itself but can follow the following steps:

Step 1: Compute $2\|\mathbf{b}\|^{-2} = (\mathbf{a}'\mathbf{a} \pm a_1(\mathbf{a}'\mathbf{a})^{1/2})^{-1}$ (computation order: n)

Step 2: Compute scalar $c = 2\|\mathbf{b}\|^{-2}(\mathbf{b}'\mathbf{x})$ (computation order: n)

Step 3: Compute $H\mathbf{x} = \mathbf{x} - c\mathbf{b}$ (computation order: n)

Because the result of Step 1 remains the same for different vectors $\mathbf{x}$, the computation of Step 1 will not be counted. Hence the computation of Algorithm A.1 using the Householder method is $\sum_{i=2 \text{ to } n} 2i^2 \approx 2n^3/3$.

Because computational reliability is more important than computational efficiency, the Householder method is very commonly used in practice and is most commonly used in this book, even though it requires twice as much computation as the Gaussian elimination method.

C. Givens Rotational Method (Givens, 1958)

$$G = G_1 G_2, \ldots, G_{n-2} G_{n-1} \tag{A.18a}$$

where

$$G_i = \left[\begin{array}{cccccccc} 1 & & \vdots & & \vdots & & & \\ & \ddots & \vdots & & \vdots & & & \\ & & 1\vdots & & \vdots & & & \\ \cdots & \cdots & \cdots\cdots & \cdots & \cdots & \cdots & \cdots & \cdots \\ & & \vdots & R_i & \vdots & & & \\ \cdots & \cdots & \cdots\cdots & \cdots & \cdots & \cdots & \cdots & \cdots \\ & & \vdots & & \vdots & 1 & & \\ & & \vdots & & \vdots & & \ddots & \\ & & \vdots & & \vdots & & & 1 \end{array}\right] \begin{array}{l} \\ \}i-1 \\ \\ \\ \}2 \\ \\ \\ \}n-i-1 \\ \\ \end{array} \tag{A.18b}$$

and

$$R_i = \begin{bmatrix} \cos\theta_i & \sin\theta_i \\ -\sin\theta_i & \cos\theta_i \end{bmatrix} \tag{A.18c}$$

Equation (A.18) shows that the component matrices G_i (or R_i) of matrix G are decided by their respective parameter θ_i ($i=n-1, n-2, \ldots, 1$). The parameter θ_i is determined by the two-dimensional vector operated by R_i. Let this vector be $\mathbf{b}_i=[x \quad y]'$. Then

$$\theta_i = \tan^{-1}\left(y/x\right) \; (= 90° \text{ if } x = 0)$$

or

$$\cos\theta_i = x/\|\mathbf{b}_i\| \quad \text{and} \quad \sin\theta_i = y/\|\mathbf{b}_i\|$$

It is easy to verify that $R_i\mathbf{b}_i = \left[\|\mathbf{b}_i\|, 0\ldots0\right]'$

The geometrical meaning of $R_i\mathbf{b}_i$ can be interpreted as the rotation of the original cartesian coordinates counterclockwise θ_i degrees so that the x-axis now coincides with $\mathbf{b}_i$. This operation is depicted in Figure A.3.

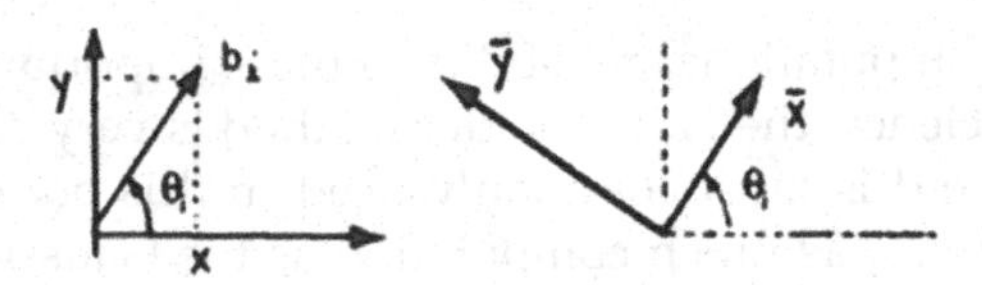

FIGURE A.3
Geometrical meaning of Givens rotational method.

The reader can refer to Example A.5 for three numerical examples of the Givens method.

It is easy to verify that according to (A.18a–c),

$$\mathbf{Ga} = [\|\mathbf{a}\|, 0 \ldots 0]' \tag{A.18d}$$

Because $R_i'R_i = I \ \forall i$, the matrix G of (A.18a–c) is a unitary matrix. Therefore, like the Householder method, the Givens rotational method is numerically stable.

It is easy to verify that the order of computation for $G\mathbf{x}$ ($\mathbf{x} \neq \mathbf{a}$) is $4n$, excluding the computation of G itself. Hence, the order of computation of Algorithm A.1 is $\sum_{i=2 \text{ to } n} 4i^2 \approx 4n^3/3$.

Although the Givens method is only half as efficient as the Householder method, it has very simple and explicit geometrical meanings. Therefore, it is still commonly used in practice and is used in Algorithm 9.2 of this book.

Finally, after Algorithm A.1 is applied and the echelon-form matrix $V^{-1}A = \bar{A}$ is obtained, we still need to compute $\mathbf{x}$ from $\bar{A}$ and $V^{-1}\mathbf{b} = \bar{\mathbf{b}}$. Eliminating the linearly dependent columns of $\bar{A}$ [see description of the echelon form of (A.10)], we have

$$\bar{A}\mathbf{x} \triangleq \begin{bmatrix} a_{11} & a_{12} & \ldots & a_{1r} \\ 0 & a_{22} & \ldots & a_{2r} \\ \vdots & & \ddots & \vdots \\ 0 & \ldots & 0 & a_{rr} \end{bmatrix} \begin{bmatrix} x_1 \\ x_2 \\ \vdots \\ x_r \end{bmatrix} = \bar{\mathbf{b}} \triangleq \begin{bmatrix} b_1 \\ b_2 \\ \vdots \\ b_r \end{bmatrix} \tag{A.19}$$

where the diagonal elements of the matrix are nonzero. It is obvious that the solution of (A.19) is

$$x_r = \frac{b_r}{a_{rr}} \qquad x_i = \frac{\left(b_i - \sum_{j=i+1}^{r} a_{ij}x_j\right)}{a_{ii}}, \quad i = r-1, r-2, \ldots, 1 \tag{A.20}$$

The computation of (A.20) is called "back substitution," whose order of computation at $r=n$ is $n^2/2$. This computation is numerically stable with respect to the Problem (A.19) itself (Wilkinson, 1965).

However, because this operation requires consecutive divisions by a_{ii} ($i = r, r-1, \ldots, 1$), the Problem (A.19) can be ill conditioned when these elements have small magnitudes. This understanding conforms with the theoretical result about the condition number $||\bar{A}||\,||\bar{A}^{-1}||$ of matrix $\bar{A}$ (2.13). In the next Section (A.28) and (A.29), we will show that $\|\bar{A}^{-1}\| = \sigma_r^{-1} \geq |\lambda_r|^{-1}$, where σ_r and $|\lambda_r| = \min_i \{|a_{ii}|\}$ are the smallest singular value and the smallest eigenvalue of $\bar{A}$, respectively. Thus, small elements a_{ii} imply large and bad condition of matrix $\bar{A}$ as well as Problem (A.19).

Comparing the resulting vector $[a_{ii}, 0 \ldots 0]'$ of the three matrix decomposition methods, both orthogonal methods (Householder and Givens) have $a_{ii} = ||\mathbf{a}_i||_2$ [see (A.17) and (A.18d)], while the Gaussian elimination method has $a_{ii} = ||\mathbf{a}_i||_\infty$ [see (A.13) and Definition 2.1]. Because $||\mathbf{a}_i||_2 \geq ||\mathbf{a}_i||_\infty$, the orthogonal methods not only are computationally more reliable than the Gaussian elimination method but also make their subsequent computation better conditioned.

A.3 Singular Value Decomposition (SVD)

Matrix singular value decomposition was proposed as early as in 1870 by Betram and Jordan. It became one of the most important mathematical tools in numerical linear algebra and linear control systems theory only in the 1970s (Klema and Laub, 1980), about a 100 years later. This is because SVD is a well-conditioned problem and because of the development of a systematic and numerically stable computational algorithm of SVD (Golub and Reinsch, 1970).

A.3.1 Definition and Existence

Theorem A.1

For any $m \times n$ dimensional matrix A, there exists an $m \times m$ and $n \times n$ dimensional unitary matrix U and V such that

$$A = U \sum V^* = U_1 \Sigma_r V_1^* \tag{A.21}$$

where

$$\sum = \begin{bmatrix} \Sigma_r & 0 \\ 0 & 0 \end{bmatrix}, \quad \Sigma_r = \text{diag}\{\sigma_1, \sigma_2, \ldots, \sigma_r\}$$

$$U = \Big[\underset{r}{U_1} \quad \vdots \quad \underset{m-r}{U_2}\Big] \quad \text{and} \quad V = \Big[\underset{r}{V_1} \quad \vdots \quad \underset{n-r}{V_2}\Big]$$

and

$$\sigma_1 \geq \sigma_2 \geq \cdots \geq \sigma_r > 0.$$

Here σ_i ($i=1, \ldots, r$) is the positive square root of the i-th largest eigenvalue of matrix A^*A and is defined as the i-th nonzero singular value of matrix A. Matrices U and V are the orthonormal right eigenvector matrices of AA^* and A^*A, respectively. In addition, there are $\min\{m,n\}-r \underline{\Delta}\, n-r$ (if $n \leq m$) zero singular values ($\sigma_{r+1}=\ldots=\sigma_n=0$) of matrix A. Equation (A.21) is defined as the singular value decomposition of matrix A.

Proof: See Stewart (1976).

A.3.2 Properties

Theorem A.2: Minimax Theorem

Let the singular values of an $m\times n$ dimensional matrix S be $\sigma_1 \geq \sigma_2 \geq \ldots \geq \sigma_n > 0$. Then,

$$\sigma_k = \min_{\dim(S)=n-k+1} \max_{\substack{\mathbf{x} \in \mathbf{S} \\ \mathbf{x} \neq 0}} \frac{\|A\mathbf{x}\|}{\|\mathbf{x}\|} \qquad k=1,2,\ldots,n \tag{A.22}$$

where the linear space $\mathbf{S}$ is spanned by the $n-k+1$ basis vectors $\{\mathbf{v}_k, \mathbf{v}_{k+1}, \ldots, \mathbf{v}_n\}$, which are the last $n-k+1$ vectors of matrix V of (A.21).

Proof

From Definition A.2, let the unitary matrix

$$V \underline{\Delta} [V_1 : V_2] \underline{\Delta} [\mathbf{v}_1 \; : \; \ldots \; : \; \mathbf{v}_{k-1} | \mathbf{v}_k \; : \; \ldots \; : \; \mathbf{v}_n]$$

Then the vectors of V_1 will span the "orthogonal complement space" $\overline{\mathbf{S}}$ of $\mathbf{S}$ such that $\overline{\mathbf{S}}'\mathbf{S} = 0$ and $\overline{\mathbf{S}} \cup \mathbf{S}$ = n-dimensional space.

Because $\mathbf{x} \in \mathbf{S}$ implies

$$\mathbf{x} = [\mathbf{v}_1 \; : \; \ldots \; : \; \mathbf{v}_{k-1} | \mathbf{v}_k \; : \; \ldots \; : \; \mathbf{v}_n] \begin{bmatrix} 0 \\ \vdots \\ 0 \\ a_k \\ \vdots \\ a_n \end{bmatrix} \begin{matrix} \}k-1 \\ \\ \}n-k+1 \end{matrix} \quad \underline{\Delta} V\mathbf{a} \tag{A.23}$$

Hence,

$$\begin{aligned}\|A\mathbf{x}\|/\|\mathbf{x}\| &= \left(\mathbf{x}^*A^*\ A\mathbf{x}/\mathbf{x}^*\mathbf{x}\right)^{1/2} \\ &= \left(\mathbf{a}^*V^*A^*\ AV\mathbf{a}/\mathbf{a}^*V^*V\mathbf{a}\right)^{1/2} \\ &= \left(\mathbf{a}^*\Sigma^2\mathbf{a}/\mathbf{a}^*\mathbf{a}\right)^{1/2} \\ &= \left[\left(a_k^2\sigma_k^2 + a_{k+1}^2\sigma_{k+1}^2 + \cdots + a_n^2\sigma_n^2\right)/\left(a_k^2 + a_{k+1}^2 + \cdots + a_n^2\right)\right]^{1/2} \\ &\leq \sigma_k\end{aligned} \tag{A.24}$$

Thus, the maximum part of the theorem is proved. On the other hand,

$$\mathbf{x} = \left[\mathbf{v}_1 \quad : \quad \ldots \quad : \quad \mathbf{v}_k \quad : \quad \mathbf{v}_{k-1} \quad : \quad \ldots \quad : \quad \mathbf{v}_n\right]\begin{bmatrix} a_1 \\ \vdots \\ a_k \\ 0 \\ \vdots \\ 0 \end{bmatrix} \tag{A.25}$$

similarly implies that $||A\mathbf{x}||/||\mathbf{x}|| \geq \sigma_k$.
The combined (A.24) and (A.25) prove Theorem A.2.

Corollary A.1 $\|\mathrm{A}\| \underline{\Delta} \max_{\mathbf{x}\neq 0} \frac{\|A\mathbf{x}\|}{\|\mathbf{x}\|} = \sigma_1$ (A.26)

Corollary A.2 $\min_{\mathbf{x}\neq 0} \frac{\|A\mathbf{x}\|}{\|\mathbf{x}\|} = \sigma_n$ (A.27)

Corollary A.3 $\sigma_1 \geq |\lambda_1| \geq \cdots \geq |\lambda_n| \geq \sigma_n$ (A.28)

where λ_i (i=1, …, n) are the eigenvalues of matrix A (if m=n) (Jiang, 1993).
Corollary A.4 If A^{-1} exists, then

$$\left\|A^{-1}\right\| \underline{\Delta} \max_{\mathbf{x}\neq 0} \frac{\left\|A^{-1}\mathbf{x}\right\|}{\|\mathbf{x}\|} = \max_{\mathbf{x}\neq 0} \frac{\left\|V_1\Sigma_r^{-1}U_1^*\mathbf{x}\right\|}{\|\mathbf{x}\|} = \sigma_n^{-1} \tag{A.29}$$

Corollary A.5 If A^{-1} exists, then

$$\min_{\mathbf{x}\neq 0} \frac{\left\|A^{-1}\mathbf{x}\right\|}{\|\mathbf{x}\|} = \sigma_1^{-1} \tag{A.30}$$

Corollary A.6

Let the singular values of two $n \times n$ matrices A and B be $\sigma_1 \geq \sigma_2 \geq \ldots \geq \sigma_n$ and $s_1 \geq s_2 > \ldots \geq s_n$, respectively, then,

$$|\sigma_k - s_k| \leq \|A - B\| \underline{\Delta} \|\Delta A\|, \quad (k = 1, \ldots, n) \tag{A.31}$$

Proof

From (A.22),

$$\sigma_k = \min_{\substack{\dim(S)=n-k+1 \\ x \in S, x \neq 0}} \max \frac{\|A\mathbf{x}\|}{\|\mathbf{x}\|} = \max_{x \in S, x \neq 0} \frac{(B + \Delta A)\mathbf{x}}{\|\mathbf{x}\|}$$

$$\leq \frac{\|B\mathbf{x}\|_{x \in S, x \neq 0}}{\|\mathbf{x}\|} + \frac{\|\Delta A\mathbf{x}\|_{x \in S, x \neq 0}}{\|\mathbf{x}\|} \tag{A.32}$$

$$\leq s_k + \|\Delta A\|$$

Similarly $s_k \leq \sigma_k + \|\Delta A\|$ (A.33)

Hence the theorem.

Corollary A.6 implies that SVD problem is well conditioned or is insensitive to the original data variation ΔA.

A.3.3 Applications

For simplicity of presentation, we let all matrices of this section be real, and we present all theorems without proof.

A. Solving of a Set of Linear Equations

$$A\mathbf{x} = \mathbf{b}, \ (\mathbf{b} \neq 0) \tag{A.34}$$

From (A.21):

$$\mathbf{x} = V_1 \Sigma_r^{-1} U_1' \mathbf{b} \tag{A.35}$$

Theorem A.3

If $\mathbf{b}$ is a linear combination of the columns of U_1, then (A.35) is an exact solution of (A.34).

This theorem proves that the columns of U_1 span the range space of A, $\mathbf{R}(A)$ (see Example A.4).

Theorem A.4

If **b** is not a linear combination of the columns of U_1, then (A.35) is the least-squares solution of (A.34). In other words, for all $\Delta\mathbf{x}\neq 0$,

$$\|A\mathbf{x}-\mathbf{b}\| \leq \|A(\mathbf{x}+\Delta\mathbf{x})-\mathbf{b}\| \tag{A.36}$$

if **x** is computed from (A.35).

Theorem A.5

If the rank of matrix A is n, then U_1 has n linearly independent columns. Thus, the necessary and sufficient condition for (A.34) to have exact solution (A.35) for all **b** is that matrix A be full rank.

Theorem A.6

The nonzero solution **x** of linear equations

$$A\mathbf{x}=0 \tag{A.37}$$

is a linear combination of the columns of V_2. In other words, the columns of V_2 span the null space of A, $\mathbf{N}(A)$.

The above result can be generalized to its dual case.

Example A.6

[See Step 2a, Algorithm 6.1.]

Let the $m\times p$ $(m\leq p)$ dimensional matrix DB be full-row rank. Then in its SVD of (A.21), $U_1=U$ and $U_2=0$. Thus, based on the duality (or transpose) of Theorem A.6, there is no nonzero solution **c** such that $\mathbf{c}DB=0$.

Based on the duality (or transpose) of Corollary A.2,
$\min||\mathbf{c}DB||=\sigma_m$, when $\mathbf{c}=\mathbf{u}_m' = (\text{the } m-\text{th column of } U)'$.

Example A.7

(See Conclusion 6.4 and its proof.)

Let the $n\times p$ $(n>p)$ dimensional matrix B be full-column rank. Then, in its SVD of (A.21), U_1 and U_2 have dimensions $n\times p$ and $n\times(n-p)$, respectively. Based on the transpose of Theorem A.6, all rows of $(n-m)\times n$ dimensional matrix T such that $TB=0$ are linear combinations of the rows of U_2'.

Now because $\mathbf{R}(U_1)\cup\mathbf{R}(U_2)=n$-dimensional space R^n and $U_1'U_2=0$, the rows of any $m\times n$ matrix C such that $[T':C']'$ is full rank must be linear combinations of the rows of U_1'. Consequently, CB must be full-column rank.

Example A.8

(See Step 4 of Algorithm 8.2 and Step 3 of Algorithm 8.3.)

Let the columns of an $n \times p$ $(n > p)$ dimensional matrix D be orthonormal. Then in its SVD of (A.21), $U = D$ and $\Sigma_r = V = I_p$. Thus, the least-squares solution (A.35) of $D\mathbf{c} = \mathbf{b}$ is

$$\mathbf{c} = D'\mathbf{b}$$

Theorem A.7

Let us define A^+ as the pseudo-inverse of matrix A such that $A^+AA^+ = A^+$, $AA^+A = A$, $(AA^+)' = AA^+$, and $(A^+A)' = A^+A$. Then,

$$A^+ = V_1 \Sigma_r^{-1} U_1'$$

Thus, from Theorems A.3 and A.4, $\mathbf{x} = A^+\mathbf{b}$ is the least-squares solution of $A\mathbf{x} = \mathbf{b}$.

B. Rank Determination

From Theorems A.3 to A.6, the rank r of an $n \times n$ dimensional matrix A determines whether the Eqs. (A.34) and (A.37) are solvable. If $r = n$, then (A.34) is solvable for all $b \neq 0$ while (A.37) is unsolvable. If $r < n$, then (A.34) may not be solvable while (A.37) has $n-r$ linearly independent solutions $\mathbf{x}$.

There are several numerical methods for rank determination. For example, Algorithm A.1 can be used to determine the number ($= r$) of linearly independent columns/rows of a matrix. The rank of a square matrix also equals the number of nonzero eigenvalues of that matrix. However, both numbers are very sensitive to the variation and uncertainty of matrix A, ΔA. Therefore, these two methods are not very reliable in rank determination.

On the other hand, the rank of matrix A also equals the number of nonzero singular values of A, and the singular values are insensitive to ΔA. Therefore, this is, so far, the most reliable method of rank determination.

Theorem A.8

If the singular values computed from a given matrix $A + \Delta A$ are $s_1 \geq s_2 \geq \ldots \geq s_n > 0$ $(r = n)$, then the necessary condition for the rank of the original matrix A to be less than n (or σ_n of $A = 0$) is $||\Delta A|| \geq s_n$, and the necessary condition for the rank of A to be less than r (or σ_r of $A = 0$) is $||\Delta A|| \geq s_r$ $(r = 1, \ldots, n)$.

Proof

Let σ_r be zero in Corollary A.6 for $r = 1, \ldots, n$, respectively.

Theorem A.8 implies that the determination of rank $= r$ (or r nonzero singular values) has an accuracy margin which is equivalent to $||\Delta A|| < \sigma_r$.

In solving the set of linear equations (A.34), the higher the determined r, the more accurate the least-squares solution (A.35), and the greater the norm of the corresponding solution because of the greater corresponding σ_r^{-1} (see the end of Section A.2). This tradeoff of accuracy and solution magnitude is studied in depth in Lawson and Hanson (1974) and Golub et al. (1976a).

From this perspective, not only the absolute magnitude of the singular values, but also the relative magnitude among the singular values should be considered in rank determination. For example, it is recommended that the r is determined so that there is a greater gap between singular values s_r and s_{r+1} than other singular value gaps.

This tradeoff between accuracy and solution magnitude (or the condition of subsequent computation) also surfaced in control systems problems. For example, such a tradeoff is involved between the condition of Eq. (4.1) and the amount of system order (or system information), as discussed at the end of Section 5.2. Such a tradeoff also appears at the Hankow matrix-based model reduction problem (Kung and Lin, 1981) and minimal order realization problem (Tsui, 1983b).

Finally, although singular values are most reliable in revealing the *total* number of linearly independent columns/rows of a matrix, they cannot reveal *which* columns/rows of that matrix are linearly independent of each other. On the other hand, *each* system matrix column or row corresponds to a certain state, a certain input, or a certain output. Hence linear dependency of *each* system matrix column/row is essential in many control problems such as controllability/observability index computation or analytical eigenvector assignment. Because the orthogonal QR matrix decomposition operation (Algorithm A.1) can reveal such linear dependency, and because this method is still quite reliable in computation (DeJong, 1975; Golub et al., 1976a; Tsui, 1983b), it is most widely used in this book.

A.4 The Applications and Solution of Matrix Equation Pair $TA-FT=LC$ and $TB=0$ (Tsui, 2004a)

In the matrix equation pair $TA-FT=LC$ (Eq. 4.1) and $TB=0$ (Eq. 4.3), parameters (A, B, C) are given plant system state space model (1.1), with dimensions $n \times n$, $n \times p$, and $m \times n$, respectively. Parameters (F, T, L) are the state space model of dynamic part of the observer (3.16a) with dimensions $r \times r$, $r \times n$, and $r \times m$, respectively, and are to be designed as the solution of this matrix equation pair.

In addition, and in practical design, the solution to satisfy this matrix equation pair must also satisfy matrix $[T':C']'$ being full row rank. This is because the purpose of Eq. (4.1) is to estimate signal $T\mathbf{x}(t)$ (Theorem 3.2). If the rows of matrix T are not linearly independent from each other and from the rows of matrix C, then this $T\mathbf{x}(t)$ signal will have redundancy with each other and with the measured signal $C\mathbf{x}(t)$.

Another desired feature for this solution is to maximize rank ($\underline{C} \equiv [T':C']'$). The reason is to minimize the constraint on the feedback control $\underline{KC}\mathbf{x}(t)$, even though a satisfactory $\underline{KC}\mathbf{x}(t)$ control signal may still be generated without a maximum possible value of rank ($\underline{C}$) in many practical situations (see Section 6.4).

To guarantee the performance of the observer/feedback controller, almost all existing design formulations require that the observer poles or the eigenvalues of matrix F be arbitrarily given. At least, these poles must be stable (see Section 5.2).

Another must-satisfied feature, not mentioned in other control literature, is that the rows of solution matrix T must be completely decoupled. Only with this feature, can the number of rows of T, or the order of the observer/feedback controller (the most important parameter of any controller system), be freely designable. It is proven in Section 4.4, Chapter 6, (Tsui, 2015), and Subsection A.4.6, that only this technical feature can make the solution to matrix equation pair (4.1) and (4.3) exist for the great majority of plant systems.

To make the rows of matrix T be completely decoupled, matrix F must be set in Jordan form (Tsui, 1987a, 1993a). Hence, solution matrix F of this matrix equation pair is fixed except its dimension.

The following is a list of five most important and direct applications of this matrix equation pair.

A.4.1 Output Feedback Controller That Can Generate $Kx(t)$ Signal (Definition 3.3)

Because Eq. (4.3) ($TB=0$) means zero input feedback to the observer (3.16a), we name such observer/controller an "output feedback compensator" – This controller takes the feedback of plant system output measurement only.

Combined with Eq. (4.1), this equation pair forms the necessary and sufficient condition for an output feedback compensator that can generate the $K\mathbf{x}(t)$ control signal (K is a constant).

The main purpose of feedback control is maintaining robust control of system output against system input disturbance and model uncertainty. Therefore, the output feedback compensator structure is really the rational feedback control structure and is by far the dominant structure of the classical control theory. From the modern control theory, this structure is the only structure that can guarantee the full realization of the critical loop transfer function and robust properties of $K\mathbf{x}(t)$ control (see Theorem 3.4).

Therefore, deriving an exact solution of this matrix equation pair and a solution general to the great majority of plant systems, is the main significance of the new design principle of this book, and is a momentous progress of the modern control theory over the past six decades.

A.4.2 Unknown Input Observer

An unknown input observer is a state observer (Definition 4.1) without taking the feedback of that plant system's unknown input (Wang et al., 1975). The necessary and sufficient condition of an unknown input observer is Eq. (4.1), (4.3) (if the given gain to the unknown input is matrix B), and rank ($\underline{C}=[T':C']'$)$=n$.

Therefore, unknown input observer is a special case of observers satisfying the matrix equation pair (4.1) and (4.3), with the special additional condition of rank ($\underline{C}$)$=n$. Because of this additional condition, unknown input observer does not exist for a great majority of plant systems, while exact solution of the matrix equation pair (4.1) and (4.3) (without the additional condition of rank ($\underline{C}$)$=n$) does exist for a great majority of plant systems (see Example 4.6 and Conclusion 6.5).

A.4.3 Robust Input Fault Detectors (Section 7.2)

The accidental plant system control input fault, is represented as an unknown input fault signal $\mathbf{d}(t)$, and appear as an additional term $B_d\mathbf{d}(t) \equiv [B_{d1}:B_{d2}]$ $[\mathbf{d}_1(t)':\mathbf{d}_2(t)']'$ in the plant system state space model (7.12). There are generally Q parts of $\mathbf{d}(t)$ ($\mathbf{d}_1(t)$ ... $\mathbf{d}_Q(t)$) in the general fault model (7.12), we limit $Q=2$ here for simplicity of presentation.

A fault detector is to detect the fault occurrence, which is the changing of $\mathbf{d}(t)$ from 0 to nonzero. This is achieved by designing the fault detector/observer with a single output $e(t)$ such that $e(t)$ changes from 0 to nonzero as $\mathbf{d}(t)$ changes from 0 to nonzero. The $e(t)$ signal is therefore named the "residual signal".

The fault isolation and diagnosis problem is much more difficult. It requires to detect not only $\mathbf{d}(t)\neq 0$, but also to identify if only one of the two $\mathbf{d}_i(t)$'s becomes nonzero then which one ($\mathbf{d}_1(t)$ or $\mathbf{d}_2(t)$). To achieve this goal, it is proved in Subsection 7.2.2 that for this fault model, two special fault detectors must work together, with each fault detector output $e_i(t)$ becomes nonzero if only if its corresponding say $\mathbf{d}_i(t)$ becomes nonzero, $i=1, 2$. This kind of special fault detector is named "robust fault detector" because its output $e_i(t)$ will remain 0 (or insensitive and robust) to the nonzero $\mathbf{d}_j(t)$'s if $j\neq i$.

It is proved in (7.14) that to satisfy this design requirement, each of the i-th robust fault detector must satisfy (4.1) in order to generate $T_i\mathbf{x}(t)$ signal, and must satisfy $T_iB_{dj}=0$ or (4.3) for $j\neq i$, and must satisfy a relatively easier condition of $0=\underline{K}_i\underline{C}_i$ for all i (which can always be satisfied by a high enough observer order or enough number of rows of $\underline{C}_i$). Therefore, equation pair (4.1) and (4.3) is the main necessary condition for the fault detection an isolation problem (Ge and Fang, 1988; Tsui, 1989).

A.4.4 Observer for Systems with Time-Delayed States

Around 2010, some people modeled the effect of time delayed states $\mathbf{x}(t-\tau)$, as an additional term $B_d\mathbf{x}(t-\tau)$ in the plant system state space model (1.1a), where the constant τ is the delay time of state $\mathbf{x}(t)$. Obviously, the matrix equation pair (4.1) and (4.3) (or $TB_d=0$) is the key design condition for this observer (Tsui, 2012b).

A.4.5 Eigen-Structure Assignment by Generalized State Feedback Control

There are two (and dual) design Algorithms 8.1 and 8.2 for computing the design solution of this assignment. Of the two main steps of these algorithms, the second main step, and the only nontrivial main step, is the same as (or the dual of) our matrix equation pair (4.1) and (4.3).

That second main step of Algorithm 8.1 is to compute the solution matrices V_q and K_q for equation pair (8.11a) and (8.11b):

$$AV_q - V_q\Lambda_q = BK_q,\ \text{and}\ T_{n-q}V_q = 0,$$

where matrix $T_{n\text{-}q}$ is already computed in Step 1 and can be treated here as the given system matrix C. Because the dual of matrix C is matrix B (see Section 1.3), the dual of (8.11a and b) is

$$T_qA' - \Lambda_q' T_q = L_qC\ \text{and}\ T_qB = 0$$

which is our matrix equation pair (4.1) and (4.3), where matrix T_q is the dual of matrix V_q as the left and right eigenvector matrices corresponding to Jordan for matrix Λ_q, while matrices L_q and K_q are the dual observer gain and state feedback control gain, respectively.

The second main step of Algorithm 8.2 is to compute solution matrices T_p and L_p for equation pair (8.16a) and (8.16b):

$$T_pA - \Lambda_pT_p = L_pC\ \text{and}\ T_pV_{n-p} = 0$$

Which is again our matrix equation pair (4.1) and (4.3), with the dimensional parameter r of (4.1) and (4.3) changed to parameter p. Here matrix V_{n-p} is already computed in Step 1 of Algorithm 8.2 and can be treated as the given system matrix B.

A.4.6 The Closed Form Solution of This Matrix Equation Pair

The above five very basic, important, and wide range applications, make the matrix equation pair (4.1) and (4.3) by far the most important mathematical requirement of the modern control system design.

Yet, surprisingly, if the dimension r of solution (F, T, L) to this matrix equation pair is flexible from 1 to n–m, then this solution is very general and very simple, as proven and demonstrated in Algorithms 5.3 and 6.1 and Conclusion 6.5, of this book. Please refer to the beginning of Chapter 5 for the publication history of this solution.

Algorithms 5.3 and 6.1 and their following analysis presented completely and explicitly, the computation and the generality of the solution of (4.1) and (4.3). For simplicity of presentation, we will assume all eigenvalues of matrix F or all poles of observer/controller be real and distinct λ_i, $i=1, \ldots, r$. Then Eqs. (4.1) and (4.3) can be expressed as

$$[\mathbf{t}_i : \mathbf{l}_i]\begin{bmatrix} A-\lambda_i I & : & B \\ -C & : & 0 \end{bmatrix} = 0, (i=1,\ldots,r) \quad \text{(A.38)}$$

where $\mathbf{t}_i$ and $\mathbf{l}_i$ are the i-th rows of solution matrices T and L corresponding to λ_i, respectively.

Equation (A.38) shows clearly that a sufficient condition for the existence of nonzero solution $[\mathbf{t}_i : \mathbf{l}_i]$ is either $m>p$, or that system (A, B, C) has at least one (or $r>0$) stable transmission zeros λ_i (Conclusion 6.1). A great majority of plant systems can satisfy either one of these two sufficient conditions (see Conclusion 6.5 and Example 4.6).

For all plant systems that do not satisfy either of the above two sufficient conditions, and for all design requirements for which some rows of solution (F, T, L) cannot satisfy (4.1) and (4.3) exactly, a solution that satisfies (4.1) exactly and (4.3) approximately, can always be easily and simply computed. See Section 6.1. This is because Eq. (4.3) ($TB=0$) is perhaps the simplest formulation, whose approximate solution has been well developed.

In other existing literature of the modern control theory, the decoupled solution of Eq. (4.1) was not derived and nor used (Tsui, 1987a, 2003a, 2015). As a result, the number of rows of solution matrix T, or the observer/controller order r, is fixed at n–m. This implies that rank ($[T':C]'$)$=n$ must be satisfied together with (4.1) and (4.3), or that the three conditions of the unknown input observer of A.4.2 must be satisfied. These three conditions are much more restrictive than our equation pair (4.1) and (4.3). Thus, the corresponding solution to those three conditions becomes very restrictive, and is nonexist for the great majority of plant systems (see Conclusions 6.4 and 6.5, and Example 4.6).

Appendix B: Design Projects and Problems

There are eight design projects listed with partial answers in this appendix. Its purpose is twofold. First, these design projects show the usefulness of the theoretical design methods of this book. Second, these design projects are the synthesized and practical exercises of the theoretical design methods of this book.

Because of the limitations on the scope of this book and of the control theory itself, only the mathematical models and mathematical design requirements, and not the physical meanings of each project, are described in this appendix. Readers are referred to the original papers for the detailed physical meanings of each project, because such understanding of the actual physical project is essential to any good design.

System 1: Airplane System (Choi and Sirisena, 1974)

$$A = \begin{bmatrix} -0.037 & 0.0123 & 0.00055 & -1 \\ 0 & 0 & 1 & 0 \\ -6.37 & 0 & -0.23 & 0.0618 \\ 1.25 & 0 & 0.016 & -0.0457 \end{bmatrix},$$

$$B = \begin{bmatrix} 0.00084 & 0.000236 \\ 0 & 0 \\ 0.08 & 0.804 \\ -0.0862 & -0.0665 \end{bmatrix}, \text{ and}$$

$$C = \begin{bmatrix} 0 & 1 & 0 & 0 \\ 0 & 0 & 1 & 0 \\ 0 & 0 & 0 & 1 \end{bmatrix}$$

(a) Using Algorithms 5.3 and 6.1, design the dynamic part of the dynamic output feedback compensator of this system, with $F=-2$.

(b) Using Algorithm 10.1, design the LQ optimal state feedback control K for $Q=I$, $R=I$. Compute the eigenvalues of the corresponding feedback system dynamic matrix $A-BK$.

(c) Using the result K of part (b), design the output part of the dynamic output $\underline{K} \triangleq \left[K_Z : K_y\right]$ of the feedback compensator of part (a) such that $K = \underline{K}\left[T' : C'\right] \triangleq \underline{K}\underline{C}$ is best satisfied.

(d) Using Algorithms 8.1 and 8.2, design K_y such that the matrix $A-BK_yC$ has the same eigenvalues of part (b).

(e) Using Algorithm 10.2, design the LQ static output feedback control $K_y C\mathbf{x}(t)$ for $Q=I$, $R=I$. The answer is:

$$K_y = \begin{bmatrix} -0.397 & -1.591 & -7.847 \\ 1.255 & 3.476 & 4.98 \end{bmatrix}$$

(f) Repeat part (e) for the generalized state feedback control $\underline{K}\underline{C}\mathbf{x}(t)$.
(g) Compare the control systems of part (c) to part (f) in terms of poles, eigenvector matrix condition number, feedback gain, and zero-input response.
(h) Design a complete fault detection/isolation/accommodation system, with poles equal to –1 and –2 and with any result selected from part (c) to part (f) as a normal (fault-free) compensator.

System 2: Four-Water-Tank System (Ge and Fang, 1988)

Figure B.1 shows a four-tank system. On the condition that $A_i = 500\,\text{cm}^2$, $S_i = 2.54469\,\text{cm}^2$ ($i=1, \ldots, 4$), and $u(t) = 1\,\text{cm}^3/\text{s}$, the state space model with the water levels h_i (cm) ($i=1, \ldots, 4$) as the four-system states is

$$A = 21.886^{-1} \begin{bmatrix} -1 & 1 & 0 & 0 \\ 1 & -2 & 1 & 0 \\ 0 & 1 & -2 & 1 \\ 0 & 0 & 1 & -2 \end{bmatrix} \quad B = \begin{bmatrix} 0.002 \\ 0 \\ 0 \\ 0 \end{bmatrix}$$

and

$$C = \begin{bmatrix} 1 & 0 & 0 & 0 \\ 0 & 0 & 1 & 0 \\ 0 & 0 & 0 & 1 \end{bmatrix}$$

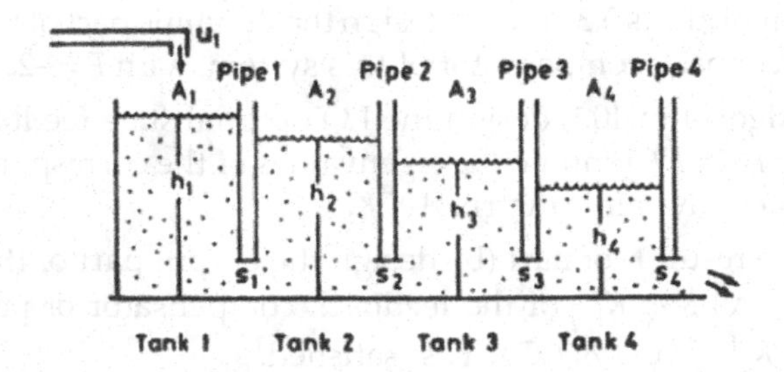

FIGURE B.1
A four-tank system.

(a) Using Algorithms 5.3 and 6.1 (especially Step 2(b) of Algorithm 6.1), design the dynamic part of a dynamic output feedback compensator of this system, with $F=-7$.

(b) Using Algorithm 7.2, design the fault detection and isolation system with $q=2$ and with robust fault detector pole (or double pole) equal to –7. The partial answer is:

$$T_1 = \begin{bmatrix} 0 & 0 & 0 & -130.613 \end{bmatrix}$$

$$T_3 = \begin{bmatrix} 0 & 16.702 & -110.655 & 0 \\ 0 & -27.144 & 135.721 & 0 \end{bmatrix}$$

$$T_5 = \begin{bmatrix} 84.582 & 0 & -84.582 & 0 \end{bmatrix}$$

$$T_6 = \begin{bmatrix} 116.517 & -18.713 & 0 & 0 \\ -114.759 & 22.952 & 0 & 0 \end{bmatrix}$$

(c) Using the duality of Algorithm 5.3, design a unique state feedback gain K to place the eigenvalues $-2.778 \pm j14.19$ and $-5.222 \pm j4.533$ in matrix $A-BK$.

(d) Compare the parameter T_0 of part (a) and T_i $(i=1, 3, 5, 6)$ of part (b) in generating the normal-state feedback K of part (c): Let $K = \underline{K}_i \underline{C}_i$, where $\underline{C}_i \triangleq \left[T_i' : C' \right]'$ $(i=0, 1, 3, 5, 6)$, and compare the accuracy of $\underline{K}_i \underline{C}_i$ and the magnitude of gain $\underline{K}_i$.

(e) Repeat Example 7.9 for the design of failure accommodation control and threshold treatment of model uncertainty and measurement noise.

System 3: A Corvette 5.7L, Multi-port, Fuel-Injected Engine (Min, 1990)

At the operating point that manifold pressure = 14.4 In-Hg, throttle position at 17.9% of maximum, engine speed = 1,730 RPM, and load torque = 56.3 ft-lb., the linearized state space model is

$$A = \begin{bmatrix} 0.779 & 0.0632 & -0.149 & -0.635 & -0.211 \\ 1 & 0 & 0 & 0 & 0 \\ 0.271 & -0.253 & 0.999 & 0 & 0.845 \\ 0 & 0 & 0 & 0 & 0 \\ 0 & 0 & 0 & 0 & 0 \end{bmatrix}$$

$$B = \begin{bmatrix} 1.579 & 0.22598 \\ 0 & 0 \\ 0 & -0.9054 \\ 1 & 0 \\ 0 & 1 \end{bmatrix} \quad \text{and} \quad C = \begin{bmatrix} 1 & 0 & 0 & 0 & 0 \\ 0 & 0 & 1 & 0 & 0 \\ 0 & 0 & 0 & 1 & 0 \end{bmatrix}$$

The five system states are change in manifold pressure, change in manifold pressure (last rotation), change in engine RPM, change in throttle position, and change in external load, respectively. The two inputs are the next rotation throttle angle change and the change of external load during the next rotation, respectively.

(a) Using Algorithms 5.3 and 6.1, design the dynamic part of an output feedback compensator for this system, with poles $-1 \pm j$.

(b) Determine the rank of matrix $[T' \colon C']'$. It should be $5 = n$.

(c) Design the fault detection, isolation, and accommodation system, with $q = 2$, and poles $= -2, -4$, and -6.

System 4: Booster Rockets Ascending Through Earth's Atmosphere (Enns, 1990)

$$A = \begin{bmatrix} -0.0878 & 1 & 0 & 0 \\ 1.09 & 0 & 0 & 0 \\ 0 & 0 & 0 & 1 \\ 0 & 0 & -37.6 & -0.123 \end{bmatrix} \quad B = \begin{bmatrix} 0 & 0 & 4.2 \times 10^{-10} \\ 0 & 0 & 1.27 \times 10^{-8} \\ 0 & 0 & 0 \\ 1 & 0 & -1.2 \times 10^{-6} \end{bmatrix}$$

$$C = \begin{bmatrix} 0 & 1 & 0 & -0.00606 \\ 0 & 0 & -37.6 & -0.123 \\ 0 & 0 & 0 & -0.00606 \end{bmatrix} \quad D = \begin{bmatrix} 0 & 1 & 0 \\ 0 & 0 & -1.2 \times 10^{-6} \\ 0 & 0 & 0 \end{bmatrix}$$

The four-system states are the angle of attack (rad.), pitch rate q (rad/s), lowest frequency elastic model deflection η, and $\dot{\eta}$, respectively. The three inputs are the error of elastic model poles v_{POLE}, error of elastic model zeros v_{ZERO}, and the thrust vectoring control u_{TVC}(lb), respectively. The three outputs are the gyro output measurement y_{GYRO}(rad/s), $\eta - v_{POLE}$, and $y_{GYRO} - q - v_{ZERO}$, respectively.

From a control theory point of view, a difficulty involved with this problem is that the third column of B is too small, while its corresponding input is the only real control input u_{TVC} (the other two inputs are artificially added to account for the errors associated with the elasticity model). In addition, an adjustment has to be made to consider the nonzero D matrix, which is assumed to be zero in this book. Nonetheless, without matrix D and by eliminating the second column of matrix B, the example becomes similar to that of System 1.

System 5: Bank-to-Turn Missile (Wise, 1990)

At the flight conditions of 16° of angle of attack, Mach 0.8 (velocity of 886.78 ft/s), and an attitude of 4,000 ft, the linearized missile rigid body airframe state space model is

$$A = \begin{bmatrix} -1.3046 & 0 & -0.2142 & 0 \\ 47.7109 & 0 & -104.8346 & 0 \\ 0 & 0 & 0 & 1 \\ 0 & 0 & -12{,}769 & -135.6 \end{bmatrix} \quad B = \begin{bmatrix} 0 \\ 0 \\ 0 \\ 12{,}769 \end{bmatrix}$$

and

$$C = \begin{bmatrix} -1{,}156.893 & 0 & 189.948 & 0 \\ 0 & 1 & 0 & 0 \end{bmatrix}$$

The four-system states are the angle of attack, pitch rate, fin deflection, and fin rate, respectively. The control input is the fin deflection command (rad.), and the two outputs are normal acceleration (ft/s²) and pitch rate (rad/s), respectively.

(a) Using Algorithms 5.3 and 6.1, design the dynamic part of the dynamic output feedback compensator of this system. Because $CB=0$, we let the compensator order $r=1$ and dynamic matrix $F=-10$.

(b) Using the duality of Algorithm 5.3, design state feedback gain K which can place each of the following four sets of eigenvalues in matrix $A-BK$ (Wilson et al., 1992):

$$\{-5.12, -14.54, -24.03 \pm j18.48\},$$

$$\{-10 + j10, -24 \pm j18\},$$

$$\{-9.676 \pm j8.175, -23.91 \pm j17.65\},$$

$$\text{and } \{-4.7 \pm j2.416, 23.96 \pm j17.65\}.$$

(c) Design the respective output part $\underline{K}$ of the dynamic output feedback compensator of part (a), for the four sets of eigenvalues of part (b).

(d) Compare the controls of parts (b) and (c), for each of the four sets of part (b). The comparison can be made in practical aspects such as the control gain (K vs. $\underline{K}$) and the zero-input response.

(e) Design a complete failure detection, isolation, and accommodation system, with $q=1$ and poles$=-14, -10\pm j10$. The normal feedback compensator can be chosen from any of the four compensators of parts (a) and (c).

System 6: Extended Medium-Range Air-to-Air Missile (Wilson et al., 1992)

At the flight condition of 10° of angle of attack, Mach 2.5 (velocity of 2,420 ft/s), and a dynamic pressure of 1,720 lb/ft², the normal roll-yaw missile airframe model is

$$A = \begin{bmatrix} -0.501 & -0.985 & 0.174 & 0 \\ 16.83 & -0.575 & 0.0123 & 0 \\ -3{,}227 & 0.321 & -2.1 & 0 \\ 0 & 0 & 1 & 0 \end{bmatrix} \text{ and } B = \begin{bmatrix} 0.109 & 0.007 \\ -132.8 & 27.19 \\ -1{,}620 & -1{,}240 \\ 0 & 0 \end{bmatrix}$$

The four-system states are sideslip, yaw rate, roll rate, and roll angle, respectively. The two control inputs are rudder position and aileron position, respectively.

(a) For each of the four sets of feedback system eigenvalues of System 5, use Algorithms 9.1 and 9.2 and the analytic decoupling rules to design the eigenvectors and the corresponding state feedback gains.

(b) Compare each of the four sets of results of part (a) with the following corresponding result of Wilson et al. (1992):

$$K = \begin{bmatrix} 1.83 & -0.154 & 0.00492 & -0.0778 \\ -2.35 & 0.287 & -0.03555 & 0.0203 \end{bmatrix}$$

$$K = \begin{bmatrix} 5.6 & -0.275 & -0.00481 & -0.989 \\ -4.71 & 0.359 & -0.00815 & 1.1312 \end{bmatrix}$$

$$K = \begin{bmatrix} 3.19 & -0.232 & 0.10718 & 0.1777 \\ -1.63 & 0.299 & -0.15998 & -0.4656 \end{bmatrix}$$

$$K = \begin{bmatrix} 1.277 & -0.172 & 0.10453 & 0.1223 \\ 0.925 & 0.2147 & -0.15696 & -0.2743 \end{bmatrix}$$

The comparison can be made in practical aspects such as feedback gain, robust stability (2.23–2.25), and zero-input response.

System 7: Chemical Reactor (Munro, 1979)

$$A = \begin{bmatrix} 1.38 & -0.2077 & 6.715 & -5.676 \\ -0.5814 & -4.29 & 0 & 0.675 \\ 1.067 & 4.273 & -6.654 & 5.893 \\ 0.048 & 4.273 & 1.343 & -2.104 \end{bmatrix} \text{ and } B = \begin{bmatrix} 0 & 0 \\ 5.679 & 0 \\ 1.136 & -3.416 \\ 1.136 & 0 \end{bmatrix}$$

(a) Repeat part (a) of System 6, but for a new eigenvalue set: {–0.2, –0.5, –5.0566, –8.6659}.

(b) Repeat part (b) of System 6, but compare the following two possible results (Kautsky et al., 1985):

$$K = \begin{bmatrix} 0.23416 & -0.11423 & 0.31574 & -0.26872 \\ 1.1673 & -0.28830 & 0.68632 & -0.24241 \end{bmatrix}$$

$$K = \begin{bmatrix} 0.10277 & -0.63333 & -0.11872 & 0.14632 \\ 0.83615 & 0.52704 & -0.25775 & 0.54269 \end{bmatrix}$$

System 8: Distillation Column (Klein and Moore, 1977)

$$A = \begin{bmatrix} -0.1094 & 0.0628 & 0 & 0 & 0 \\ 1.306 & -2.132 & 0.9807 & 0 & 0 \\ 0 & 1.595 & -3.149 & 1.547 & 0 \\ 0 & 0.0355 & 2.632 & -4.257 & 1.855 \\ 0 & 0.00227 & 0 & 0.1636 & -0.1625 \end{bmatrix} \text{ and}$$

$$B = \begin{bmatrix} 0 & 0 \\ 0.0638 & 0 \\ 0.0838 & -0.1396 \\ 0.1004 & -0.206 \\ 0.0063 & -0.0128 \end{bmatrix}$$

(a) Repeat part (a) of System 6, but for a new set of eigenvalues {–0.2, –0.5, –1, $-1 \pm j$}.

(b) Repeat part (b) of System 6, but compare the following result of Kautsky et al. (1985):

$$K = \begin{bmatrix} -159.68 & 69.844 & -165.24 & 125.23 & -45.748 \\ -99.348 & 7.9892 & -14.158 & -5.9382 & -1.2542 \end{bmatrix}$$

References

Anderson B D O. 1979. *Optimal Filtering.* Englewood Cliffs: Prentice Hall.

Anderson B D O, Moore J B. 1971. *Linear Optimal Control.* Englewood Cliffs, NJ: Prentice Hall.

Athanassiades M. 1963. Optimal control for linear time-invariant plants with time, fuel, and energy constraints. *IEEE Trans. Appl. Ind.*, 81: 321–325.

Athans M, Falb P L. 1966. *Optimal Control.* New York: McGraw Hill.

Bachelier O, Bosche J, Mehdi D. 2006. On pole assignment via eigenstructure assignment approach. *IEEE Trans. Automatic Control,* AC 51: 1554–1558.

Bachelier O, Mehdi D. 2008. Non-iterative pole placement technique: A step further. *J. Franklin Inst.*, 345: 267–281.

Balakrishnan A V. 1984. *Kalman Filtering Theory.* New York: Optimization Software Inc.

Bjorck, Hammaling S. 1983. A Schur method for the square root of a matrix. *Linear Algebra Ann.*, 52/53: 127–140.

Brand L. 1968. The companion matrix and its properties. *Ameri. Math. Monthly,* 75: 146–152.

Bryson Jr A E, Ho Y C. 1969. *Applied Optimal Control.* Waltham: Blaisdell Publishing Co.

Byers R. 1983. Hamiltonian and Symplectic Algorithms for the Algebraic Riccati Equation (Ph D Thesis). Cornell: Cornell University.

Byers R. 1987. Solving the algebraic Riccati equation with the matrix sign function. *Linear Algebra Ann.*, 85: 267–279.

Byers R. 1990. A Hamiltonian Jacobi algorithm. *IEEE Trans. Automatic Control,* AC-35: 566–570.

Chang S S L. 1961. *Synthesis of Optimal Control.* New York: McGraw Hill.

Chen C T. 1984. *Linear System Theory and Design.* 2nd Edition. New York: Holt, Rinehart and Winston.

Chen C T. 1993. *Analog & Digital Control System Design: Transfer function, State Space, & Algebraic Methods.* Philadelphia: Sauders College Publishing.

Chen B M, Saberi A, Sannuti P. 1991. A new stable compensator design for exact and approximate loop transfer recovery. *Automatica,* 27: 257–280.

Cho D, Paolella P. 1990. Model-based failure detection and isolation of automotive powertrain systems, *Proc. 9th American Control Conf.*, 3: 2898–2905.

Choi S S, Sirisena H R. 1974. Computation of optimal output feedback gains for multivariable systems. *IEEE Trans. Automatic Control,* AC-19: 257.

Chu D. 2000. Disturbance decoupled observer design for linear time invariant systems: A matrix pencil approach. *IEEE Trans. Automatic Control,* AC-45: 1569–1575.

Chu E W E. 1993a. Approximate pole assignment. *Int. J. Control,* 58: 471–484.

Chu E.W.E. 1993b. Letter to the author.

Dahleh M A, Pearson Jr. B. 1987. L1 optimal compensators for continuous time systems. *IEEE Trans. Automatic Control,* 32: 889–895.

Darouach M. 2000. Existence and design of functional observers for linear systems. *IEEE Trans. Automatic Control,* AC-45(5): 940–943.

Davison E J. 1975. The numerical solution of $X=A_1X+A_2X+D$, $X(0)=C$. *IEEE Trans. Automatic Control*, AC-20: 566–567.
Davison E J. 1976. Remarks on multiple transmission zeros of linear multivariable systems. *Automatica*, 12: 195.
Davison E J. 1978. An algorithm for the calculation of transmission zeros of the system (C, A, B, D) using high gain output feedback. *IEEE Trans. Automatic Control*, AC-23: 738–741.
Davison E J, Wang S H. 1974. Properties and calculation of transmission zeros of linear multivariable systems. *Automatica*, 10: 643–658.
DeJong L S. 1975. Numerical Aspects of Realization Algorithms in Linear Systems Theory (Ph D Thesis). Netherland: Eindhoven University.
Dickman A. 1987. On the robustness of multivariable linear feedback in state space representation. *IEEE Trans. Automatic Control*, AC-32: 407.
Dongarra et al. 1979. "LINPACK" User's Guide, SIAM, Philadelphia.
Dorato P. 1987. A historical review of robust control. *IEEE Control Systems Magazine*, 7(2): 44–47.
Doyle J. 1978. Guaranteed margins for LQG regulators. *IEEE Trans. Automatic Control*, AC-23: 756–757.
Doyle J, Francis B A, Tannenbaum A R. 1992. *Feedback Control Theory*. New York: MacMillan Publishing Company.
Doyle J, Glover K, Khargoneker P P, Francis B A. 1989. State-space solutions to standard H_2 and H_2H_∞ control problems. *IEEE Trans. Automatic Control*, AC-34: 831–847.
Doyle J, Stein G. 1979. Robustness with observers. *IEEE Trans. Automatic Control*, AC-24: 607–611.
Doyle J., Stein G. 1981. Multivariable feedback design: Concepts for a classical/modern synthesis. *IEEE Trans. Automatic Control*, AC-26: 4–16.
Duan G R. 1993a. Solution to matrix equation AV+VF=BM and their application to eigen-structure assignment in linear systems. *IEEE Trans. Automatic Control*, AC-38: 276–280.
Duan G R. 1993b. Robust eigen-strcuture assignment via dynamic compensator. *Automatica*, 29: 469–474.
Dullerud G E, Francis B A. 1992. L_1 analysis and design of sampled-data systems. *IEEE Trans. Automatic Control*, AC-37: 436–446.
Emami-Naeini A, Akhter M, Rock S. 1988. Effect of model uncertainty on fault detection, the threshold selector. *IEEE Trans. Automatic Control*, AC-33: 1106–1115.
Enns D E. 1990. Structured singular value synthesis design example: rocket stabilization, *Proc. 9th ACC*, 3: 2514.
Fahmy M, O'Reilly J. 1982. On eigen-structure assignment in linear multivariable systems. *IEEE Trans. Automatic Control*, AC-27: 690–693.
Fernando T L, Trinh H M, Jennings L. 2010. Functional observability and the design of minimum order linear functional observers. *IEEE Trans. Automatic Control*, AC-55(5): 1269–1273.
Fletcher L R, Magni J F. 1987. Exact pole assignment by output feedback part 1,2. *Int. J. Control*, 45: 1995–2019.
Fortmann T E, Williamson D. 1972. Design of low order observers for linear feedback control laws. *IEEE Trans. Automatic Control*, AC-17: 301–308.
Fowell R A, Bender D J, Assal F A. 1986. Estimating the plant state from the compensator state. *IEEE Trans. Automatic Control*, AC-31: 964–967.

Fox L. 1964. *An Introduction to Numerical Linear Algebra.* London: Oxford University Press.

Francis B A. 1987. *A Course in H∞ Control Theory, 88 in Lecture Notes in Control and Information Sciences.* New York: Springer Verlag.

Francis J G F. 1961, 1962. The QR transformation, Parts I and II. *Computer J.*, 4: 265–271, 332–345.

Frank P M. 1990. Fault diagnosis in dynamic systems using analytical and knowledge based redundancy - A survey and some new results. *Automatica,* 26: 459–474.

Friedland B. 1962. A minimum response time controller for amplitude and energy constraints. *IEEE Trans. Automatic Control,* AC-7: 73–74.

Friedland B. 1986. *Control System Design—An Introduction to State Space Methods.* New York: McGraw Hill.

Friedland B. 1989. On the properties of reduced order Kalman filters. *IEEE Trans. Automatic Control,* AC-34: 321–324.

Fu M Y. 1990. Exact, optimal, and partial loop transfer recovery, *Proc. 29th IEEE Conf. on Decision and Control*: 1841–1846.

Gantmacher F R. 1959. *The Theory of Matrices, 1 and 2.* New York: Chelsea.

Ge W, Fang C Z. 1988. Detection of faulty components via robust observation. *Int. J. Control,* 47: 581–600.

Geromel J C, Peres P L D, Souza S R. 1993. H∞ robust control by static output feedback, *Proc. 12th American Control Conf.*, 1.

Gertler J. 1991. Analytical redundancy methods I fault detection and isolation, survey and synthesis, *Proc. IFAC Safe Process Symposium.*

Givens W. 1958. Computation of plane unitary rotations transforming a general matrix to triangular form. *J. Soc. Industr. Appli. Math.*, 6: 26–50.

Golub G H, Klema V, Stewart G. 1976a. Rank degeneracy and least squares solutions. STAN-CS, 76: 559.

Golub G H, Nash S, Van Loan C. 1979. A Hessenberg Schur method for the problem AX+XB=C. *IEEE Trans. Automatic Control,* AC-24: 909–913.

Golub G H, Reinsch C. 1970. Singular value decomposition and least square problems. *Numer. Math.*, 14: 403.

Golub G H, Van Loan C F. 1989. *Matrix Computations.* 2nd Edition. Baltimore: Johns Hopkins Univ. Press.

Golub G H, Wilkinson J H. 1976b. Ill conditioned eigensystems and the computation of the Jordan form, SIAM Rev., 18: 578.

Grace A, Laub A J, Little J N, Thompson C. 1990. MATLAB User's Guide. The Math Works, Inc., South Natick, MA.

Graham D, Lathyop R C. 1953. The synthesis of optimum response: Criteria and standard forms. *AIEE,* 72, Part II: 273–288.

Gupta R D, Fairman F W, Hinamoto T. 1981. A direct procedure for the design of single functional observers. *IEEE Trans. Circuits Systems,* CAS-28: 294–300.

Helmke U, Moore J B. 1992. L2 sensitivity minimization of linear system representations via gradient flows. *J. Mathematical Systems Control Theory.*

Horisberger H P, Belanger P R. 1974. Solution of the optimal constant output feedback problem by conjugate gradients. *IEEE Trans. Automatic Control,* AC-19: 434–435.

Hou M, Muller P C. 1992. Design of observers for linear systems with unknown inputs. *IEEE Trans. Automatic Control,* AC-37: 871–875.

Householder A S. 1958. Unitary triangularization of a nonsymmetric matrix. *J. Ass. Comp.*, 5: 339–342.

Hu T, Lin Z. 2001. *Control Systems with Actuator Saturation: Analysis and Design*. Boston: Birkhuser.

Huang Y S, MacFarlane A G J. 1982. *Multivariable Feedback: A Quasi Classical Approach, 40, in Lecture Notes in Control and Information Sciences*. New York: Springer Verlag.

Jiang E X. 1993. Bounds for the smallest singular value of a Jordan block with an application to eigenvalue perturbation. *Proc. SIAM Conf. in China*.

Juang Y T, Kuo T S, Hsu C F. 1986. Stability robustness analysis for state space models. *Proc. IEEE Conf. on Decision and Control*, 745.

Kaileth T. 1980. *Linear Systems*. Englewood Cliffs, NJ: Prentice Hall.

Kalman R E. 1960. Contributions to the theory of optimal control. *Bol. Soc. Mat. Mex.*, 5: 102–119.

Kautsky J, Nichols N K, Van Dooren P. 1985. Robust pole assignment in linear state feedback. *Int. J. Control*, 41: 1129–1155.

Khargoneker P P, Peterson I R, Rotea M A. 1988. H∞ optimal control with state feedback. *IEEE Trans. Automatic Control*, AC-33: 786–788.

Kimura H. 1975. Pole assignment by gain output feedback. *IEEE Trans. Automatic Control*, AC-20: 509–516.

Klein G, Moore B C. 1977. Eigenvalue generalized eigenvector assignment with state feedback. *IEEE Trans. Automatic Control*, AC-22: 140–141.

Klema V, Laub A. 1980. The singular value decomposition: Its computation and some applications. *IEEE Trans. Automatic Control*, AC-25: 164–176.

Konigorski U. 2012. Pole placement by parameter output. *System & Control Lett.*, 61: 292–297.

Kouvaritakis B, MacFarlane A G J. 1976. Geometrical approach to analysis and synthesis of system zeros. *Int. J. Control*, 23: 149–166.

Kudva P, Viswanadham N, Ramakrishna A. 1980. Observers for linear systems with unknown inputs. *IEEE Trans. Automatic Control*, AC-25: 113–115.

Kung S Y, Lin D W. 1981. Optimal Hankel norm model reductions: Multivariable systems. *IEEE Trans. Automatic Control*, AC-26: 832–852.

Kwakernaak H. 1993. Robust control and H∞ optimization—Tutorial paper. *Automatica*, 29: 255–273.

Kwakernaak H, Sivan R. 1972. *Linear Optimal Control Systems*. New York: Wiley Intersciences.

Kwon B H, Youn M J. 1987. Eigenvalue generalized eigenvector assignment by output feedback. *IEEE Trans. Automatic Control*, AC-32: 417–421.

Laub A J. 1979. A Schur method for solving algebraic Riccati equations. *IEEE Trans. Automatic Control*, AC-24: 913–921.

Laub A J. 1985. Numerical linear algebra aspects of control design computations. *IEEE Trans. Automatic Control*, AC-30: 97–108.

Laub A J, Linnemann A. 1986. Hessenberg and Hessenberg/triangular forms in linear system theory. *Int. J. Control*, 44: 1523.

Laub A J, Moore B C. 1978. Calculation of transmission zeros using QZ techniques. *Automatica*, 14: 557–566.

Lawson C L, Hanson R J. 1974. *Solving Least Square Problems*. Englewood Cliffs: Prentice Hall.

Lay D C. 2006. *Linear Algebra and Its Applications*. 3rd Edition. Boston: Pearson Addison-Wesley.

Lehtomati N A, Sandell Jr N R, Athans M. 1981. Robustness results in linear quadratic gaussian based multivariable control designs. *IEEE Trans. Automatic Control,* AC-26: 75–92.

Levine W S, Athans M. 1970. On the determination of optimal constant output feedback gains for linear multivariable systems. *IEEE Trans. Automatic Control,* AC-15: 44–48.

Levis A H. 1987. Research directions for control community: Report of September 1986 workshop. *IEEE Trans. Automatic Control,* AC-32: 275–285.

Lewkowicz L, Sivan R. 1988. Maximal stability robustness for state equations. *IEEE Trans. Automatic Control,* AC-33: 297–300.

Liu Y, Anderson B D O. 1990. Frequency weighted controller reduction methods and loop transfer recovery. *Automatica,* 26: 487–497.

Liubakka M K. 1987. Application of Failure Detection and Isolation Theory to Internal Combustion Engines (MS Thesis). Ann Arbor: University of Michigan.

Luenberger D G. 1967. Canonical forms for linear multivariable systems. *IEEE Trans. Automatic Control,* AC-12: 290–293.

Luenberger D G. 1971. An introduction to observers. *IEEE Trans. Automatic Control,* AC-16: 596–603.

MacFarlane A G J, Karcanias N. 1976. Poles and zeros of linear multivariable systems: A survey of the algebraic, geometric and complex variable theory. *Int. J. Control,* 24: 33–74.

Magni J F. 1987. Exact pole assignment by output feedback, Part 3. *Int. J. Control,* 45: 2021–2033.

Miminis G S, Paige C C. 1982. An algorithm for pole assignment of time invariant linear systems. *Int. J. Control,* 34: 341–345.

Min P S. 1990. Validation of controller inputs in electronically controlled engines, *Proc. 9th ACC,* 3: 2887.

Misra P, Patel R V. 1989. Numerical algorithms for eigenvalue assignment by constant and dynamic output feedback. *IEEE Trans. Automatic Control,* AC-34: 579–588.

Moler C B, Stewart G W. 1973. An algorithm for generalized matrix eigenvalue problems. *SIAM J. Numerical Analysis,* 10: 241–256.

Moore B C. 1976. On the flexibility offered by state feedback in multivariable systems beyond closed loop eigenvalue assignment. *IEEE Trans. Automatic Control,* AC-21: 689–692.

Moore J B, Tay T T. 1989. Loop transfer recovery via H_∞/H_2 sensitivity recovery. *Int. J. Control,* 49: 1249–1271.

Munro N. 1979. *Proc. Instn. Elect. Engrs.,* 126: 549.

Niemann H H, Sogaard Andersen P, Stoustrup J. 1991. Loop transfer recovery for general observer architectures. *Int. J. Control,* 53: 1177–1203.

O'Reilly J. 1983. *Observers for Linear Systems.* London: Academic Press.

Paige C C, Van Loan C F. 1981. A Schur decomposition for Hamiltonian matrices. *Linear Algebra Annuals,* 41: 11–32.

Patel R V. 1978. On transmission zeros and dynamic output feedback. *IEEE Trans. Automatic Control,* AC-23: 741–742.

Patel R V. 1981. Computation of matrix fraction descriptions of linear time invariant systems. *IEEE Trans. Automatic Control,* AC-26: 148–161.

Petkov P, Christov N, Konstantinov M. 1984. Computational algorithm for pole assignment of linear multiinput systems. *IEEE Trans. Automatic Control,* AC-29: 1044–1047.

Pontryagin L S, Boltyanskii V, Gankrelidze R, Mishehenko E. 1962. *The Mathematical Theory of Optimal Processes.* New York: Interscience Publishers.

Postlethwaite I, MacFarlane A G J. 1979. *A Complex Variable Approach to the Analysis of Linear Multivariable Feedback Systems, 12 in Lecture Notes in Control and Information Sciences.* New York: Springer Verlag.

Qiu L, Davison E J. 1986. New perturbation bounds for the robust stability of linear state space models. *Proc. IEEE Conf. on Decision and Control,* 751.

Rosenbrock H H. 1973. The zeros of a system. *Int. J. Control,* 18: 297–299.

Rosenbrock H H. 1974. *Computer Aided Control System Design.* London: Academic Press.

Rosenthal J, Wang X C. 1992. The mapping degree of the pole placement map. *SIAM Conf. on Control and Its Applications,* CP9 Linear Systems I.

Rothschild D, Jameson A. 1970. Comparison of four numerical algorithms for solving the Lyapunov matrix equations. *Int. J. Control,* 11: 181–198.

Saberi A, Chen B M, Sannuti P. 1991. Theory of LTR for nonminimum phase systems, recoverable target loops, and recovery in a subspace. *Int. J. Control,* 53: 1067–1115, 1116–1160.

Saberi A, Chen B M, Sannuti P. 1993. *Loop Transfer Recovery: Analysis and Design.* New York: Springer Verlag.

Saberi A, Sannuti P. 1990. Observer design for loop transfer recovery and for uncertain dynamic systems. *IEEE Trans. Automatic Control,* AC-35: 878–897.

Saeki M. 1992. H_∞/LTR procedure with specified degree of recovery. *Automatica,* 28: 509–517.

Safonov M G, Athans M. 1977. Gain and phase margin for multiloop LQG regulators. *IEEE Trans. Automatic Control,* AC-22: 173–179.

Sage A P, White III C C. 1977. *Optimum Systems Control.* 2nd Edition. Englewood Cliffs: Prentice Hall.

Shaked U, Soroka E. 1985. On the stability robustness of the continuous time LQG optimal control. *IEEE Trans. Automatic Control,* AC-30: 1039–1043.

Sinswat V, Patel R V, Fallside F. 1976. A method of computing invariant zeros and transmission zeros of invertible systems. *Int. J. Control,* 23: 183–196.

Sobel K M, Shapiro E Y, Rooney R H. 1984. Synthesis of direct lift control laws via eigenstructure assignment. *Proc. Nat. Aero. Elect. Conf. Ohio*: 570–575.

Sogaard Andersen L. 1986. Issues in robust multivariable observer based feedback design (Ph D Thesis). Denmark: Technical University of Denmark.

Sogaard Andersen L. 1987. Comments on 'On the loop transfer recovery'. *Int. J. Control,* 45: 369–374.

Spurgeon S K. 1988. An assignment of robustness of flight control systems based on variable structure techniques (Ph D Thesis). England: University of York.

Stein G, Athans M. 1987. The LQG/LTR procedure for multivariable feedback control design. *IEEE Trans. Automatic Control,* AC-32: 105–114.

Stewart G W. 1976. *Introduction to Matrix Computations.* London: Academic Press.

Stoustrup J, Niemann H H. 1993. The general H_∞ problem with static output feedback. *Proc. 12th American Control Conf.,* 1.

Syrmos V L. 1993. Computational observer design techniques for linear systems with unknown inputs using the concept of transmission zeros. *IEEE Trans. Automatic Control,* AC-38: 790–794.

Syrmos V L, Abdallah C, Dorato P. 1994. Static output feedback: A survey, *Proc. 33rd IEEE CDC*: 837–842.

Syrmos V L, Lewis F L. 1993. Output feedback eigen-structure assignment using two Sylvester equations. *IEEE Trans. Automatic Control,* AC-38: 495–499.

Tahk M, Speyer J. 1987. Modeling of parameter variations and asymptotic LQG synthesis. *IEEE Trans. Automatic Control,* AC-32: 793–801.

Tits A L, Yan Y. 1996. Globally convergent algorithm for robust pole assignment by state feedback. *IEEE Trans. Automatic Control,* AC-41: 1432–1452.

Toivonen H T. 1985. A globally convergent algorithm for the optimal constant output feedback problem. *Int. J. Control,* 41: 1589–1599.

Truxal J G. 1955. *Automatic Feedback Control System Synthesis.* New York: McGraw-Hill.

Tsui C C. 1983b. Computational Aspects of Realization and Design Algorithms in Linear Systems Theory (Ph D Thesis). Stony Brook: State University of New York.

Tsui C C. 1985. A new algorithm for the design of multi-functional observers. *IEEE Trans. Automatic Control,* AC-30: 89–93.

Tsui C C. 1986a. An algorithm for computing state feedback in multi-input linear systems. *IEEE Trans. Automatic Control,* AC-31: 243–246.

Tsui C C. 1986b. On the order reduction of linear function observers. *IEEE Trans. Automatic Control,* AC-31: 447–449.

Tsui C C. 1986c. Comments on 'New technique for the design of observers'. *IEEE Trans. Automatic Control,* AC-31: 592.

Tsui C C. 1987a. A complete analytical solution to the equation TA – FT = LC and its applications. *IEEE Trans. Automatic Control,* AC-32: 742–744.

Tsui C C. 1987b. On preserving the robustness of an optimal control system with observers. *IEEE Trans. Automatic Control,* AC-32: 823–826.

Tsui C C. 1988a. A new approach of robust observer design. *Int. J. Control,* 47: 745–751.

Tsui C C. 1988b. On robust observer compensator design. *Automatica,* 24: 687–691.

Tsui C C. 1989. On the solution to the state failure detection problem. *IEEE Trans. Automatic Control,* AC-34: 1017–1018.

Tsui C C. 1990. A new robustness measure for eigenvector assignment, *Proc. 9th American Control Conf.*: 958–960.

Tsui C C. 1992. Unified output feedback design and loop transfer recovery, *Proc. 11th American Control Conf.*: 3113–3118.

Tsui C C. 1993a. On the solution to matrix equation TA – FT = LC and its applications. *SIAM. J. on Matrix Analysis,* 14: 33–44.

Tsui C C. 1993b. Unifying state feedback/LTR observer and constant output feedback design by dynamic output feedback. *Proc. IFAC World Congress,* 2: 231–238.

Tsui C C. 1993c. A general failure detection, isolation and accommodation system with model uncertainty and measurement noise. *Proc. IFAC World Congress,* 6: 341–348.

Tsui C C. 1994a. A new robust stability measure for state feedback systems. *Systems & Control Letters,* 23: 365–369.

Tsui, C.C. 1994b. A general failure detection, isolation an accommodation system with model uncertainty and measurement noise. *IEEE Trans. Automatic Control,* AC-39: 2318–2321.

Tsui C C. 1996a. A new design approach of unknown input observers. *IEEE Trans. Automatic Control,* AC-41: 464–468.

Tsui C C. 1996b. Author's reply to 'Comments on the loop transfer recovery'. *IEEE Trans. Automatic Control,* AC-41: 1396.

Tsui C C. 1997. The design generalization and adjustment of a failure isolation and accommodation system. *Int. J. Systems Science*, 28: 91–107.

Tsui C C. 1998a. What is the minimum function observer order? *J. Franklin Institute*, 35B/4: 623–628.

Tsui C C. 1998b. The first general output feedback compensator that can implement state feedback control. *Int. J. Systems Science*, 29: 49–55.

Tsui C C. 1999a. A design algorithm of static output feedback control for eigen-structure assignment, *Proc. 1999 IFAC World Congress*, Q: 405–410.

Tsui C C. 1999b. A fundamentally novel design approach that defies separation principle, *Proc. 1999 IFAC World Congress*, G: 283–288.

Tsui C C. 1999c. High performance state feedback, robust, and output feedback stabilization control—A systematic design algorithm. *IEEE Trans. Automatic Control*, AC-44: 560–563.

Tsui C C. 2000a. The applications and a general solution of a fundamental matrix equation pair, *Proc. 3rd Asian Control Conf.*, 3035–3040.

Tsui C C. 2000b. What is the minimum function observer order? *Proc. 3rd World Congress on Intelligent Control and Artificial Intelligence*: 2811–2816.

Tsui C C. 2001. A design algorithm of static output feedback control for eigen-structure assignment, *Proc. 2001 American Control Conf.*: 1669–1674, Arlington, VA.

Tsui C C. 2002. A new state feedback control design approach which guarantees the critical realization, *Proc. 4th World Congress on Intelligent Control and Automation*: 199–205.

Tsui C C. 2003a. What is the minimum function observer order? *Proc. 2003 European Control Conf.* Session Observer 1.

Tsui C C. 2003b. The applications and a general solution of a fundamental matrix equation pair, *Proc. 2003 European Control Conf.* Session Robust Control 5.

Tsui C C. 2004a. An overview of the applications and solutions of a fundamental matrix equation pair. *J. Franklin Inst.*, 341/6: 465–475.

Tsui C C. 2004b. Six-dimensional expansion of output feedback design for eigen-structure assignment, *Proc. 2004 Chinese Control Conf.*, WA 4.

Tsui C C. 2004c. *Robust Control System Design—Advanced State Space Techniques*. 2nd Edition. New York: Marcel Dekker.

Tsui C C. 2005. Six-dimensional expansion of output feedback design for eigen-structure assignment. *J. Franklin Inst.*, 342/7: 892–901.

Tsui C C. 2006. Eight irrationalities of basic state space control system design, *Proc. 6th World Congress on Intelligent Control and Automation*: 2304–2307.

Tsui C C. 2012. Overcoming eight drawbacks of the basic separation principle of state space control design, *Proc. 2012 Chinese Control Conference*: 213–218.

Tsui C C. 2012a. The best possible theoretical result of minimal order linear functional observer design. *J. Univ. of Sci. & Technol. China*, 42/7: 603–608.

Tsui C C. 2012b. Observer design for systems with time-delayed states. *Int. J. Automat. Comput.*, 9(1): 105–107.

Tsui C C. 2014.The Theoretical Part of Linear Functional Observer Design Problem is Solved, *Proc. 2014 Chinese Control Conf.*: 3456–3461.

Tsui C C. 2015. Observer design – A survey. *Int. J. Automat. Comput.*, 12(1): 50–61.

Tsui C C, Chen C T. 1983a. An algorithm for companion form realization. *Int. J. Control*, 38: 769–779.

Van Dooren P. 1981. The generalized eigenstructure problem in linear system theory. *IEEE Trans. Automatic Control*, AC-26: 111–129.

Van Dooren P. 1984. Reduced order observers: A new algorithm and proof. *Systems & Control Lett.*, 4: 243–251.
Van Dooren P, Emaimi Naeini, Silverman L. 1978. Stable extraction of the Kronecker structure of pencils, *Proc. IEEE 17th Conf. on Decision and Control*: 521–524.
Van Loan C F. 1984. A symplectic method for approximating all the eigenvalues of a Hamiltonian matrix. *Linear Algebra Ann.*, 61: 233–251.
Vidyasagar M. 1984. The graphic metric for unstable plants and robustness estimates for some systems. *IEEE Trans. Automatic Control*, AC-29: 403.
Vidyasagar M. 1985. *Control System Synthesis: A Factorization Approach*. Cambridge: MIT Press.
Wang J W, Chen C T. 1982. On the computation of the characteristic polynomial of a matrix. *IEEE Trans. Automatic Control*, AC-27: 449–451.
Wang S H, Davison E J, Dorato P. 1975. Observing the states of systems with unmeasurable disturbances. *IEEE Trans. Automatic Control*, AC-20: 716–717.
Wang X A. 1996. Grassmannian, central projection, and output feedback control for eigenstructure assignment. *IEEE Trans. Automatic Control*, AC-41: 786–794.
Wang X A, Konigorski U. 2013. On linear solutions of the output feedback pole assignment problem. *IEEE Trans. Automatic Control*, AC-58(9): 2354–2359.
Weng Z X, Shi S J. 1998. H_∞ loop transfer recovery synthesis of discrete time systems. *Int. J. Robust & Nonlinear Control*, 8: 687–697.
Wilkinson J H. 1965. *The Algebraic Eigenvaluc Problem*. London: Oxford University Press.
Willems J C. 1995. Book review of mathematical systems theory: The influence of R. E. Kalman. *IEEE Trans. Automatic Control*, AC-40: 978–979.
Wilson R F, Cloutier J R, Yedavali R K. 1992. Control design for robust eigen-structure assignment in linear uncertain systems. *Control Systems Magazine*, 12(5): 29–34.
Wise K A. 1990. A Comparison of six robustness tests evaluating missile autopilot robustness to uncertain aerodynamics, *Proc. 9th ACC*, 1: 755.
Wolovich W.A. 1974. Linear Multivariable Systems. New York: Springer-Verlag.
Xu H.G. 1991. Solving Algebraic Riccati Equations via Skew-Hamilton matrices. Ph.D. Thesis, Fudan University, China.
Yan W Y, Teo K L, Moore J B. 1993. A gradient flow approach to computing LQ optimal output feedback gains, *Proc. 12th American Control Conf.*, 2: 1266–1270.
Yeh H H, Banda S S, Chang B C. 1992. Necessary and sufficient conditions for mixed H_2 and H_∞ optimal control. *IEEE Trans. Automatic Control*, AC-37: 355–358.
Youla D C, Bongiorno J J, Lu C N. 1974. Single loop feedback stabilization of linear multivariable dynamic systems. *Automatica*, 10: 159–173.
Zames G. 1981. Feedback and optimal sensitivity: Model reference transformations, multiplicative seminorms, and approximate inverse. *IEEE Trans. Automatic Control*, AC-26: 301–320.
Zheng D Z. 1989. Some new results on optimal and suboptimal regulators of LQ problem with output feedback. *IEEE Trans. Automatic Control*, AC-34: 557–560.
Zhou K M. 1992. Comparison between H_2 and H_∞ controllers. *IEEE Trans. Automatic Control*, AC-37: 1442–1449.
Zhou K M, Doyle J C, Glover K. 1995. *Robust and Optimal Control*. New Jersey: Prentice.

Index

adaptive fault control 163–166
  in coordination with robust control 43, 164, 167
algorithms
  algebraic Riccati equation 226
  block-observable Hessenberg form 101
  Lyapunov equation $TA-A'T=C$ 229
  matrix equation $TA-FT-LC$ 111
  matrix equation $TB=0$ (and $TA-FT=LC$) 118
  minimal order observer design 146
  observable Hessenberg form 98
  pole assignment by output feedback 182
  quadratic optimal output feedback 229
  QR matrix decomposition 244
  rank-one eigenvector assignment 205
  rank-two eigenvector assignment 206
  robust failure detection observers 159
  Schur triangularization 226
asymptotic LTR 82
asymptotic stability 27

back substitution 251
bandwidth 29
basis 241
  of feedback system eigenvectors 107, 180
  orthogonal/orthonormal 242
bidiagonal form matrix 9
block-controllable Hessenberg form 180
block-observable canonical form 16, 103
block-observable Hessenberg form 99
  computation of 101
blocking zeros 19

canonical form system matrix
  controllable 15
  Jordan 9
  observable 14
characteristic polynomial 10

classical control theory 1
  in measuring performance 28
  in measuring robust stability 41
closed-loop system 50
companion form matrix 15
complete response 26
condition number
  of back substitution 251
  of computational problem 36
  of eigenvalue assignment problem 198
  of general eigenvalue pattern 178, 254
  of matrix 37
  of matrix equation $TA-FT=LC$ 112
  of triangular matrix 251
controllability 11
  index 181
  of state feedback system 57
  of static output feedback system 62
coupling 41

decoupling
  of feedback system eigenstructure 212
  in observer dynamic part 85, 135, 145
  in sensitivity 41
  of the solution of $TA-FT=LC$ 41, 76, 85, 113, 145
defective eigenvector *see* generalized eigenvector
detectable fault level 169
disturbance 50
duality 12
dynamic equation 2
  of observer 64
  of output feedback compensator 69
dynamic output feedback 69
  state feedback implementation 87
dynamic part of the system 2

echelon-form matrix
  computation of 246
  lower 100

eigenstructure assignment 198, 203
eigenvalue 9
  multiple 9
  sensitivity of 37
eigenvalue assignment
  by state feedback 179
  by static output feedback 181
  condition of 198
eigenvector
  of companion form matrix 108, 211
  generalized, defective 9
  left/right 9
eigenvector assignment
  freedom of 198
  in static output feedback 182, 199
exp(At)
  of Jordan form matrix 26
  properties of 27

failed-state component 158
fault
  accommodation 163
  detectable fault level 170
  detection and isolation 157
  detection residual threshold 169
  with model uncertainty and noise 167
  signal 158
fault detector 158
  existence of 159
  requirement of 158–159
feedback system 49
  dynamic output 69
  observer 65
  state 56
  static output 60
final value theorem 32, 178
Frobenius norm 35
frequency response 29
freedom of
  matrix equation *TA–FT=LC* 112
  eigenvector assignment 198
function observer 76

Gaussian eliminated method 247
generalized eigenvector 9
  to sensitivity of eigenvalues 212
generalized state of feedback control 61, 86
  probability of stabilization 93, 95, 137
Gershgorin theorem 41

H_∞ problem 65
H_∞–H_2 problem 65, 232
Hamiltonian matrix 225
  order reduction of decomposition of 227
Hessenberg form
  block-controllable 180
  block-observable 99
  computation of 98
  lower, upper 98
  observable 98
Householder method 248

identity matrix 5, 242
ill conditioned 36
input 2
  disturbance 50
  fault signal 158
  reference 50
irreducible system 11

Jordan block 9
Jordan form 9
  in controllability/observability test 12
  in exp(At) 26
  in the solution of *TA–FT=LC* 113

Kalman filter 74, 82
Kalman inequality 58
Kronecker product 109

L_1 problem 232
Laplace transform 5
  final value theorem 32
least-square solution 256
linear dependence/independence 237
linearity 5
linear quadratic problem *see* quadratic optimal control
linear space 240
  basis of 241
  dimension of 240
linear time-invariant system 1

linear transformation 242
    change of basis 242
    orthogonal/orthonormal 242
logic fault isolation operation 159
loop gain 55
loop transfer function
    of feedback system 50
    of observer feedback system 66
    of open-loop system 50
    of quadratic optimal control
        system 58
    of state feedback system 57
    of static output feedback
        system 60
loop transfer recovery (LTR) 69, 82
low sensitivity 25
    of feedback system 49
    of performance 37
    of quadratic optimal control
        system 59
    of stability 41
    of state feedback control 57
    relation to high performance 25
Lyapunov equation 114, 228
    solution of 229

matrix
    bidiagonal 9
    block-Hessenberg form 99, 180
    companion form 15
    condition number of 37
    echelon form 100
    Hamiltonian 225
    Hessenberg form 98
    Jordan form 9
    orthogonal/orthonormal 242
    polynomial 17
    pseudo-inverse 257
    rank of 241
    rational polynomial 10
    Schur triangular form 226
    state matrix 2
    trace of 36
    unitary 243
    Vandermonde 106
MIMO system 2
minimal order observer 76
    advantage of 143, 157
    design formulation of 146
    design history of 76
Minimax theorem 253
minimum-phase 81, 82
    strictness of 92, 94
model
    state space 1
    transfer function 6
    uncertainty 51
modern control theory *see* state space
        control theory

noise 54
norm
    of matrix 35
    of vector 35
normalized vector 243
null space 241
numerical linear algebra 34
Nyquist stability test 42, 58

observability 11
    index 17, 102
    of state feedback system 57
    of static output feedback system 62
observer
    function 76
    general 62
    minimal-order 86
    state 74
    unknown input 81
open-loop system 50
order
    of linear computation 248, 249
    of minimal-order observer
        149, 155
    of observer 62
    of output feedback compensator 88
    of robust failure detector 160
    of system 11
orthogonal
    linear transformations 242
    vectors 239
orthonormal basis vectors 242
output 2
    feedback system 87
    measurement noise 54
    part of the system 2

performance 28
  indication of 28
  relation to robustness 25, 178
perturbation analysis, of eigenvalue sensitivity 38
pole 18
projection 239
pseudo-inverse matrix 257

QR decomposition 244 quadratic optimal control 223
  by generalized state feedback 227
  robust stability of 58, 59
  by state feedback 225

range space 241
rank
  column/row 241
  computation of 257
  full column/row 241
  of matrix 257
realization 18
reducible system 11
reference input 50
residual signal 158
response 5
  steady state 27
  transient 28
  zero-input 5
  zero-state 5
Riccati equation 225
robust control *versus* adaptive control 143, 169
robust failure detector 159
robustness *see* Low sensitivity
robust performance *see* Low sensitivity, of performance
robust stability *see* Low sensitivity, of stability
Routh-Hurwitz stability test 42

Schur triangularization 226
sensitivity
  of computation (*see* Condition number)
  to disturbance 54
  of eigenvalues 37
  of Jordan-form matrix 37
  to model uncertainty 52
  of multiple eigenvalues 178, 212
  of performance 37
  of stability 41
separation property
  of dynamic output feedback system 86
  necessary condition 79
  sufficient condition 77
similarity 10
similarity transformation 10
simulation 5
  of eigenstructure assignment result 217
singular value decomposition (SVD) 252
SISO system 2
special coordinate basis 136
spectrum norm 35
stability 27
state 2
state equation *see* dynamic equation
state feedback control 56
  generalized 61, 86
state matrix 2
state observer 74
  full-order 74
  reduced-order 75
  unknown input 81
state space control theory 1
static output feedback control 60
steady state response 27
strong stabilization 137
Sylvester equation $TA-FT=C$ 114
system
  feedback 50
  irreducible 11
  linear time-invariant 1
  MIMO 2
  open-loop 50
  reducible 11
  single 25, 49
  SISO 2

$TA-FT=LC$ matrix equation
  in eigenstructure assignment 179
  in fault detectors 158
  in general observers 63
  in observer feedback system poles 77
  versus other basic equations 113
  versus Sylvester equation 114

threshold treatment
  of effect of measurement noise 169
  of effect of model uncertainty 169
time constant 7
trace of matrix 36
tractability of design 217
transfer function 6
  loop 50
transformation
  linear 242
  orthogonal/orthonormal 242
  similarity 10
  symplectic 227
transient response 28
transmission zero 19
  computation of 20
  of output feedback compensator
    design 20
  of output feedback system 20

unitary matrix 243
unknown input observer 81

Vandermonde matrix 108
vector
  basis 241
  norm 35

weighting factors
  of eigenvector assignment 207
  of quadratic optimal criteria 223–224
  of robust stability measure 44
well-conditioned problem 36

zero-input response 5
zeros 19
  blocking 19
  transmission 19
zero state response 5

Printed in the United States
by Baker & Taylor Publisher Services

Printed in the United States
by Baker & Taylor Publisher Services